Luis L. Bonilla and
Stephen W. Teitsworth
Nonlinear Wave Methods for Charge Transport

Related Titles

Whitham, GB

Linear & Nonlinear Waves (Cloth)

1974
ISBN 978-0-471-94090-6

Luis L. Bonilla and Stephen W. Teitsworth

Nonlinear Wave Methods for Charge Transport

WILEY-VCH Verlag GmbH & Co. KGaA

The Authors

Luis L. Bonilla
Gregorio Millan
Institute of Fluid Dynamics,
Nanoscience and Industrial Mathematics
Universidad Carlos III de Madrid
Leganes
Spain
bonilla@ing.uc3m.es

Stephen W. Teitsworth
Department of Physics
Duke University
Durham
USA
teitso@phy.duke.edu

Cover

Cover illustration shows numerical simulation of charge density as a function of space (horizontal direction) and time (vertical direction) for a GaAs/AlAs superlattice, depicting periodic formation and propagation of charge accumulation (red–yellow) and depletion (blue–purple) layers. Figure courtesy of Huidong Xu.

Library of Congress Card No.: applied for
British Library Cataloguing-in-Publication Data:
A catalogue record for this book is available from the British Library.
Bibliographic information published by the Deutsche Nationalbibliothek
The Deutsche Nationalbibliothek lists this publication in the Deutsche Nationalbibliografie; detailed bibliographic data are available on the Internet at <http://dnb.d-nb.de>.

Typesetting le-tex publishing services GmbH, Leipzig
Printing Strauss GmbH, Mörlenbach
Binding Litges & Dopf GmbH, Heppenheim
Cover Design Adam-Design, Weinheim

Printed in the Federal Republic of Germany
Printed on acid-free paper

ISBN 978-3-527-40695-1

Contents

Nonlinear Wave Methods for Charge Transport. Luis L. Bonilla and Stephen W. Teitsworth

ISBN: 978-3-527-40695-1

Preface

What is the Rationale of this Book?

The primary theme of this book is to introduce and develop *mathematical* techniques for the treatment of nonlinear wave phenomena and associated singular perturbation methods, especially those that are useful for treating nonlinear electronic transport. The practice of implementing these techniques is largely illustrated through their application to charge transport problems in bulk semiconductors and semiconductor superlattices. Broader connections to other nonlinear wave problems, for example, dislocation dynamics in crystals and the dynamics of nerve pulse conduction, are also discussed. This book is also designed to provide a self-contained introduction to the *physics* of nonlinear electronic transport phenomena in solid material, especially in the form of drift-diffusion models. The treatment is targeted towards scientists, mathematicians, and engineers as well as graduate students in these fields. The level of discussion is designed to be accessible to those who have completed an undergraduate degree in mathematics, the physical sciences, or most engineering fields.

Of course, there are many excellent books already available that approach nonlinear wave methods and related techniques from the perspective of applied or pure math. However, we have found that most or all of these require a high level of commitment and comprehension due to their length and the depth of coverage. While this book is less comprehensive in its coverage than some of the classic texts [1–4], it is designed to allow the reader to quickly gain proficiency with a subset of techniques that the authors have found to be very useful.

Additionally, there are excellent books on electronic transport ranging from solid state texts to nonlinear phenomena. The current one is designed to be self-contained in the treatment of this piece of physics, though appropriate references are provided for those readers who want to achieve a greater breadth in their understanding of electronic transport phenomena. Regarding the references included in this book, it should be noted that the authors have not attempted to be exhaustive, but rather, the focus has been placed on including a smaller list of references that we have found to be most helpful.

Nonlinear Wave Methods for Charge Transport. Luis L. Bonilla and Stephen W. Teitsworth

ISBN: 978-3-527-40695-1

What are the Prerequisites?

We assume that the reader has taken a set of math courses that are typically required as part of a physics, engineering, or math undergraduate degree. This includes, most importantly, introductory courses in linear algebra, and ordinary and partial differential equations. Additionally, it would be helpful if the student has also had courses on nonlinear dynamics and asymptotic methods. For those who have not explored such material, we have included references to basic texts covering that material. Readers may wish to consult those when background questions arise. We do assume that the reader has had at least a one year introductory physics course. Particularly important, is material from electrostatics, circuits, electrical current flow, and electromagnetic waves. It will also be helpful if the student has taken elementary courses on quantum mechanics and solid state physics though, again, we attempted to follow a self-contained treatment with references supplied for additional background material.

References

1 Bender, C.M. and Orszag, S.A. (1978) *Advanced Mathematical Methods for Scientists and Engineers*, McGraw Hill, N.Y.

2 J. Kevorkian and J. Cole (1996) *Multiple Scale and Singular Perturbation Methods*, Springer, N.Y.

3 P. A. Lagerstrom (1988) *Matched asymptotic expansions*, Springer, N.Y.

4 G.B. Whitham (1974 *Linear and Nonlinear Waves*, Wiley, N.Y.

Acknowledgments

This book would not have been possible were it not for years of collaboration and teaching on the subject during the past two decades. First of all, we would like to thank family, colleagues, and friends for their supportive and understanding role during this large and occasionally frustrating process. While we will not attempt to name everyone, we would like to single out the following colleagues who played an important role in our research: (alphabetically) Mike Bergmann, Ana Carpio, Manuel Carretero, Ramón Escobedo, Jorge Galán, Holger Grahn, Pedro Hernando, Francisco Higuera, Jörg Kastrup, Joe Keller, Manuel Kindelan, Slava Kochelap, Amable Liñán, Roberto Merlin, Miguel Moscoso, John Neu, Álvaro Perales, Gloria Platero, Inma Rodríguez Cantalapiedra, Marco Rogozia, David Sánchez, Carlos Velasco, Stephanos Venakides, and Bob Westervelt. ST thanks Eckehard Schöll and Andreas Amann for helpful conversations concerning material in Section 9.2. Among current and former students, we would like to thank: Huidong Xu, Kevin Brown, Martin Heinrich, and Ignacio Arana.

Also, we would like to thank Duke undergrad, Jeffrey Peyser, for preparing a number of the figures.

We would also like to acknowledge our editor at WILEY-VCH, Anja Tschörtner, for her helpful and persistent encouragement.

Nonlinear Wave Methods for Charge Transport. Luis L. Bonilla and Stephen W. Teitsworth

ISBN: 978-3-527-40695-1

1
Introduction

1.1
Overview of Nonlinear Wave Phenomena

Examples of nonlinear wave phenomena abound in the physical, biological, and social sciences. One of the most striking instances is the solitary wave in shallow water in a narrow channel. In 1844, the Scottish naval engineer, John Scott Russell recalled an event some 10 years prior in which he observed a single humped wave (which he called a "wave of translation") that preserved its shape over a great distance as he followed it on horseback, arguably the first reported experimental observation of a nonlinear solitary wave [1]. The wave he observed had a height of about 0.3 m and a width of about 10 m, and was created by the sudden stopping of a barge in a narrow canal. He was so clearly struck by this observation that he described this experience as his "first chance interview with that singular and beautiful phenomenon". Initial attempts to explain this remarkable phenomena using approaches based on the then well-known linear wave equation failed. While Boussinesq and Lord Rayleigh made considerable theoretical progress in the 1870s, it was not until 1895 (and some 50 years after Russell's intitial experimental report) with the treatment of the Korteweg–de Vries (KdV) nonlinear wave equation, that the solitary shallow water wave received its conclusive explanation [2]. The KdV equation can be written in the form:

$$2\frac{\partial \psi}{\partial t} + 3\psi\frac{\partial \psi}{\partial x} + \frac{1}{3}\frac{\partial^3 \psi}{\partial x^3} = 0\,, \tag{1.1}$$

where $\psi(x, t)$ denotes the height of the fluid surface at position x along a straight narrow channel, and t denotes time. Take note of the presence of the quadratic nonlinearity in the second term. This equation can be derived from basic laws of fluid dynamics. One finds a family of single-humped solutions to this equation that move *without dispersing*. Furthermore, one finds that *the speed of the wave is dependent on its amplitude*. Both of these properties cannot be captured by using a standard linear wave equation. For instance, an initial shape in the solitary wave form is made up of several wavelengths and must, therefore, quickly disperse and devolve into its constituent modes.

Nonlinear Wave Methods for Charge Transport. Luis L. Bonilla and Stephen W. Teitsworth

ISBN: 978-3-527-40695-1

Another more recent example, and one with which many are familiar, is the dynamics of traffic flow, including the causes of traffic jams. One can construct a simple and remarkably descriptive nonlinear wave model with the following two simple ingredients: 1) the law of "mass" conservation, in this case car conservation, which can be written for one lane of traffic flow in the following one- dimensional form as

$$\frac{\partial \rho}{\partial t} + \frac{\partial j}{\partial x} = 0 , \tag{1.2}$$

where $\rho(x, t)$ denotes the number of cars per unit length at space-time point (x, t), and $j(x, t)$ denotes the rate at which the cars pass the point x per unit time at time t, in other words, the vehicle "current"; and 2) a constitutive relationship between j and ρ, which takes the form $j = \int \nu(\rho)\, d\rho$ where $\nu(\rho)$, denotes the *density-dependent* velocity of the vehicles, which may be obtained by empirical observation [3]. Clearly, if the density is very high, everyone drives more slowly and at some point of high enough density, the average velocity tends to zero. Incorporating this density-dependent velocity into the vehicle conservation equation gives rise to a nonlinear wave equation of the following form:

$$\frac{\partial \rho}{\partial t} + \nu(\rho)\frac{\partial \rho}{\partial x} = 0 . \tag{1.3}$$

Among other things, this equation reveals *shock wave* solutions moving at a precise speed and in which the density jumps discontinuously between relatively high and low values for a wide class of functional forms for $\nu(\rho)$. If a diffusive term is added in the right hand side of Eq. (1.3), the discontinuity in the shock wave is somewhat smoothed: the wave profile has a steep gradient in a narrow region whose width is determined by the diffusion coefficient. This system illustrates yet another feature of nonlinear wave systems that are not possible to find in solutions of a linear wave equation, namely, shock-like boundary layers between well-defined values with a shape that exhibits no dispersion as the shock moves through its medium.

A prominent example from biology concerns the propagation of nerve impulses in humans and animals. The effective transmission line in this case is the axon and the wave is of an electrochemical nature involving the lateral diffusion of ions across the membrane boundary of the axon. This type of nonlinear wave can be described by the following set of equations known as the FitzHugh–Nagumo model [4, 5]:

$$\frac{dU}{dt} = f(U) - W + I + D\frac{\partial^2 U}{\partial x^2} , \tag{1.4}$$

$$\frac{dW}{dt} = U - BW , \tag{1.5}$$

where $U(x, t)$ corresponds to the lateral electric potential across the axon membrane, $f(U)$ is a nonlinear function which describes regenerative self-excitation in the system, $W(x, t)$ corresponds to an outward-flowing ion current, I denotes a stimulus current, D is an effective diffusion constant for the membrane potential, and B is a constant. The solutions of this system exhibit several features associated with nerve pulses, for example, a traveling solitary wave structure in which the waves have a particular amplitude, time-dependence of the variables that typically possess regions of rapid and then gradual change which suggests the role of *multiple time scales*, and refractory behavior in which there is a certain period of time during which it is not possible to excite the medium. It should also be noted that none of these features is captured by a linear wave equation approach.

Chemical reactions provide yet another example. There is a class of autocatalytic chemical reactions (e. g., Belousov–Zhabotinsky reaction) which show temporal oscillatory behavior in the constituent molecular concentrations when they are well-stirred in order to keep them spatially homogeneous [6]. When they are not stirred, these same systems can show propagating spiral wave patterns of great complexity. This behavior can be modeled by a reaction-diffusion system:

$$\frac{\partial n_i}{\partial t} = \nabla \cdot (D \nabla n_i) + R_i(\{n_j\}) , \tag{1.6}$$

where the $n_i(r, t)$ (with $i = 1, 2, \ldots, N$) denote the space and time-dependent concentrations of different molecular species in the reacting mixture, and R_i is the reaction rate for the i-th species. Typically, R_i is a nonlinear function of the concentrations because the reactions involve two or more molecules per reaction. For example, in a simple binary reaction of the form $\mathrm{A} + \mathrm{B} \to \mathrm{C}$, the corresponding reaction rate R_C for species C includes a term that is proportional to $n_\mathrm{A} n_\mathrm{B}$ according to the mass action law. For specific reactions, this system exhibits nonlinear spiral waves in two spatial dimensions. If one neglects the reaction term, one has the usual linear diffusion equation and thus, the underlying parabolic partial differential equation is very different than the wave equation. The reaction terms may give rise to *bistability* of two different time-independent solutions, another common feature that distinguishes nonlinear from linear wave phenomena.

In this book, we will focus on the development and application of nonlinear wave methods for problems involving the transport of electric charge in solid materials. The realization that nonlinear charge waves occur in condensed matter is relatively young and is largely traceable back to the discovery of the Gunn effect in GaAs in the early 1960s [7]. In a seminal paper, Gunn found that when doped GaAs samples were subjected to sufficiently high voltage, they emitted strong microwave radiation. He subsequently verified in a series of capacitive probe measurements that the microwave radiation was due to periodically cycling domains of high electric field propagating along the samples in the direction of current flow. In the next section, we review the basics of classical charge transport in solid media and point out, in simple terms, where it is that the crucial nonlinearities may arise.

1.2
Nonlinear Waves and Electronic Transport in Materials

Here, we review the essentials of electronic transport in solids at a level similar to that of a typical undergrad course in electricity and magnetism. We also introduce the drift-diffusion form of electrical current. We start by recalling that conservation of charge is expressed by the following well-known continuity equation

$$\nabla \cdot \boldsymbol{j} + \frac{\partial \rho}{\partial t} = 0 , \tag{1.7}$$

where $\boldsymbol{j}$ now denotes the three-dimensional *electrical current density* (SI units are $\mathrm{A/m^2}$) and $\rho(\boldsymbol{r}, t)$ denotes the three-dimensional charge density. We also have the basic electrostatic relationship between electric field and charge density, one of the Maxwell equations expressed as

$$\nabla \cdot \boldsymbol{E} = \frac{\rho(\boldsymbol{r}, t)}{\kappa \epsilon_0} . \tag{1.8}$$

Finally, we have a constitutive relation between the electric field and current density that flows in the material. The simplest and most common assumption is the local version of Ohm's law,

$$\boldsymbol{j} = \sigma \boldsymbol{E} , \tag{1.9}$$

that is, a linear proportionality between the applied electric field and the current that flows, where σ denotes the electrical conductivity. Even though this is an empirical relationship, it works quite well for many materials and applications. It can be derived in a number of ways, from the elementary Drude argument to sophisticated quantum mechanical calculations that take full account of energy band structure and scattering mechanisms [8, 9].

Let us briefly recall the Drude form of the dc conductivity,

$$\sigma = \frac{n e^2 \tau}{m^*} , \tag{1.10}$$

where n denotes the volume density of mobile charge carriers (e. g., electrons), e is the electron charge, m^* is the effective mass of the charge carrier in the material of interest [8], and τ denotes the scattering time. In both semiconductors and metals, the mobile charge carriers (electrons and/or holes) are compensated by a background of fixed charge (associated with ionic constituent atoms of the underlying crystal lattice) of opposite sign. If we denote the density of these fixed charges by N, then we have for total charge density the following relationship:

$$\rho(\boldsymbol{r}, t) = e\,(n(\boldsymbol{r}, t) - N) , \tag{1.11}$$

where we have assumed that n can vary in space, but that N is constant.

If we now combine Poisson's Law and Ohm's Law, Eqs. (1.8) and (1.9), respectively, we have:

$$\nabla \cdot \boldsymbol{j} = \sigma \nabla \cdot \boldsymbol{E} = \frac{\sigma}{\kappa \epsilon_0} \rho(\boldsymbol{r}) . \tag{1.12}$$

This equation can be brought into a form depending only on $\rho(\boldsymbol{r}, t)$ by using the charge continuity equation (1.7)

$$\frac{\partial \rho}{\partial t} = -\frac{\sigma}{\kappa \epsilon_0} \rho(\boldsymbol{r}, t) \equiv -\frac{\rho(\boldsymbol{r}, t)}{\tau_d} , \tag{1.13}$$

where we have defined the *dielectric relaxation time*: $\tau_d \equiv \kappa\epsilon_0/\sigma$. Equation (1.13) is often referred to as the dielectric relaxation equation. Imagine an initial condition in which the charge density at a point $\boldsymbol{r}$ has a value $\rho_0 \neq 0$ which might occur due to a thermal fluctuation or a rapid voltage pulse applied to the system. Then at subsequent times, the charge density decays exponentially to zero as mobile charges from other parts of the material flow into the region at point $\boldsymbol{r}$ to neutralize the overall charge density

$$\rho(\boldsymbol{r}, t) = \rho_0 \exp\left[-\frac{t}{\tau_d}\right] . \tag{1.14}$$

This result leads to the immediate conclusion that we do *not* expect to find nonlinear charge waves in transport systems that are well-described by Ohm's Law; the state in which $\rho \to 0$ in the interior of a homogeneous material is always stable. However, Eq. (1.14) does yield an additional insight, namely, that instability might occur if the conductivity σ were somehow negative. Referring back to the elementary dc conductivity expression, Eq. (1.10), we can see that this is unlikely since the constituent's parameters are all positive. However, if we allow for a nonlinear constitutive relationship between current density and applied field, we can recognize the possibility to achieve a *differential* conductivity that is negative for a certain range of field values. This book is largely centered on such systems in nature.

Returning to Ohm's Law, cf. Eq. (1.9), we note that it can be written in the form $j = env$ where v denotes the drift velocity, that is, the average velocity of all the carriers in an applied electric field E. Regarding the Ohm's Law case, it is stated that $v = e\tau E/m^*$, that is, a linear relationship between drift velocity and applied field. We can generalize this to consider a nonlinear dependence of drift velocity on field which we denote by $v(E)$. Additionally, a study of statistical transport theories shows that, in general, the current will take the form of a gradient expansion in the density of the mobile charges [10, 11]. If we restrict ourselves to only one spatial dimension (call it the *x*-direction), j_x can be expanded in a series of terms proportional to $\partial^i n/\partial x^i$. For charge transport problems (even nonlinear ones), it generally suffices to retain the first two terms of such an expansion (i. e., $i = 0$ and 1) which gives the drift-diffusion expression for current density:

$$j_x = e\left[nv(E) - D(E)\frac{\partial n}{\partial x}\right] , \tag{1.15}$$

where $D(E)$ denotes the (possibly field-dependent) diffusion constant for the charge carriers. Let us now repeat the above steps that previously lead to the dielectric relaxation equation, though now using the drift-diffusion form of the current. Firstly, we note that we use the Poisson equation to write the charge continuity equation in the form

$$\frac{\partial j_x}{\partial x} + \kappa\epsilon_0\frac{\partial^2 E}{\partial x\partial t} = 0 , \tag{1.16}$$

which can be immediately integrated with respect to x to yield

$$j_x + \kappa\epsilon_0\frac{\partial E}{\partial t} = J(t) . \tag{1.17}$$

In this form of Ampère's law, the total current density $J(t)$ is the sum of the electrical current density j_x and the displacement current $\kappa\epsilon_0 \partial E/\partial t$. Multiplied by the cross section, it denotes the boundary current that flows into and out of the structure. At this point, we replace the current density j_x by the drift-diffusion expression to write:

$$e n v(E) - e D(E)\frac{\partial n}{\partial x} + \kappa\epsilon_0 \frac{\partial E}{\partial t} = J(t)\,. \tag{1.18}$$

However, this equation involves both variables n and E, and it is desirable to write a dynamical equation entirely in terms of one of these. We can eliminate n in favor of E using the Poisson Equation, cf. Eq. (1.8), and also recalling the relationship between n and ρ, cf. Eq. (1.11):

$$\frac{\partial E}{\partial t} + v(E)\frac{\partial E}{\partial x} - D(E)\frac{\partial^2 E}{\partial x^2} = \frac{J(t) - e N v(E)}{\kappa\epsilon_0}\,. \tag{1.19}$$

We can immediately see a qualitative analogy with the traffic flow equation, cf. Eq. (1.3). The field-dependent drift velocity of carriers is analogous to the density-dependent vehicle velocity. For appropriately shaped $v(E)$, one may also expect the formation of shock waves. Additionally, we have diffusion and source terms to consider that are similar in form to those in the first equation from the FitzHugh–Nagumo model, cf. Eq. (1.4). Therefore, we may also expect to see oscillatory and excitatory behavior that is reminiscent of nerve propagation. As we shall see, Eq. (1.19) forms the basis of an effective model of the Gunn effect studied in Chapter 6.

Another way in which nonlinearity may enter into the charge transport picture in a very basic and widespread manner is through the inclusion of trapping dynamics, whereby the mobile charge carriers may become trapped and liberated from trapping sites. This will be discussed in detail in Chapter 7, and thus, we mention only the simplest aspect of this behavior here. Let $n_t(\boldsymbol{r}, t)$ denote the density of charge carriers that are trapped on trapping sites. If the total density of traps is N, then the local charge density should be written as $\rho(\boldsymbol{r}, t) = e\,(n(\boldsymbol{r}, t) + n_t(\boldsymbol{r}, t) - N)$. Let us consider the form of local rate equation for the n_t. Clearly, n_t will be increased when a free charge carrier is trapped on an available site, a process known as *capture*. This process must be proportional to the density of free carriers and the density of available sites, that is, proportional to the product $n(N - n_t)$. Similarly, we expect n_t to decrease due to excitation of trapped charge carriers, for example, by external illumination of an appropriate frequency or by the absorption of energy from other excitations in the material, for example, phonons [8]. This process is known as *generation* and is only proportional to the density of trapped charges n_t. Another process that decreases n_t occurs when a sufficiently energetic free carrier collides with a trap carrier and imparts sufficient energy to liberate it. This process is known as *impact ionization* and is proportional to both n and n_t. We can put these ingredients together in order to write a rate equation of the form:

$$\frac{\partial n_t}{\partial t} = -G n_t + R n(N - n_t) - K n n_t\,, \tag{1.20}$$

where the quantities G, R, and K are respectively called the coefficients of generation, capture, and impact ionization. We can immediately see that this rate equation is nonlinear in a qualitatively similar form as is found in the chemical reactions discussed in the previous section, cf. Eq. (1.6). It is also interesting to note that an additional feature not found in the chemical reaction literature is that the kinetic coefficients may have a strong dependence on electric field. As we shall see in Chapter 7, on trap-controlled space-charge instabilities, this feature plays a central role.

1.3 Structural Outline of the Book

We conclude this chapter by providing a road map of the presentation of material in this book. Chapters 2–4 are devoted to introducing key mathematical techniques that are especially useful for analyzing nonlinear wave phenomena in electronic transport systems. We begin in Chapter 2 by reviewing basic concepts and facts from nonlinear dynamics, including a summary of common bifurcations that are often encountered. Much of this material will be familiar to those who have previously studied nonlinear dynamics and, in this case, a fast reading is possible with the understanding that it can be referred to as necessary. In the second half of Chapter 2, we analyze bifurcations using a multiple scales approach known as the Chapman–Enskog method. This method has certain advantages in bifurcation analysis over the well-known normal form approach, especially for models of electronic transport. In Chapter 3, we introduce basic concepts of nonlinear waves in spatially continuous excitable media, illustrated using the FitzHugh–Nagumo and related models. Among the concepts presented are co-moving frame analysis, nonlinear wave speed determination methods, and stability analysis. In Chapter 4, we look at the application of these nonlinear wave methods to spatially discrete systems, especially periodic arrays where the new phenomena of wave pinning is expected to emerge.

Chapter 5 reviews the quantum mechanical underpinnings of the drift-diffusion transport models that are the focus of this book. After a discussion of fundamental quantum transport, we apply these methods to derive drift-diffusion currents for *strongly-coupled* semiconductor superlattices. Chapter 6 covers the theory of the Gunn effect in GaAs in which space charge instability is due to the peculiar nonlinear dependence of the drift velocity $v(E)$ on electric field. This chapter also begins with a self-contained review of the essential semiconductor physics underlying the Gunn effect. The concept of a greatly simplified asymptotic model for long samples is also introduced. In Chapter 7, we turn to the case where the instability is a result of field-dependent dynamics relative to trap states, also called trap-controlled instabilities. The treatment is largely parallel to that of the Gunn effect, though there are some striking differences between these two phenomena. In Chapter 8, we turn to the nonlinear dynamics of weakly-coupled semiconductor superlattices which are found moving as well as static (equivalently, pinned) electric field domains. It is possible to understand many observed phenomena by applying singular perturbation methods to the limit $N_{SL} \to \infty$, where N_{SL} is the number of periods, or sequential layers of quantum wells, in the superlattice. Finally, in Chapter 9, we conclude with a brief and admittedly biased survey of some other systems in which the methods and tools developed here also provide useful insight.

References

1 Russell, J.S. (1845) Report on Waves, in Report of the 14th meeting of the British Association for the Advancement of Science, York, Sept. 1844, London 1845, pp. 311–390.

2 Korteweg, D.J. and de Vries, G. (1895) On the Change of Form of Long Waves Advancing in a Rectangular Canal, and on a New Type of Long Stationary Waves, *Philosophical Magazine*, **39**, 422–443.

3 Witham, G.B. (1974) *Linear and Nonlinear Waves*, John Wiley & Sons, Inc, New York.

4 Keener, J. and Sneyd, J. (1998) *Mathematical Physiology*, Springer, New York.

5 FitzHugh, R. (1961) Impulses and physiological states in theoretical models of nerve membrane. *Biophysical Journal*, **1**, 445–466.

6 Zhabotinsky, A.M. (1991) A history of chemical oscillations and waves. *Chaos*, **1**, 379–386.

7 Gunn, J.B. (1965) Instabilities of current and of potential distribution in GaAs and InP. *Proceedings of Symposium on Plasma Effects in Solids* (ed. J. Bok), Dunod, Paris, pp. 199–207.

8 Ashcroft, N.W. and Mermin, N.D. (1976) *Solid State Physics*, Brooks-Cole, New York.

9 Grahn, H.T. (1999) *Introduction to Semiconductor Physics*, World Scientific, Singapore.

10 Chapman, S. and Cowling, T.G. (1970) *The Mathematical Theory of Non-uniform Gases*, 3rd edn, Cambridge University Press, Cambridge.

11 Cercignani, C. (2000) *Rarefied Gas Dynamics: From Basic Concepts to Actual Calculations*, Cambridge, New York.

12 Cercignani, C., Illner, R. and Pulvirenti, M. (1994) *The Mathematical Theory of Dilute Gases*, Springer, New York.

2
Dynamical Systems, Bifurcations, and the Chapman–Enskog Method

2.1
Introduction

In this chapter, we review the basic concepts of dynamical systems theory. We commence with a brief discussion of what constitutes a dynamical system and some examples. Then, we review the simplest and most common types of bifurcations that can occur and also present normal forms for these. Such bifurcations are most commonly treated using various types of center manifold reductions. However, in the final section, we discuss bifurcations from the point-of-view of a multiple scales approach known as the Chapman–Enskog method, a less common approach. This method is shown to have significant advantages over other approaches in analyzing bifurcations for certain types of dynamical systems. While there are many excellent books and reviews on dynamical systems theory, we offer a brief review of the terminology used throughout this book. Those readers who are familiar with dynamical systems theory may wish to move directly to Section 2.3.

2.2
Review of Dynamical Systems Concepts

Briefly, a dynamical system consists of a set of possible states and a rule for time evolution which determines uniquely present states in terms of past states. If the rule is applied at discrete times, we have a discrete-time dynamical system. In this book, we shall consider continuous-time dynamical systems, whose general form is a set of differential equations. For the systems we are treating here, these equations are *nonlinear* and, additionally, they typically depend on some adjustable parameter or set of such parameters. If the space of possible states is designated as $\mathbb{U}$, the general form of the dynamical system is

$$\frac{du}{dt} = f(u; \alpha) , \tag{2.1}$$

where $u \in \mathbb{U}$, and $\alpha \in \mathbb{R}^m$. Here, we have assumed that the parameters can be expressed as a set of m real numbers. In this book, the state space $\mathbb{U}$ will generally

Nonlinear Wave Methods for Charge Transport. Luis L. Bonilla and Stephen W. Teitsworth

ISBN: 978-3-527-40695-1

correspond to either $\mathbb{R}^n$ or to a suitable space of functions of the variables position x and time t defined on either infinite, semi-infinite, or bounded intervals in x and t.

Examples of nonlinear dynamical systems abound in all fields of science. A relatively simple and classic example of a nonlinear dynamical system is provided by the simple pendulum in which the state space is two-dimensional. In its most basic form, we consider an ideal frictionless pendulum. Depending upon the initial conditions, or equivalently on the total system energy E, the motion may consist of a perfect periodic swinging of the pendulum mass for lower values of E, and the rotational motion of the pendulum in either a clockwise or counter-clockwise direction for larger values of E. For a specific value of E between these two extremes, one has motion between successive local (unstable) maxima that takes infinite time. This is an example of a dynamical system in which energy is a constant of the motion.

One may also consider a closely related but more complicated system in which energy is *not* conserved and a sinusoidal driving torque is added to the mix, the so-called damped, driven pendulum. This is an example of a dissipative dynamical system, the presence of which implies that after initial transients run their course, the motion settles down to certain attracting forms of motion. For the damped, driven pendulum, attracting states typically consist of small-amplitude limit cycles for small amplitude sinusoidal drive, while for larger amplitude drive, the response may be a type of chaotic behavior in which, for example, the time response does not settle down to a purely periodic or purely rotational motion. The models of nonlinear electronic transport discussed in this book are certainly of this latter type. With the exception of superconductors, the transport of charge through matter always involves dissipation of driving energy as a result of collisions of the charge carriers with one another as well as the host material. For this reason, we focus our discussion on dissipative dynamical systems and commence with a discussion of basic types of attracting states that can be expected.

2.2.1
Attractors

In this section, we classify some of the most common attracting states. These are, respectively, fixed points, limit cycles, and chaotic attractors. As mentioned above, attractors are the states or sets of states to which nearby trajectories flow as $t \to \infty$. This somewhat qualitative definition serves our purposes well and we leave to in-depth texts on dynamical systems, such as [1, 2], to present the underlying mathematical concepts more rigorously by examining, for example, the properties of stable invariant sets in the underlying systems.

2.2.1.1 Steady States – Fixed Points

The simplest and most prevalent attractors are certainly the *fixed points* of the dynamical system, defined as the solutions of

$$f(u;\alpha) = 0\,. \tag{2.2}$$

Because the function $f(u;\alpha)$ is nonlinear in the argument u, there may exist multiple real-valued solutions of this equation for a single value of α, and we can denote these solutions as $u^{(j)}(\alpha)$, where the index j counts the number of distinct fixed points. This notation also serves as an indication that the fixed points are generally smooth functions of the parameter α. As a simple example, consider the following one-dimensional ordinary differential equation:

$$\frac{du}{dt} = u(2-u)(u-1) - \alpha_1 \,, \tag{2.3}$$

which is closely related to the FitzHugh–Nagumo model that is studied extensively in Chapter 3. First, consider the case where $\alpha_1 = 0$. Then, by inspection, we see immediately that there are three fixed points, that is, $u^{(1)} = 0$, $u^{(2)} = 1$, and $u^{(3)} = 2$. As α_1 increases slightly from zero, it is easy to show that the fixed points evolve smoothly and approximately as $u^{(1)}(\alpha_1) = -\alpha_1/2$, $u^{(2)}(\alpha_1) = 1 + \alpha_1/2$, and $u^{(3)}(\alpha_1) = 2 - \alpha_1/2$. Graphically, these fixed points are simply the intersection points of the f curve with the horizontal line at α_1, as shown in Figure 2.1. Notice that for $\alpha_1 > 1$, there is only one fixed point. Similarly for $\alpha_1 < -1$.

Of course, not all fixed points are attractors. In order to be attracting, the flow induced by the dynamical system must be such that all initial conditions that are near to the fixed point must flow towards it. More precisely, we have the following condition. Let u^* denote a fixed point. Then, u^* is an attractor (equivalently, u^* is asymptotically stable) provided that there exists a neighborhood $B \subseteq U$ of u^* such that for all $u(t=0) \in B$, we have $u(t \to \infty) = u^*$. A less stringent condition is stability of u^*. u^* is stable if there exists a neighborhood $B \subseteq U$ of u^* such that for all $u(t=0) \in B$, we have $u(t) \in B$ for all $t > 0$. Fixed points which are *not* stable are termed *unstable* points.

By linearizing the dynamical system about a fixed point, one can easily determine its stability type and, therefore, whether or not it is an attractor. Linearizing Eq. (2.1)

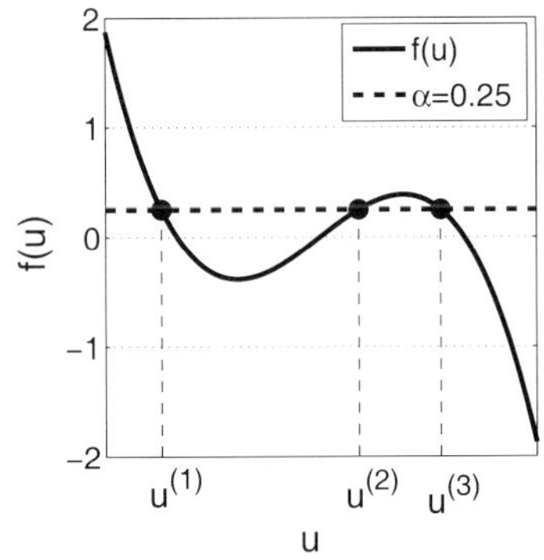

Fig. 2.1 Graphical determination of fixed points for the one-dimensional system, Eq. (2.3). The fixed points coincide with the intersection points of the curve $\nu = u(2-u)(u-1)$ with the horizontal line $\nu = \alpha$.

about a fixed point u^* yields

$$\dot{\delta u} = f(u^* + \delta u; \alpha) \simeq D f|_{u^*} \cdot \delta u \,, \tag{2.4}$$

where the overdot denotes time differentiation, $D f|_{u^*}$ denotes the Jacobian matrix evaluated at u^*, and δu is the assumed small deviation of the flow $u(t)$ from the fixed point u^*. Equation (2.4) represents a standard linear system of ordinary differential equations with constant coefficients. Then, *according to the linear system* (2.4), it is clear that the fixed point u^* is an attractor provided that all eigenvalues of $D f|_u^*$ have a negative real part. Letting n denote the dimension of the state space $\mathbb{U}$, there are n such eigenvalues (some of which may be degenerate), and we denote these as $\lambda_i(\alpha)$ where $i = 1, 2, \ldots, n$. Then, we have the following condition for the stability of a fixed point.

Fixed point stability criterion The fixed point u^* is asymptotically stable provided that the eigenvalues of the Jacobian matrix $D f|_{u^*}$ satisfy $\mathrm{Re}[\lambda_i(\alpha)] < 0$ for all i. On the other hand, the fixed point is *unstable* when at least one of the eigenvalues has a positive real part.

Returning to the example of Eq. (2.3) with $\alpha_1 = 0$, we see that the fixed points at $u^{(1)} = 0$ and $u^{(3)} = 2$ are asymptotically stable since $f'(0) = -2 < 0$ and $f'(2) = -1 < 0$. Furthermore, the fixed point $u^{(2)} = 1$ is unstable since $f'(1) = +1 > 0$. It is interesting to note that the presence of more than one stable fixed point in this example provides a simple illustration of *bistability*. Namely, depending on the initial conditions, the system evolves either to $u^{(1)}$ or $u^{(3)}$. The set of initial conditions in the state space $\mathbb{U}$ that lead to a particular attractor is referred to as the *basin of attraction* for that attractor. In the one-dimensional example above, the basin of attraction for $u^{(1)}$ (respectively, $u^{(3)}$) is the interval $[-\infty, u^{(2)})$ (respectively, $(u^{(2)}, +\infty]$). In other words, the two basins are separated by the unstable *fixed* point $u^{(2)}$, which is said to comprise the *basin boundary*.

Finally, we note that the situation in which one or more of the eigenvalues has a *zero* real part and all others have a negative real part, generally indicates that the system is at a bifurcation point with respect to variation in the control parameter α. Below, we will treat some of the classic bifurcations associated with fixed points. But, first, let us look at two other types of attracting states.

2.2.1.2 Limit Cycles

In many nonlinear dynamical systems, one finds an attracting state which is oscillatory in nature. This means that there exists a minimum period $T > 0$ such that $u_{\mathrm{lc}}(t) = u_{\mathrm{lc}}(t + T)$ for arbitrary t. However, $u_{\mathrm{lc}}(t) \neq u_{\mathrm{lc}}(t + \tau)$ for all $T > \tau > 0$. For the autonomous equations (2.1), f does not depend explicitly on t and therefore $u_k(t + c)$ with constant c is also a periodic solution representing the same orbit as $u_k(t)$. If all nearby initial conditions evolve towards this orbit as $t \to \infty$, the limit cycle is a stable attractor. A simple example of a *stable* limit cycle is provided by the following two-dimensional dynamical system:

$$\dot{x} = x - y - x(x^2 + y^2) \,, \tag{2.5}$$

$$\dot{y} = x + y - y(x^2 + y^2) \,. \tag{2.6}$$

In fact, this limit cycle can be determined analytically by writing the dynamical system in polar coordinates, that is, $r = (x^2 + y^2)^{1/2}$ and $\theta = \arctan(y/x)$, as follows:

$$\dot{r} = r(1 - r^2) , \tag{2.7}$$

$$\dot{\theta} = 1 . \tag{2.8}$$

Clearly, there is a limit cycle when $\dot{r} = 0$ at $r = 1$, and it is stable since $r(1 - r^2)$ is positive for r smaller than 1, and negative for r greater than 1. A plot of this limit cycle and typical nearby trajectories is shown in Figure 2.2. This figure also shows the time trace of $x(t)$ associated with this limit cycle.

A simple example of an *unstable* limit cycle is obtained by flipping the sign of the r-equation as follows:

$$\dot{r} = r(r^2 - 1) , \tag{2.9}$$

$$\dot{\theta} = 1 . \tag{2.10}$$

Now, one can see that $\dot{r}$ is negative for r just less than unity and positive for r just greater than unity. Nearby trajectories are driven away from the unstable limit cycle at $r = 1$, either to the stable fixed point at $r = 0$ or to the attracting fixed point at $r = \infty$.

In higher dimensional systems (i. e., $n > 2$), one has the analogue of a Jacobian condition to determine the stability properties of limit cycles. Briefly, one linearizes the system about a limit cycle solution, that is, $u(t) = u_{\text{lc}}(t) + \delta u(t)$, and substitutes this form into the dynamical system. By retaining only linear terms, one arrives at the following dynamical equation for $\delta u(t)$:

$$\dot{\delta u} = f(u_{\text{lc}}(t) + \delta u; \alpha) \simeq Df|_{u_{\text{lc}}(t)} \cdot \delta u \equiv A(t) \cdot \delta u , \tag{2.11}$$

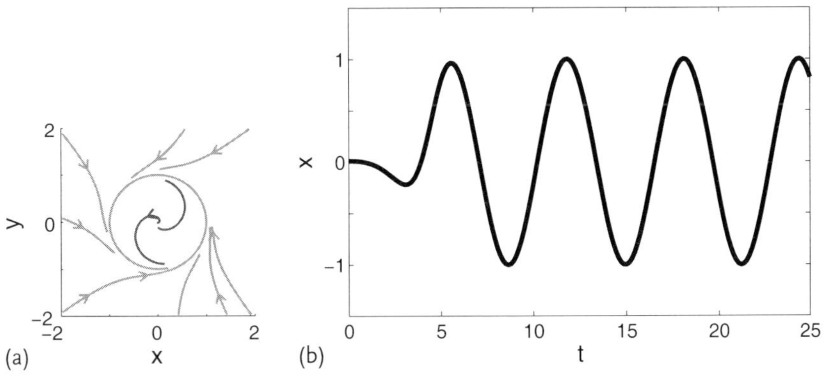

Fig. 2.2 (a) Phase plane portrait of limit cycle for Eqs. (2.5) and (2.6), and nearby trajectories are also shown. (b) $x(t)$ associated with the limit cycle, including an initial transient.

where we have defined the matrix $A(t)$ which has the property that $A(t) = A(t+T)$. The solution is $\delta u(t) = M(t) \cdot \delta u(0)$ where the fundamental matrix $M(t)$ satisfies $\dot{M} = A(t) \cdot M$ with initial condition $M(0) = I_n$, where I_n denotes the identity matrix in $\mathbb{R}^n$. Since $A(t) = A(t+T)$, $M(t+T)$ is also a fundamental matrix and therefore, an invertible matrix exists with constant entries such that $M(t+T) = M(t)C$, and thus $C = M(T)$. The matrix $M(T)$ after one period is known as the *monodromy* matrix, and we have $\delta u(T) = M(T) \cdot \delta u(0)$ and $\delta u(nT) = M(T)^n \cdot \delta u(0)$. The stability of the limit cycle is determined by the property that $\delta u(nT)$ tends to zero as $n \to \infty$, which leads to the criterion that follows [1].

Limit cycle stability criterion Let the eigenvalues of the monodromy matrix $M(T)$, be denoted by $1, \mu_1, \mu_2, \ldots, \mu_{n-1}$. Then, the limit cycle is linearly (locally) stable provided that $|\mu_i| < 1$ for all $i = 1, 2, \ldots, n-1$. If any of the $|\mu_i|$ is greater than unity, then the limit cycle is unstable.

The eigenvalues μ_j are called *Floquet multipliers*. Since the monodromy matrix is invertible, we can write it as $M(T) = e^{BT}$, where B is a matrix with possibly complex coefficients. Its eigenvalues α_j are called *Floquet exponents* and are related to the Floquet multipliers by $\mu_j = e^{\alpha_j T}$, $j = 1, \ldots, n-1$ and $\alpha_0 = 0$. The previous stability theorem is equivalent to saying that the limit cycle is linearly stable if all the Floquet exponents with indices $j = 1, \ldots, n-1$ have negative real parts. If any one has positive real parts, the limit cycle is unstable.

It is immediate to check that the matrix $P(t) = M(t)e^{-Bt}$ has period T, and therefore, the fundamental matrix $M(t) = P(t)e^{Bt}$. Then, the solutions of Eq. (2.11) are linear combinations of products of $e^{\alpha t}$ times T-periodic functions of time. This is the *Floquet theorem*, a one-dimensional version of the Bloch theorem for a Schrödinger equation with a periodic potential.

For higher dimensional systems ($n \gg 2$), the computation of the monodromy matrix $M(T)$ in practice is likely to be numerically complex. For further details on this approach, the reader may consult, for example, [1]. Practically speaking, one can numerically identify a stable limit cycle by utilizing the property that all nearby trajectories flow towards the limit cycle. If one chooses an initial condition for the limit cycle in the neighborhood of the limit cycle, it will converge to the limit cycle as time progresses. Similarly, one could locate a completely unstable limit cycle, that is, one for which all of the eigenvalues $|\mu_i|$ are greater than unity. In that case, one runs time backwards in the numerical simulation. Of course, limit cycles of saddle type (in which some of the $|\mu_i|$ are greater and some are lesser than unity) cannot be simply located.

2.2.1.3 Chaotic Attractors

There are many excellent treatments of chaotic attractors currently available, see for example [2, 3]. For this reason, we give only a brief summary of the salient properties that define a chaotic attractor and that are useful to recall in later sections of this book. The reader who has no prior familiarity with concepts of chaotic dynamics may wish to consult one of the introductory texts.

Returning to our nonlinear dynamical system, Eq. (2.1), we can define a *flow mapping* ϕ_t which describes the action of f over finite times; thus, given an initial condition u_0 at starting time $t = 0$, we have

$$u(t) = \phi_t(u_0)\,, \tag{2.12}$$

where the function ϕ_t is continuous and differentiable in its arguments provided that f satisfies mild continuity conditions [4]. Then, small initial deviations δu_0 evolve according to a *tangent mapping* defined by

$$\delta u(t) \simeq T_t(\delta u_0; u_0) \equiv \partial\phi_t/\partial u_0|_{u_0} \cdot \delta u_0\,, \tag{2.13}$$

which is a linear operator on the small deviations δu_0, though generally a nonlinear function of u_0.

Since the dynamical systems of interest for electronic transport are dissipative and bounded, we expect that motion will approach an attracting set A where

$$A = \phi_{t\to\infty}(P)\,, \tag{2.14}$$

where P is an open subset of the state space $\mathbb{U}$. We have already discussed two types of attractors, namely, fixed points and limit cycles. A third type is the *chaotic* (or strange) attractor. In this case, the attracting flow is still bounded, but exhibits nonperiodic behavior and sensitive dependence on initial conditions. Three useful quantitative techniques for distinguishing and characterizing chaotic behavior are described in what follows.

Lyapunov exponents The tangent mapping T_t is a real linear operator and transforms a small initial n-sphere about u_0 in an evolving n-dimensional ellipsoid at subsequent points $u(t)$ along the flow. If the time-dependent lengths of the principal axes of the ellipsoid are denoted by $\delta u_i(t)$ for $i = 1, 2, \ldots, n$, then we define the Lyapunov exponents of the attractor A as

$$\Lambda_i \equiv \lim_{t\to\infty} \frac{1}{t} \ln \left\| \frac{\delta u_i(t)}{\delta u_i(0)} \right\|\,, \qquad i = 1, 2, \ldots, n\,. \tag{2.15}$$

Since the principal axes of the ellipsoid are eigenvectors of the transformation T_t, and the eigenvalues give the corresponding rates of stretching or contraction, we may also write for the *set* of Lyapunov exponents

$$\{\Lambda_i\} = \lim_{t\to\infty} \frac{1}{t} \ln[\mathrm{spec}(T_t)]\,, \tag{2.16}$$

where $\mathrm{spec}(T_t)$ denotes the set of all n eigenvalues of the transformation T_t. It is customary to rank order the Lyapunov exponents according to $\Lambda_1 > \Lambda_2 > \ldots > \Lambda_n$.

Let us note that the three types of attractors mentioned above are characterized by the following Lyapunov spectra:

1. fixed point: $\Lambda_i < 0$ for all i;
2. limit cycle: $\Lambda_1 = 0$ and $\Lambda_i < 0$ for all $i > 1$; and
3. chaotic attractor: $\Lambda_1 > 0$ and the remaining Λ_i may be positive, negative, or zero.

Since the dynamical system is assumed to be dissipative, the sum of all Lyapunov exponents is always negative, that is, $\Sigma_{i=1}^{n} \Lambda_i < 0$. We also note a higher dimensional analogue of limit cycle behavior is possible in which more than one exponent is zero with the remaining exponents negative. This can be thought of as a *limit torus* and is associated with *quasiperiodic* behavior.

Fractal dimension Another important property of chaotic attractors is that they possess *fractal structure*, that is, an intricate geometrical structure on all length scales in the state space $\mathbb{U}$. A useful way to quantify this structure is by computing the fractal dimension of the attracting set A. In contrast to the Lyapunov exponents above, this method does not make explicit use of the *dynamics* on the attractor, only its geometry. A myriad of different definitions are possible; we present here one of the simplest, namely, the Mandelbrot or *box* dimension [2, 5]. One way to compute the box dimension is as follows. one covers or tiles the state space with n-dimensional hypercubes of side ϵ. Then, count the number of such cubes that contain elements of A and call this $N_A(\epsilon)$. Then, the fractal or box dimension of the attractor A is defined as the following limit

$$d_A = \lim_{\epsilon \to 0} \left[\frac{\ln N_A(\epsilon)}{\ln \epsilon} \right] . \tag{2.17}$$

In practice, the box dimension may not be the easiest or most reliable to numerically compute [2, 3], and it is common to utilize other dimension definitions such as the correlation dimension [6] or the multifractal spectrum "$f(\alpha)$" [7].

Let us note that the three types of attractors mentioned above are characterized by the following fractal dimensions:

1. Fixed point: $d_A = 0$. this follows since only one cube is required to cover the fixed point, regardless of length ϵ. Therefore, $N_A(\epsilon) = 1 = \epsilon^0$ and the dimension is zero.
2. Limit cycle: $d_A = 1$ since the limit cycle is a simple closed curve in the state space. Thus, the number of cubes required to cover A scales linearly with cube length ϵ, implying $N_A(\epsilon) \propto \epsilon^1$ and the dimension is 1.
3. Chaotic attractor: d_A generally possesses a *non-integer* value. If the dimension of the state space is much greater than unity, that is, $n \gg 1$, and $d_A = O(1)$, and one refers to the chaotic behavior as *low-dimensional*. On the other hand, if one finds that $d_A \gg 1$, one calls the attractor *high-dimensional*.

Finally, let us note that a limit torus (as defined above) has an *integer* dimension p with $1 < p \leq n$.

Power spectrum Both of the methods described above are capable of providing relatively rigorous characterization of chaotic attractors. A typical drawback, especially for higher dimensional systems, $n \gg 1$, is that their numerical implementation can be very challenging. This caveat also extends to their application to experimental data.

The *power spectrum* provides a useful, if somewhat less rigorous, means of characterizing the attractors of a dynamical system. This is true of both numerical simulations as well as experimental measurement. The use of power spectra in connection with chaotic attractors is suggested by the fact that a chaotic signal has unpredictability that is reminiscent of stochastic processes, for example, noise in electronic transport systems. In the latter case, a delta-correlated random process, for example, shot noise in a diode, yields a power spectrum that is flat over several decades of frequency. Power spectra of chaotic, but nonetheless deterministic, processes can be expected to have broadband features in their power spectra as well.

We focus on one particular scalar component of the dynamical system trajectory $u(t)$ (cf. Eq. (2.1)) which, for notational simplicity, we denote as $x(t)$. First of all, we define a truncated signal, $x_\tau(t) \equiv x(t)[\theta(t+\tau) - \theta(t-\tau)]$ where $\theta(t)$ denotes the unit step function, that is, $\theta(t) = 0$ for $t < 0$ and $\theta(t) = 1$ for $t > 0$. Now, consider the Fourier transform of x_τ:

$$\widetilde{X}(\omega) = \int_{-\infty}^{+\infty} x_\tau(t) e^{i\omega t} dt . \tag{2.18}$$

Then the power spectrum of $x(t)$ is defined as

$$S_x(\omega) \equiv \lim_{\tau\to\infty} \frac{1}{2\tau} |\widetilde{X}(\omega)|^2 . \tag{2.19}$$

Let us note that the three types of attractors mentioned above are characterized by the qualitative features in their power spectra:

1. Fixed point: $S_x(\omega) \propto \delta(\omega)$, since the system tends to a time-independent value.
2. Limit cycle: $S_x(\omega) = \Sigma_{j=1}^{\infty} a_j \delta(\omega - 2\pi j/T)$ where T is the fundamental period of the limit cycle and the harmonic amplitudes a_j are positive and decreasing with index j.
3. Chaotic attractor: $S_x(\omega)$ possesses broadband features frequently punctuated by delta-function peaks [8].

Finally, let us note that a limit torus (as defined above) has multiple incommensurate frequencies. In contrast, the delta function peaks are not all multiples of a single frequency as for the simple limit cycle above, but rather sums and differences of two or more incommensurate frequencies [9].

2.2.2 Bifurcations – Basic Definitions and Types

Generally speaking, a *bifurcation* in the behavior of a dynamical system (cf. Eq. (2.1)) refers to a fundamental change in the nature of the attractor(s) as the control parameter α is varied through a specific value α^*, often referred to as a *bifurcation* or *critical* point. Many bifurcations of interest involve a small variation in α such that a fixed point changes its stability type. Such bifurcations are referred to as *local* bifurcations and they may be characterized by studying the eigenvalues of the Jacobian matrix of the linearized system, cf. Eq. (2.4). Additionally, different bifurcations are characterized by the minimal number of different components of control parameter vector α that are needed to completely described all of the nearby flow types that are involved. We begin our discussion of local bifurcation by examining some common local bifurcations that require the variation of only one component of control parameter which, for simplicity, we call α. Bifurcations that have this property are called *codimension-1* [1].

We may define a local bifurcation point denoted by $\alpha^* = 0$, such that the Jacobian of the linearized system possesses at least one eigenvalue with zero real part. Furthermore, we may assume that for $\alpha < 0$ the fixed point is stable (i. e., $\text{Re}[\lambda_i(\alpha)] < 0$ for all i), while for $\alpha > 0$, there is at least one eigenvalue with a real part greater than zero, indicating that the original fixed point has lost stability. Without loss of generality, let us assume that the eigenvalues are arranged according to the value of the real part so that $\text{Re}[\lambda_i(\alpha)] \geq \text{Re}[\lambda_{i+1}(\alpha)]$ for all i. Note that the equality is possible when there are: 1) complex conjugate pairs of eigenvalues, or 2) higher order degeneracy. The most common program for describing bifurcations in real dynamical systems is based on the *Center Manifold Theorem* [1], which implies that it is always possible through smooth, though *nonlinear*, coordinate transformations to bring a dynamical system into an appropriate normal form. Such normal forms can be viewed as paradigm dynamical systems for particular bifurcations that are in some sense, the simplest dynamical systems one can write that exhibit the sought-after bifurcation. For high-dimensional dynamical systems, the identification of the appropriate coordinate transformations can be a very complex task. However, another approach, which is discussed in the final section of this chapter, is based on multiple scales methods [10]. Such methods take advantage of the fact that in the neighborhood of the bifurcation point, there is always a small parameter, namely, the ratio of the small real part(s) of the eigenvalue(s) that cross zero to the real parts of the remaining eigenvalues. For example, in a two-dimensional system undergoing a saddle-node bifurcation (discussed below), one has $|\lambda_1|/|\lambda_2| \ll 1$. A potential advantage of methods based on multiple scales is that one is using the natural dynamics of the system to identify the correct normal form for the bifurcating system. We return to this point in Section 2.3.

However first, we discuss the essential features of bifurcations where precisely one eigenvalue crosses the imaginary axis in the λ-plane and, afterward, turn to the case in which a complex conjugate pair crosses the imaginary axis.

2.2.2.1 Saddle-Node Bifurcation (Local)

The *saddle-node* bifurcation (also called a *fold* or *tangent* bifurcation) corresponds to the collision of a stable and an unstable fixed point, and provides an important mechanism for the creation and annihilation of fixed points. This bifurcation is also characterized by the property that the spectrum of Jacobian about the stable fixed point has one real eigenvalue that approaches zero through negative values. Near the saddle-node bifurcation point, we can write $\lambda_1 \propto -\alpha^{1/2} + o(\alpha^{1/2})$ where the bifurcation point is taken to be $\alpha^* = 0$. By little "$o(\alpha^{1/2})$" we mean that the leading order correction term goes to zero "faster" than $\alpha^{1/2}$ as $\alpha \to 0$. In contrast to this, big "$O(\alpha)$" would have meant that the absolute value of the leading order correction is smaller than some unspecified positive constant times α as $\alpha \to 0$. The normal coordinate along which the saddle-node structure is visible is associated with the eigenvector $\hat{U}_1$ corresponding to eigenvalue λ_1. By transforming the full dynamical system into a $\hat{U}_i$ eigenvector basis such that the fixed point for $\alpha = 0$ in this basis is the zero vector, one can show that the $\hat{u}$ such that $u(t) \sim \hat{u}(t)\,\hat{U}_1 + \ldots$ obeys the following normal form equation:

$$\frac{d\hat{u}}{dt} = \hat{\alpha} + \hat{u}^2 + O(\hat{u}^3) \equiv \hat{f}(\hat{u}, \hat{\alpha})\,, \tag{2.20}$$

where the transformed parameter $\hat{\alpha}$ is a smooth, monotone increasing function of α such that $\hat{\alpha} = 0$ at the bifurcation point. In the neighborhood of the stable fixed points that coalesce at $\alpha = 0$, we can write $\hat{f}(\hat{u}, \hat{\alpha}) \sim \hat{\alpha} + \hat{u}^2$. Figure 2.3 shows the function $\hat{f}$ versus coordinate $\hat{u}$ as the control parameter crosses the bifurcation point. There are two key points to note as α crosses zero: 1) two fixed points exist on one side of this bifurcation while none exist on the other side, and 2) the eigenvalue associated with the Jacobian, that is, $|\hat{f}'(\hat{u} = 0, \hat{\alpha} = 0)|$, vanishes. Returning to a slightly more complicated example shown in Figure 2.1, one can see that there are saddle-node bifurcations in the system Eq. (2.3) for values $\alpha = 1$ and $\alpha = -1$. Interestingly, this system reveals another important common feature associated with saddle-node bifurcations. When the pair of fixed points annihilate, there is typically another distant fixed point (or other type of attractor) to which all initial conditions now flow. In the context of nonlinear models of electronic transport, this

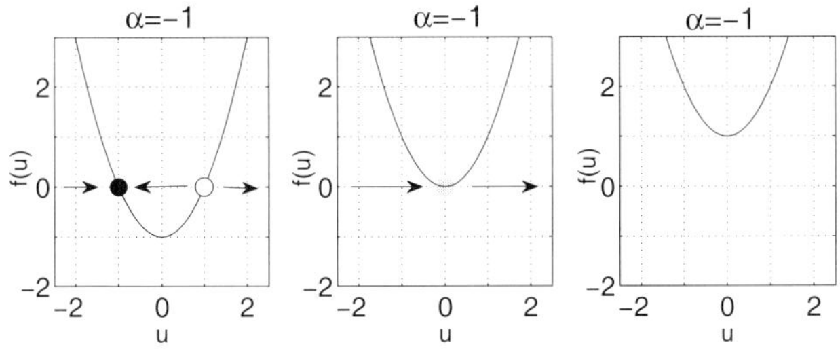

Fig. 2.3 Illustration of the saddle-node bifurcation for Eq. (2.17). On one side of this bifurcation, there are two fixed points (one stable and the other unstable), while on the other side, there are no fixed points in the neighborhood.

feature frequently implies the presence of *hysteresis* in observed dynamical behavior as control parameter is ramped up and down.

2.2.2.2 **Transcritical and Pitchfork Bifurcations (Local)**

As with the saddle-node bifurcation, transcritical and pitchfork bifurcations are also associated with a single eigenvalue of the Jacobian matrix of the linearized system crossing the imaginary axis in the λ-plane. In some sense, they are less generic than the saddle-node case and they typically reflect physical constraints placed on the system. In some systems, for example, fixed points cannot appear and disappear out of "thin air", and yet they can exchange stability type with one another. The normal form for *transcritical* bifurcation captures this behavior written as

$$\frac{d\hat{u}}{dt} = \hat{\alpha}\hat{u} - \hat{u}^2 + O(\hat{u}^3), \tag{2.21}$$

where $\hat{u}$ denotes the coordinate along which the bifurcation is revealed. In this case, the fixed point $u = 0$ (resp., $u = \alpha$) is stable for $\alpha < 0$ (resp. $\alpha > 0$) and unstable for $\alpha > 0$ (resp., $\alpha < 0$). Thus, the two fixed points change continuously with α and reveal an *exchange of stability*. This is conveniently illustrated in a bifurcation diagram in which the two evolving fixed points are plotted versus control parameter α, as shown in Figure 2.4a.

In systems that possess underlying symmetry, it is plausible that fixed points may only appear and disappear in symmetric pairs. This behavior gives rise to the following normal form for the *supercritical pitchfork* bifurcation written as

$$\frac{d\hat{u}}{dt} = \hat{\alpha}\hat{u} - \hat{u}^3 + O(\hat{u}^4), \tag{2.22}$$

where we again use $\hat{u}$ to denote the coordinate along which the bifurcation is revealed. In this case, the fixed point at $\hat{u} = 0$ is stable for $\hat{\alpha} < 0$ and unstable when $\hat{\alpha} > 0$. For $\hat{\alpha} > 0$, a symmetric pair of stable fixed points emerge from the origin with coordinates $\hat{u} = \pm\hat{\alpha}^{1/2}$. The corresponding bifurcation diagram is shown in Figure 2.4b. Interestingly, there is a closely related, yet distinct, normal form for the *subcritical* pitchfork bifurcation given by

$$\frac{d\hat{u}}{dt} = \hat{\alpha}\hat{u} + \hat{u}^3 + O(\hat{u}^4). \tag{2.23}$$

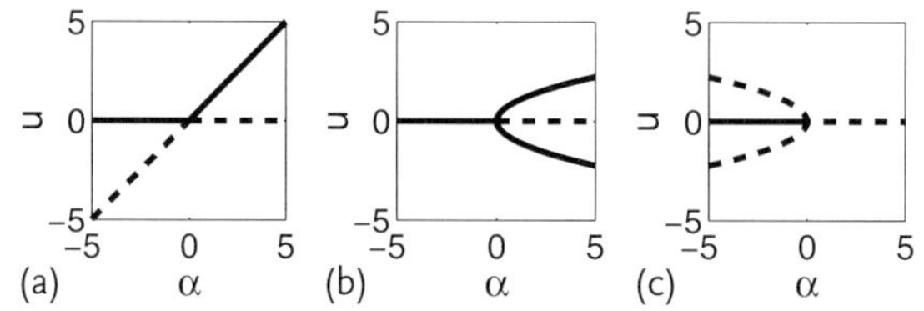

Fig. 2.4 (a) Transcritical bifurcation diagram for Eq. (2.18) revealing an exchange of stability. The stable (unstable) fixed point is depicted with a solid (dashed) line. (b) pitchfork bifurcation diagram for supercritical case, cf. Eq. (2.19), (c) Pitchfork bifurcation diagram for subcritical case, cf. Eq. (2.20).

In this case, the fixed point at $\hat{u} = 0$ is again stable for $\hat{\alpha} < 0$ and unstable when $\hat{\alpha} > 0$. However, we now have the appearance of a symmetric pair of *unstable* fixed points that emerge from the origin with coordinates $\hat{u} = \pm(-\hat{\alpha})^{1/2}$ for $\hat{\alpha} < 0$. The corresponding bifurcation diagram is shown in Figure 2.4c. Subcritical bifurcations are frequently associated with the appearance of hysteresis in the overall behavior of a dynamical system.

2.2.2.3 **Hopf Bifurcation (Local)**

Unlike the previous bifurcation scenarios where a stable fixed point loses its stability by colliding with an unstable fixed point (saddle-node case) or exchanges stability type with another fixed point, there is another local mechanism by which a fixed point changes stability type. In particular, as the control parameter is varied, the fixed point may collide with a branch of stable or unstable limit cycles. A key signature of this type of bifurcation is that the Jacobian of the dynamical system linearized about the fixed point has a *complex conjugate pair* of eigenvalues that cross the imaginary axis in the λ-plane, with all other eigenvalues having negative real parts. If the bifurcation point is denoted by $\alpha^* = 0$ and the fixed point is stable for $\alpha < 0$, then we have that $\lambda_1(\alpha) \sim \alpha\lambda_1{}'(0) + i\omega_0$, and $\lambda_2(\alpha) = \overline{\lambda_1}(\alpha)$, where $\mathrm{Re}[\lambda_1{}'(0)] > 0$ and the prime denotes differentiation with respect to α.

The normal forms for this bifurcation are inherently two-dimensional (unlike the previous examples) and one of them is given by

$$\dot{x} = \alpha x - y - x(x^2 + y^2), \tag{2.24}$$

$$\dot{y} = x + \alpha y - y(x^2 + y^2). \tag{2.25}$$

As in our earlier discussion of limit cycles, this system is most easily analyzed by transforming to polar coordinates, yielding

$$\dot{r} = r(\alpha - r^2), \tag{2.26}$$

$$\dot{\theta} = 1. \tag{2.27}$$

For $\alpha < 0$, there is only one fixed point of the r equation at $r = 0$, that is, a fixed point at the origin. Furthermore, a simple linear stability analysis shows that this fixed point is stable provided that $\alpha < 0$. For $\alpha > 0$, there is a second solution at $r_{\mathrm{lc}} = \sqrt{\alpha}$ which corresponds to a limit cycle of amplitude α. Note that for $\alpha > 0$, the fixed point at $r = 0$ is now unstable while the limit cycle is stable. The stability of the limit cycle is easily seen by linearizing Eq. (2.26) about the limit cycle solution. By denoting the small difference $\delta r \equiv r - \sqrt{\alpha}$, we have to leading order $\dot{\delta r} = -2\alpha\delta r$, and the limit cycle is stable. A phase portrait of this system below and above the bifurcation point is shown in Figure 2.5. This type of Hopf bifurcation is referred to as *supercritical*: it leads to the creation of a branch of *stable* limit cycles.

There is also a *subcritical* Hopf bifurcation which is manifested by the creation of a branch of *unstable* limit cycles. The normal form for the subcritical Hopf bifurca-

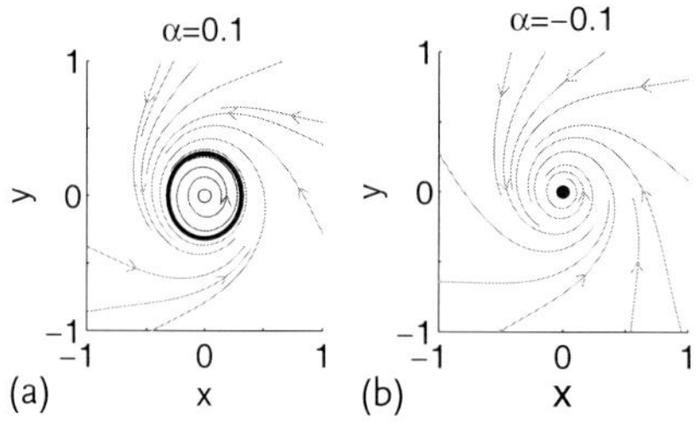

Fig. 2.5 Phase portraits corresponding to just above (a) and below (b) a supercritical Hopf bifurcation diagram for Eqs. (2.24) and (2.25).

tion is given by

$$\dot{x} = \alpha x - y + x(x^2 + y^2) , \tag{2.28}$$

$$\dot{y} = x + \alpha y + y(x^2 + y^2) . \tag{2.29}$$

Note that the only difference with the supercritical normal form (cf. Eqs. (2.24) and (2.25)) is a sign change on the nonlinear terms. As before, the subcritical form is profitably written in polar form as

$$\dot{r} = r(\alpha + r^2) , \tag{2.30}$$

$$\dot{\theta} = 1 . \tag{2.31}$$

It is immediately evident that limit cycles only exist for $\alpha < 0$. Furthermore, linear stability analysis shows that these limit cycles are unstable. The fixed point at $r = 0$ is stable for $\alpha < 0$ and unstable for $\alpha > 0$, similar to the supercritical case. A phase portrait of this system below and above the bifurcation point is shown in Figure 2.6. It should also be noted that the subcritical Hopf bifurcation is often associated with hysteretic behavior since the branch of unstable limit cycles eventually collides with some other stable entity (e. g., a branch of large amplitude limit cycles). An example of this behavior is discussed for a simple nonlinear electrical circuit model in [11].

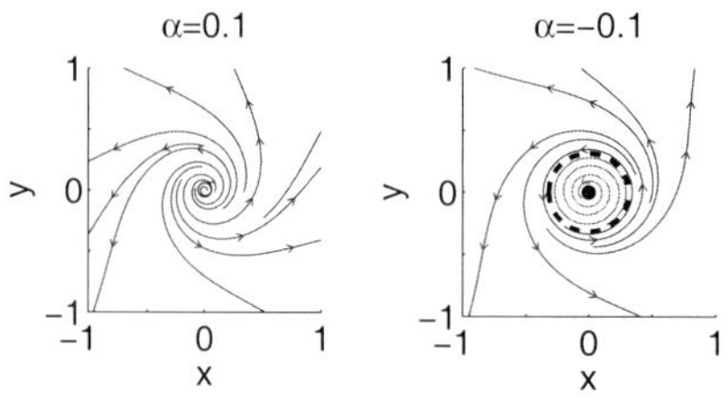

Fig. 2.6 Phase portraits corresponding to just above (a) and below (b) a subcritical Hopf bifurcation diagram for Eq. (2.30).

It is relatively straightforward to identify the presence of a Hopf bifurcation in arbitrary dynamical systems. On the other hand, it can be computationally complex to determine whether the bifurcation is sub- or supercritical. As an example, consider a general two-dimensional dynamical system undergoing a Hopf bifurcation at $\alpha = 0$. Then, at the bifurcation point, $\alpha = 0$, the system can always be written in the form

$$\dot{x} = -\omega_0 y + f(x, y) , \tag{2.32}$$

$$\dot{y} = \omega_0 x + g(x, y) . \tag{2.33}$$

Then, the following condition may be derived [4] to determine the nature of the bifurcation. Define a such that

$$a = \frac{1}{16}\left[f_{xxx} + f_{xyy} + g_{xxy} + g_{yyy}\right] + \frac{1}{16\omega_0}\left[f_{xy}(f_{xx} + f_{yy})\right.$$
$$\left. - g_{xy}(g_{xx} + g_{yy}) - f_{xx}g_{xx} + f_{yy}g_{yy}\right] , \tag{2.34}$$

where all derivatives are to be evaluated at the fixed point, $(x = 0, y = 0)$. Then, the bifurcation is subcritical (resp. supercritical) provided that $\mathrm{Sgn}(a) > 0$ (resp., $\mathrm{Sgn}(a) < 0$). One may use center manifold techniques to identify analogous criteria for higher dimensional systems (see, e. g., [1]), but the complexity of these methods generally increases with system dimension.

2.2.2.4 **Degenerate Hopf and Takens–Bogdanov Bifurcations (Local, Co-dimension 2)**

Up to this point, we have discussed bifurcations that require the variation of only one control parameter in order to be fully mapped. These are the so-called co-dimension 1 bifurcations. However, there are important bifurcation phenomena that require the independent variation of two or more control parameters. We focus here on two prominent examples of co-dimension 2 bifurcations, namely, the *degenerate Hopf* bifurcation (also known as the Bautin bifurcation), and the *Takens–Bogdanov* bifurcation. One advantage of these two is that the normal forms can still be written in the form of a two-dimensional dynamical system. When one examines systems that exist in higher dimensional phase space, additional possibilities open up. We then refer the reader to [1] for an up-to-date introduction of this subject.

The degenerate Hopf bifurcation is most easily motivated by considering what happens when the a parameter of Eq. (2.34) is precisely 0. Then, we cannot say whether the Hopf bifurcation is sub- or supercritical. Note that in order to reach this state, two independent parameters have to be adjusted. We must have $a = 0$ and we must also have a complex conjugate pair of eigenvalues of the Jacobian with zero real part. A normal form to describe this situation is written in polar coordinates as

$$\dot{r} = r\left(\alpha_1 + \alpha_2 r^2 + s r^4\right) , \tag{2.35}$$

$$\dot{\theta} = 1 , \tag{2.36}$$

where s is a real parameter. Consider the case $s = -1$. Then, the stationary values of Eq. (2.35) are given by $r = 0$, corresponding to the fixed point at the origin and $r^2 = (\alpha_2 \pm \sqrt{\alpha_2^2 + 4\alpha_1})/2$. The latter term, provided it is real and positive, describes one or two coexisting limit cycles. There are no limit cycle solutions when $\alpha_2 < 2\sqrt{-\alpha_1}$ and $\alpha_1 < 0$ and there are two limit cycle solutions when $\alpha_2 > 0$ and $-\alpha_2^2/4 < \alpha_1 < 0$. Otherwise, there is one limit cycle solution. The stability of the limit cycles is examined by linearizing Eq. (2.35) about the appropriate solution. The top panels of Figure 2.7 illustrates the three distinct types of phase portraits that are possible in this system. One of these simply has a stable fixed point at the origin. Another has two nested limit cycles of different stability type surrounding a locally stable fixed point. The final portrait consists of a single stable limit cycle surrounding an unstable fixed point.

The lower panel of Figure 2.7 shows the bifurcation diagram as a function of the two control parameters, α_1 and α_2. The contours labeled H_+ and H_- indicate lines of sub- and supercritical Hopf bifurcations, while the contour labeled as P indicates a bifurcation in which the stable and unstable limit cycles collide and annihilate one another. This is referred to as a fold or saddle-node bifurcation for limit cycles. Referring back to the discussion of criteria for the stability of limit cycles based on the monodromy matrix $M(T)$, one sees that that this occurs when one of

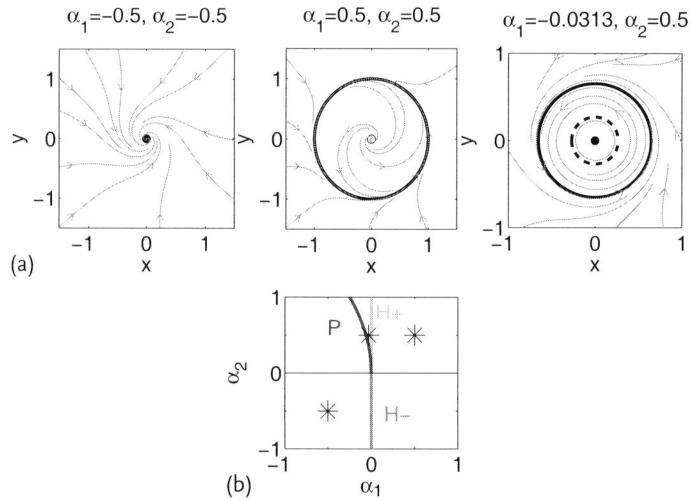

Fig. 2.7 (a) The different topological flow structures in the neighborhood of a degenerate Hopf bifurcation point. (b) Bifurcation phase diagram for different behaviors in the $\alpha_1 - \alpha_2$ parameter space.

the eigenvalues of $M(T)$ approaches unit magnitude, that is, $|\mu_i| \to 1-$ for some i. This is one of two generic ways in which a *large* amplitude limit cycle is annihilated.

Next, we turn to the Takens–Bogdanov (TB) bifurcation. This bifurcation is characterized by the fact that there are two eigenvalues that are identically zero and, hence, it is sometimes called a *double-zero* bifurcation. Therefore, one can view a Takens–Bogdanov point both as a limiting case for simple Hopf bifurcation in which the frequency ω_0 is tending to zero. One can also view the TB point as a limiting case in which two saddle-node bifurcations are somehow coalescing. The normal form for a TB bifurcation can be written as

$$\dot{x} = y\,, \tag{2.37}$$

$$\dot{y} = \alpha_1 + \alpha_2 x + x^2 + sxy\,. \tag{2.38}$$

Again, we consider the case $s = -1$. The different possible phase portraits *are* now somewhat richer than for the degenerate Hopf bifurcation just discussed. The

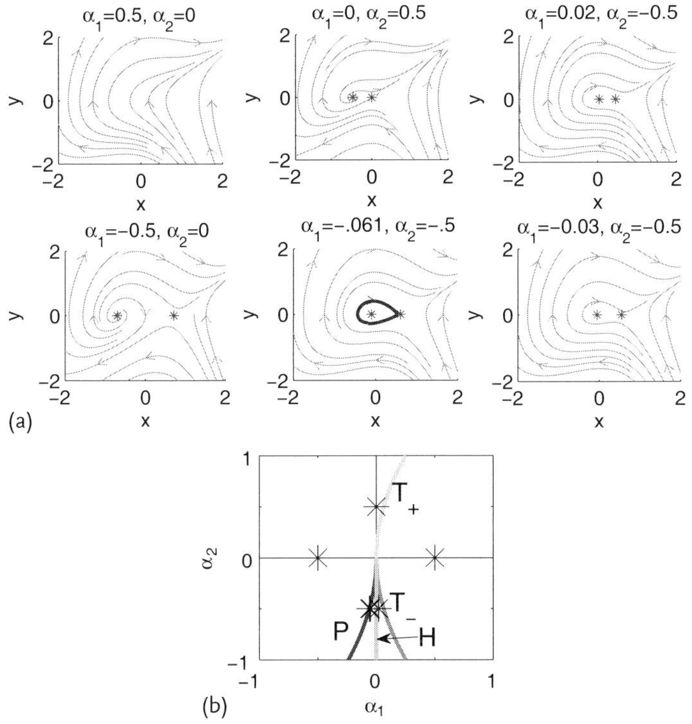

Fig. 2.8 (a) The different topological flow structures in the neighborhood of a TB bifurcation point. The middle panel of the second row shows a homoclinic orbit which exists along a particular trajectory in the $\alpha_1 - \alpha_2$ parameter space. (b) Bifurcation phase diagram for different behaviors in the $\alpha_1 - \alpha_2$ parameter space.

six panels of Figure 2.8a illustrate the distinct types of phase portraits that are possible in this system. The left panel of the first row indicates a smooth flow with no fixed points or limit cycles. The right panel of the first row and the left panel of the second row indicate a pair of unstable (one is a saddle and the other an unstable node) fixed points. The middle panel of the second row indicates the presence of a *homoclinic orbit* that emanates from one fixed point along the local unstable manifold, circulates around the other fixed point, and then returns to the original fixed point along the stable manifold. Finally, the phase plane depicted in the right panel of the second row contains two unstable fixed points and a small stable limit cycle (not shown) encircling the leftmost fixed point.

The bifurcation diagram is shown in Figure 2.8b. The contours labeled T denote saddle-node bifurcations in which pairs of fixed points annihilate, contours labeled H indicate Hopf bifurcations, and the contour marked P denotes a saddle-homoclinic bifurcation in which a large amplitude limit cycle collides with a fixed point and creates the homoclinic connection. The contour of homoclinic bifurcations constitutes a second generic means by which a large amplitude limit cycle is annihilated.

2.2.2.5 Heteroclinic and Homoclinic Connections as Examples of Nonlocal Bifurcations

We now turn briefly to discuss nonlocal (equivalently, global) bifurcations. By definition, these are bifurcations that cannot be neatly characterized by looking at the local neighborhoods of fixed points or limit cycles. Two nonlocal bifurcations that play an important role in the construction of nonlinear waves are heteroclinic and homoclinic bifurcations. First of all, let us recall that a *heteroclinic connection* refers to the existence in state space of a trajectory that connects one fixed point to a distinct one. Clearly, at least one of these fixed points must be of saddle type, that is, with a stable and an unstable manifold. More precisely, the unstable manifold of one of the fixed points must coincide with the stable manifold of the other fixed point. In some cases, these heteroclinic connections only exist for a precise value of control parameter, while in other cases, they may be robust (e. g., consider the simple example of Eq. (2.3)). In the former case, one has a *heteroclinic bifurcation*. A simple, two-dimensional paradigm dynamical system that exhibits the heteroclinic bifurcation can be written as

$$\dot{x} = 1 - x^2 - \alpha x y \,, \tag{2.39}$$

$$\dot{y} = x y + \alpha(1 - x^2) \,. \tag{2.40}$$

This system has two fixed points (both of saddle type) at coordinates $(-1, 0)$ and $(+1, 0)$, independent of α. In Figure 2.9, we show the phase portraits for $\alpha < 0$, $\alpha = 0$, and $\alpha > 0$. In particular, one notes the heteroclinic connection from $(-1, 0)$ to $(+1, 0)$ for $\alpha = 0$.

Homoclinic bifurcations are associated with the appearance of homoclinic connections or "orbits" in the state space. A *homoclinic connection* refers to the existence of a trajectory that connects one fixed point back to itself. By definition then, the

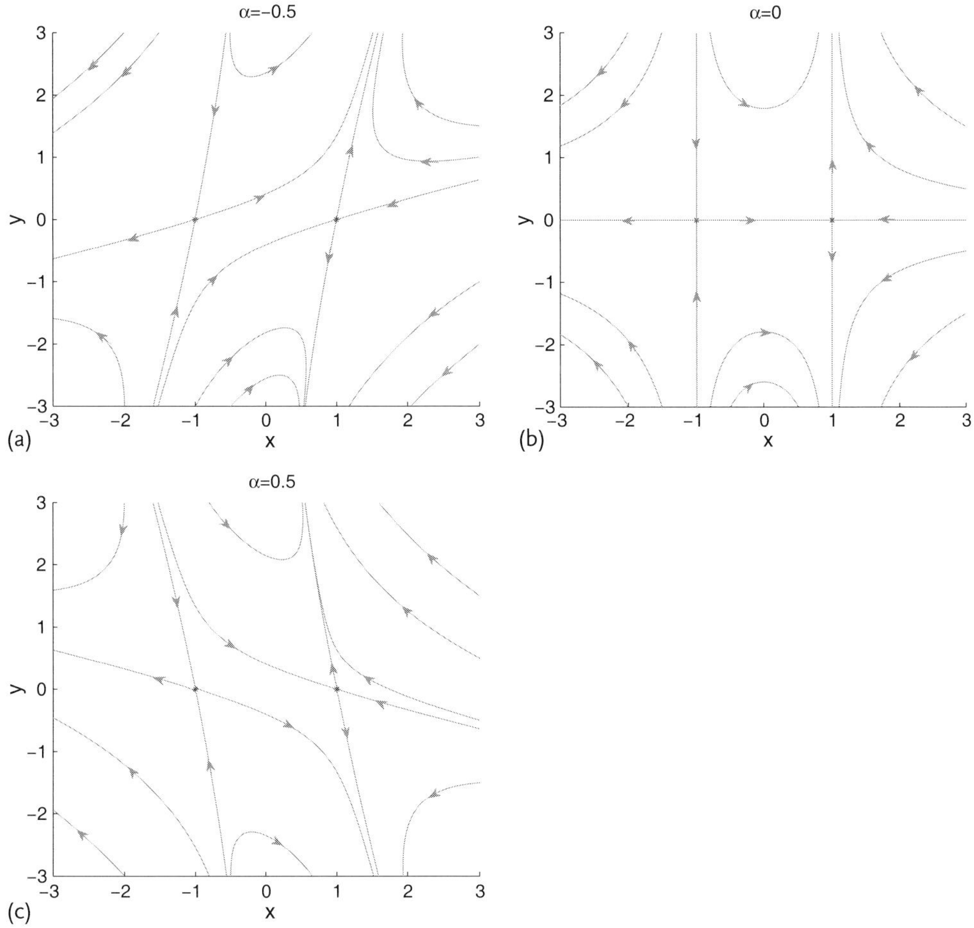

Fig. 2.9 Sequence of phase portraits showing a heteroclinic bifurcation for the dynamical system of Eqs. (2.39) and (2.40): (a) $\alpha = -0.5 < 0$, (b) $\alpha = 0$, and (c) $\alpha = +0.5 > 0$.

fixed point must be of saddle type as the orbit departs along the unstable manifold, makes an excursion around a second fixed point, and then returns to the original fixed point along its stable manifold. The existence of a homoclinic connection is typically dependent on the control parameter having a precise value. We have already encountered homoclinic connections and bifurcations in the discussion of the Takens–Bogdanov point (cf., the middle panel of the second row in Figure 2.8a).

2.3
Analysis of the Hopf Bifurcation: An Introduction to the Chapman–Enskog Method

2.3.1
Multiple Scales and Chapman–Enskog Methods

How does one determine which normal form is the appropriate one to use in a bifurcation that one encounters with some dynamical system under study? One approach is to identify the center manifold, that is, the phase space directions along which the bifurcations occur, and then to make multi-dimensional nonlinear changes of coordinates to bring the system into a recognizable normal form. While this procedure is, in principle, straightforward, it can become dauntingly cumbersome as the state space dimensionality increases. Nonetheless, there are excellent expositions of this approach in the literature (see, e. g., [1] and references therein). In the remainder of this chapter, we develop an alternative method, the Chapman–Enskog method, that takes advantage of natural separation in time scales near the bifurcation point [12]. We illustrate this approach with some detailed examples. These can be initially skimmed over during a first reading and then returned to as necessary.

The Chapman–Enskog method (CEM) has long been used to derive hydrodynamic equations of a parabolic type from kinetic equations of the Boltzmann type [13]. The complexity of these problems to which the method was first applied has perhaps obscured the realization that the CEM is a powerful singular perturbation method that can be used as a viable and attractive alternative to normal form calculations and regular multiple scales methods. Below, we provide a compact alternative to normal form for the Hopf bifurcation that is completely general. Then, we apply the method to a particular example of a two-dimensional system that describes population dynamics.

2.3.2
General Formulation of the Hopf Problem Using CEM

In nonlinear dynamical systems, we often consider bifurcation problems in which one or more solutions bifurcate from a given one as one parameter crosses a certain threshold value. These dynamical systems often can be reduced to study a lower dimensional system which describes a center manifold to which the flow of the complete system is attracted after a short transient. The calculation and solution of the reduced dynamical system on the center manifold can be carried out efficiently by the CEM in the general case. We start from a complete description of the linear stability problem about the specific solution whose bifurcations are to be studied. This solution is linearly stable for values of the bifurcation parameter below a threshold. As time goes to infinity, the solution of the linear stability problem evolves towards some constant or a function of time which contains a number of arbitrary constants which we generically call the *amplitude*. The time evolution of the amplitude describes the reduced dynamics on the center manifold. The idea is to use the amplitude of the bifurcating solution as the small expansion parameter ϵ, to expand the bifurcation parameter(s) and the bifurcating solution in powers of ϵ, to expand the time derivatives of the amplitude in powers of ϵ and to calculate the coefficients and coefficient functions in the bifurcation parameter and the amplitude equations in such a way that the successive terms of the bifurcating solution are bounded. Once we have calculated a sufficient number of terms in the amplitude equations and the control parameter expansion, we analyze the resulting lower dynamical system to determine the bifurcating solutions and their stability.

We introduce the CEM by considering a dynamical system of completely general form in the neighborhood of a point where a Hopf Bifurcation is known to occur [12]. Specifically, let us consider the system:

$$\frac{du}{dt} = f(u; \alpha) , \tag{2.41}$$

where $u(t)$ denotes an n-dimensional state vector in $\mathbb{R}^n$ and α denotes the bifurcation parameter. Suppose further that the steady state $u = 0$ is a stable fixed point of this system for $\alpha < 0$ and is unstable when $\alpha > 0$. Also, all eigenvalues of the system Jacobian $L(\alpha) \equiv \partial f(0, \alpha)/\partial u$, except for a single complex conjugate pair denoted as $\lambda(\alpha)$ and its twin $\overline{\lambda(\alpha)}$, possess negative real part in the neighborhood of $\alpha = 0$. We further suppose that the active eigenvalue has the form $\lambda(\alpha) \sim i\omega_0 + \alpha\lambda'(0)$ with $\omega_0 > 0$ and $\operatorname{Re}\lambda'(0) > 0$ when α is close to zero. (Note that $\lambda(\alpha) \sim i\omega_0 + \alpha\lambda'(0)$ is equivalent to $\lambda(\alpha) = i\omega_0 + \alpha\lambda'(0) + o(\alpha)$, as $\alpha \to 0$.)

Now, consider the linearization of Eq. (2.41) with respect to the steady state $u = 0$ at $\alpha = 0$,

$$\frac{du}{dt} \approx L(0)u . \tag{2.42}$$

By recognizing that there is a small-amplitude limit cycle in the neighborhood of $\alpha = 0$, we look for solutions to Eq. (2.42) of the form $u = A_0\,\phi_0\,e^{i\omega_0 t} + \text{cc}$, plus terms that decay exponentially fast as time increases. Here, A_0 is a complex constant and ϕ_0 denotes the eigenvector corresponding to eigenvalue $i\omega_0$, and solves the following linear equation, $L(0)\phi_0 = i\omega_0\phi_0$. Similarly, $\phi_0^\dagger$ denotes the eigenvector corresponding to the adjoint problem, $L(0)^\dagger\phi_0^\dagger = i\omega_0\phi_0^\dagger$. For a finite dimensional system, the determination of the eigenvectors ϕ_0 and $\phi_0^\dagger$ is a straightforward exercise in matrix algebra. An explicit example of this procedure in two dimensions is presented in the next section. On the other hand, if the underlying system is a partial differential equation, then we must solve the adjoint problem with homogeneous boundary conditions. We mention a few points about this generalization at the end of this section. Such an extension is used in semiconductor problems, for example, in the analysis of the Gunn effect, as shown in Chapter 6.

Our application of the CEM assumes the following expansions for the key variables of the problem:

$$\alpha = \alpha_1\epsilon + \alpha_2\epsilon^2 + O(\epsilon^3) , \tag{2.43}$$

$$u = \epsilon\,A(t;\epsilon)\,\phi_0\,e^{i\omega_0 t} + \text{cc} + \sum_{n=2}^{\infty} \epsilon^n\,u^{(n)}(t; A, \overline{A}) , \tag{2.44}$$

where the amplitude A obeys the ansatz:

$$\frac{dA}{dt} = \sum_{n=1}^{\infty} \epsilon^n F_n(A, \overline{A}) . \tag{2.45}$$

The first of these equations defines ϵ and we can see that as ϵ traverses through zero, we simply cross the bifurcation. This parameter is determined by finding the coefficients α_i. The second equation expands the solution in terms of powers of ϵ and uses the fact that the limit cycle amplitude approaches zero as $\epsilon \to 0$. These first two equations are also used in the common multiple scales expansion in which the functions $u^{(n)}$ depend on the fast time scale t, and on the slow time scale $\tau = \alpha t$, [14]; see also [10]. The third equation forms the core of the CEM: the coefficients α_i and the functions F_i are determined in such a way that the functions $u^{(n)}$ are bounded as $t \to \infty$, for A fixed. The reason for the third equation is to use the fact that $A(t; \epsilon)$, the complex envelope of the oscillation with frequency ω_0, evolves in a slow time scale ($\epsilon^2 t$ as shown below) and its rate may contain terms of different order in ϵ. The functions $u^{(n)}$ depend on the fast scale corresponding to oscillations with frequency ω_0 as well as on the slow time scale through their dependences on A. All terms in Eq. (2.44) that decrease exponentially in time are neglected. As it is also standard in multiple scales analysis, the α_n and polynomial functions F_n are determined so that the solutions $u^{(n)}$ are bounded as $t \to \infty$. This is a form of the nonresonance condition.

Substituting Eqs. (2.43)–(2.45) into Eq. (2.41) yields the following hierarchy of linear equations written as

$$[L - i\omega_0]\,\phi_0 = 0 ,$$

$$\frac{du^{(2)}}{dt} - Lu^{(2)} = -F_1\phi_0 e^{i\omega_0 t} + \text{cc} + \alpha_1 L'\phi_0 A e^{i\omega_0 t} + \text{cc} + \frac{1}{2} f^2\left(\left(A\phi_0 e^{i\omega_0 t} + \text{cc}\right)^2\right) , \tag{2.46}$$

$$\frac{du^{(3)}}{dt} - Lu^{(3)} = \alpha_1(\ldots) + F_1(\ldots) - F_2\phi_0 e^{i\omega_0 t} + \text{cc} + \alpha_2 L'\phi_0 A e^{i\omega_0 t} + \text{cc} + f^2\left(u^{(2)}, A\phi_0 e^{i\omega_0 t} + \text{cc}\right) + \frac{1}{6} f^3\left(\left(A\phi_0 e^{i\omega_0 t} + \text{cc}\right)^3\right) , \tag{2.47}$$

and so on. The first equation is satisfied automatically by the eigenvector ϕ_0 which corresponds to eigenvalue $i\omega_0$. Here, $f^k(v^1, \ldots, v^k)$ denote the symmetric k-linear form with components

$$f_i^k(v^1, \ldots, v^k) = \sum_{a_1=1}^{n} \cdots \sum_{a_k=1}^{n} \frac{\partial^k f_i(0; 0)}{\partial u_{a_1} \ldots \partial u_{a_k}} v_{a_1}^1 \ldots v_{a_k}^k , \tag{2.48}$$

so that $f^1 v = L(0)v \equiv Lv$. Similarly, we define the ith component of the symmetric k-linear form:

$$f_i^{k,l}(v^1,\ldots,v^k) = \sum_{a_1=1}^{n}\cdots\sum_{a_k=1}^{n} \frac{\partial^{k+l} f_i(0;0)}{\partial\alpha^l \partial u_{a_1}\ldots\partial u_{a_k}} v^1_{a_1}\ldots v^k_{a_k}\,, \tag{2.49}$$

so that $f^{1,1} = L'(0) \equiv L'$. In case f depends on two parameters α and β, we define another symmetric k-linear form as

$$f_i^{k,l,m}(v^1,\ldots,v^k) = \sum_{a_1=1}^{n}\cdots\sum_{a_k=1}^{n} \frac{\partial^{k+l+m} f_i(0;0,0)}{\partial\alpha^l \hat{E}\partial\beta^m \partial u_{a_1}\ldots\partial u_{a_k}} v^1_{a_1}\ldots v^k_{a_k}\,. \tag{2.50}$$

If, in the previous definitions, j of the entries ($2 \leq j \leq k$) correspond to the same vector so that $v^1 = \ldots = v^j$, we have simplified the notation by writing $f^k((v^1)^j, v^{j+1},\ldots,v^k) = f^k(v^1,\ldots,v^1,v^{j+1},\ldots,v^k)$.

The non-homogeneous equations in the hierarchy should be solved imposing that the $u^{(n)}$ be bounded as $t \to \infty$, thereby requiring that nonresonance conditions be satisfied. This, in turn, determines the α_i and F_i. We should also impose that the $u^{(n)}$ do not contain terms that are proportional $B\phi_0 e^{i\omega_0 t}$ + cc. The reason for this is that, by hypothesis, all such terms are already contained in $A(t;\epsilon)$. One way to ensure that this condition is satisfied is to impose $\int_{-\pi/\omega_0}^{\pi/\omega_0} e^{-i\omega_0 t}\langle\phi_0^\dagger, u^{(n)}\rangle\, dt = 0$, where $\phi_0^\dagger$ is the eigenvector corresponding to the adjoint problem

$$L^\dagger(0)\phi_0^\dagger = i\omega_0\phi_0^\dagger\,,$$

and the scalar product is the standard vector product in which the row vector $\phi_0^\dagger$ multiplies the column vector ϕ_0.

Equation (2.46) yields $F_1 + \alpha_1\lambda'(0)A = 0$, where

$$\lambda'(0) = \frac{\left\langle\phi_0^\dagger, L'(0)\phi_0\right\rangle}{\left\langle\phi_0^\dagger, \phi_0\right\rangle}\,, \tag{2.51}$$

and there is no contribution from the quadratic terms in Eq. (2.46). We select

$$\alpha_1 = 0\,, \quad F_1 = 0\,. \tag{2.52}$$

$$u^{(2)} = \frac{A^2}{2} e^{i2\omega_0 t}\left[2i\omega_0 I - L\right]^{-1} f^2\left(\phi_0^2\right) + \mathrm{cc} - |A|^2 L^{-1} f^2\left(\phi_0, \overline{\phi_0}\right)\,, \tag{2.53}$$

where I denotes the identity matrix. Solvability of Eq. (2.47) implies

$$F_2 = \alpha_2\lambda'(0)A - \mu\, A|A|^2\,, \tag{2.54}$$

where

$$\mu = \left\langle \phi_0^\dagger, \phi_0 \right\rangle^{-1} \left\{ \frac{1}{2} \left\langle \phi_0^\dagger, f^2\left(\overline{\phi_0}, \left([L - i2\omega_0 I]^{-1} f^2\left((\phi_0)^2\right)\right)\right) \right\rangle \right.$$
$$\left. + \left\langle \phi_0^\dagger, f^2\left(\phi_0, L^{-1} f^2\left(\phi_0, \overline{\phi_0}\right)\right) \right\rangle - \frac{1}{2} \left\langle \phi_0^\dagger, f^3\left((\phi_0)^2, \overline{\phi_0}\right) \right\rangle \right\} . \quad (2.55)$$

Substituting these results into Eq. (2.45), the following amplitude equation is found:

$$\frac{dA}{dt} = \epsilon^2 \alpha_2 \lambda'(0)\, A - \epsilon^2 \mu\, A\,|A|^2 + O(\epsilon^3) . \quad (2.56)$$

These formulas for the bifurcation equation and its coefficients have been obtained many times before using normal form techniques: see [15] for an explicit determination of the parameter μ, whose real part sign determines whether the bifurcation is sub- or supercritical. In fact, a straightforward evaluation of μ for the general two-dimensional system (cf. Eqs. (2.32) and (2.33)) shows that it gives the same criteria as was stated in Eq. (2.34). If $\mathrm{Re}(\mu) = 0$, then we must calculate higher order terms in the Chapman–Enskog expansion, Eq. (2.44). The systematic way in which the CEM yields such terms is a great advantage with respect to other methods such as standard multiple scales.

Similar ideas can be used to develop amplitude equations for pattern forming systems governed by partial differential equations [16]. In such cases, we must rescale space variables appropriately and assume that the F_n also depend on spatial derivatives of A!

2.3.2.1 An Example from Physiology

Let us consider the FitzHugh–Nagumo (FHN) model of nerve conduction in a space-clamped membrane such as the outer membrane of a squid giant axon threaded with a metallic conductor.

$$\frac{1}{c}\frac{du}{dt} = g(u) - v + I_a , \quad (2.57)$$

$$\frac{dv}{dt} = u - Bv , \quad (2.58)$$

with $g(u) = u(u-a)(2-u)$ and $0 < a < 2$. The excitation variable u is a membrane potential and the recovery variable v is an outward ion current. u and v are independent of space because the axon is threaded with a metallic conductor. The cubic source $g(u)$ is an ionic current and I_a is an applied current. c is the ratio of the time constants of u and v. For fixed I_a, we consider the values of B for which there may be three stationary solutions of the FHN system. The line $v = u/B$ is tangent to $v = g(u) + I_a$ if $B = B_o = 1/g'(u_o)$ such that $g(u_o) - u_o g'(u_o) + I_a = 0$. For $I_a = 0$, we obtain $B_o = B_1 = 4/(2-a)^2$, whereas the definition of u_o shows that it is a decreasing function of I_a. Then, B_o increases with I_a and for all $B < B_1$, we are sure that there is only one stationary solution for any positive value of I_a (for $B > B_1$ there may be three stationary solutions depending on the value of

I_a). For $0 < I_a < I_1$ and $I_a > I_3$, where $I_j = [B^{-1} + 4a/3 - (2+a)u_j/3]u_j$, $u_j = [2 + a + (-1)^j\sqrt{4 + a^2 - 2a}]/3$, $j = 1, 2$, the only stationary solution of the FHN system, $(u, v) = (u_*, u_*/B)$, has $g'(u_*) < 0$, whereas for $I_1 < I_a < I_2$, $g'(u_*) > 0$. We shall study the FHN system in Chapter 3 in more detail. Here, we shall show that for a fixed $B < B_1$ and an applied current in the interval $I_1 < I_a < I_2$, there is a Hopf bifurcation at a threshold value of I_a.

In fact, the linear stability of the stationary solution is determined by the eigenvalues of the matrix:

$$L(I_a, B) = \begin{pmatrix} c\,g'(u_*(I_a, B)) & -c \\ 1 & -B \end{pmatrix}, \tag{2.59}$$

which are

$$\lambda = \frac{cg'_* - B}{2} + i\Omega\,, \quad \Omega = \sqrt{c\,(1 - Bg'_*) - \left(\frac{cg'_* - B}{2}\right)^2}\,, \tag{2.60}$$

and its complex conjugate. The FHN system can be written as

$$\frac{d\boldsymbol{u}}{dt} = L(I_a, B)\boldsymbol{u} + \frac{1}{2}\,f^2((\boldsymbol{u} - \boldsymbol{u}_*)^2) + \frac{1}{6}\,f^3((\boldsymbol{u} - \boldsymbol{u}_*)^2)\,, \tag{2.61}$$

where

$$f^2(\boldsymbol{\varphi}, \boldsymbol{\psi}) = c f''_* \varphi_u \psi_u \begin{pmatrix} 1 \\ 0 \end{pmatrix}, \quad f^3(\boldsymbol{\varphi}, \boldsymbol{\psi}, \boldsymbol{\zeta}) = -6c\varphi_u\psi_u\zeta_u \begin{pmatrix} 1 \\ 0 \end{pmatrix}, \tag{2.62}$$

$$\boldsymbol{u} = \begin{pmatrix} u \\ v \end{pmatrix}, \quad \boldsymbol{\varphi} = \begin{pmatrix} \varphi_u \\ \varphi_v \end{pmatrix}, \ldots \tag{2.63}$$

Let us adopt I_a as our bifurcation parameter. At the critical value I_{ac}, we have $cg'(u_*(I_{ac}, B)) = B$ according to Eq. (2.60). Then, $\lambda(I_{ac}) = i\omega_0$ with $\omega_0 = \sqrt{c - B^2}$. In addition, $\lambda'(I_{ac}) = c^2 g''(u_*)^2 B(1 - iB/\omega_0)/(2\omega_0^2)$, where we have used $cg'_* = B$ with $g'_* = g'(u_*)$. At the critical current, the eigenvectors of L and $L^\dagger$ corresponding to $i\omega_0$ are

$$\phi_0 = \begin{pmatrix} 1 \\ \frac{B - i\omega_0}{c} \end{pmatrix}, \quad \phi_0^\dagger = \begin{pmatrix} 1 \\ -B + i\omega_0 \end{pmatrix}, \tag{2.64}$$

respectively. Then the complex coefficient μ Eq. (2.55) in the amplitude equation (2.56) is

$$\mu = \frac{3c}{2} - \frac{c^2 Bg''(u_*)^2}{4\omega_0^2} + i\frac{c}{2\omega_0}\left(\frac{cg''(u_*)^2(3B^2 + 2c)}{6\omega_0^2} - 3B\right). \tag{2.65}$$

The bifurcation is supercritical if $\operatorname{Re}\mu > 0$ and subcritical otherwise. It turns out that $\operatorname{Re}\mu = 0$ for $B = B_* = c\left[4 + a^2 - 2a - \sqrt{(4 + a^2 - 2a)^2 - 9/c}\right]/3$ and $I_a = I_{a*} = -g(u_*) + u_*/B_*$ with $u_* = \left[2 + a - ((4 + a^2 - 2a)^2 - 9/c)^{1/4}\right]/3$. At the point (I_{a*}, B_*) of the parameter space, we cannot draw the bifurcation diagram with the results thus far obtained. We must calculate higher order terms in the CEM in order to depict the bifurcation diagram.

Moreover, if $u_* = (2+a)/3$, then $g''_* = 0$ and $\mathrm{Re}\,\lambda'(I_{ac}) = 0$, with $I_{ac} = (2+a)[B^{-1} + 2a - 2(2+a)^2/9]/3$. This happens for $B = B_c = c[-2a + (2+a)^2/3]$. The point (I_{ac}, B_c) corresponds to a degenerate Hopf bifurcation and its bifurcation diagram can be determined by carrying out the CEM with a different scaling. We shall now consider these cases of degenerate Hopf bifurcation.

2.3.3
Utility of the CEM for Higher Order Bifurcations

There are two standard cases of the Hopf bifurcation that require calculating more than one term in the amplitude equation. Both cases can be contemplated as a two-parameter bifurcation and we assume that the $u = 0$ is a solution of Eq. (2.41), so that $f(0; \alpha, \beta) = 0$.

Firstly, let us assume that there is another parameter β such that the curve of Hopf bifurcations in the (α, β) plane, denoted by $\beta_0(\alpha)$, has a maximum or a minimum at $(0, 0)$: $\beta_0(\alpha) \sim \pm\kappa\alpha^2$ as $\alpha \to 0$. Assume the trivial solution is unstable inside the parabola. If we move α at fixed $\beta = 0$, the real part of the eigenvalue $\lambda(\alpha, 0)$ is negative for all $\alpha \neq 0$, so that $\mathrm{Re}\,\lambda(\alpha, 0)$ has a maximum at $\alpha = 0$. Then, we have $\mathrm{Re}\,\partial_\alpha\lambda(0,0) = 0$, $\mathrm{Re}\,\partial^2_\alpha\lambda(0,0) < 0$ and we consider $\mathrm{Re}\,\partial_\beta\lambda(0,0) \neq 0$. We will call this the case of a degenerate simple eigenvalue. In the second case, let us assume that $\mathrm{Re}\,\mu = 0$ for $\beta = 0$, so that $\mathrm{Re}\,\mu \sim \mu_1\beta$, $\mu_1 \neq 0$ as $\beta \to 0$. This second case will be called a degenerate Hopf bifurcation. We shall now show how to calculate the amplitude equation in both cases using the CEM.

2.3.3.1 Degenerate Simple Eigenvalue
In this case,

$$\lambda(\alpha, \beta) \sim i\omega_0 + i\alpha\omega_1 + \beta\lambda_\beta + \frac{\alpha^2}{2}\lambda_{\alpha\alpha} + \alpha\beta\lambda_{\alpha\beta} + \frac{\beta^2}{2}\lambda_{\beta\beta}\,, \tag{2.66}$$

where $\omega_1 = \mathrm{Im}\,\partial_\alpha\lambda(0,0) \neq 0$, and $\lambda_\alpha = \partial_\alpha\lambda(0,0)$, $\lambda_{\alpha\beta} = \partial_\alpha\partial_\beta\lambda(0,0)$, and so on. To analyze this case as a two-parameter Hopf bifurcation, we shall assume that the parameters can be expanded in series of powers of ϵ (the bifurcation amplitude), so that, in addition to Eqs. (2.43)–(2.45), we have

$$\beta = \beta_1\epsilon + \beta_2\epsilon^2 + O(\epsilon^3). \tag{2.67}$$

Inserting Eqs. (2.43)–(2.45) and (2.67) in (2.41), we obtain the following hierarchy of equations:

$$[L(0,0) - i\omega_0]\,\phi_0 = 0\,,$$

$$\left[\frac{\partial}{\partial t} - L\right] u^{(2)} = -F_1\phi_0 e^{i\omega_0 t} + \mathrm{cc} + [\alpha_1 L_\alpha + \beta_1 L_\beta]\left(Ae^{i\omega_0 t}\phi_0 + \mathrm{cc}\right) + \frac{1}{2} f^2\left(\left(A\phi_0 e^{i\omega_0 t} + \mathrm{cc}\right)^2\right)\,, \tag{2.68}$$

$$\left[\frac{\partial}{\partial t} - L\right] u^{(3)} = \alpha_1 L_\alpha u^{(2)} + \beta_1(\ldots)$$
$$+ \left[\alpha_2 L_\alpha + \beta_2 L_\beta + \frac{\alpha_1^2}{2} L_{\alpha\alpha}\right] \left(A e^{i\omega_0 t} \phi_0 + \text{cc}\right)$$
$$- \left(F_1 \partial_A u^{(2)} + F_2 \phi_0 e^{i\omega_0 t}\right) + \text{cc}$$
$$+ f^2 \left(u^{(2)}, A\phi_0 e^{i\omega_0 t} + \text{cc}\right) + \frac{1}{2}\partial_\alpha f^2 \left(\left(A\phi_0 e^{i\omega_0 t} + \text{cc}\right)^2\right)$$
$$+ \frac{1}{6} f^3 \left(\left(A\phi_0 e^{i\omega_0 t} + \text{cc}\right)^3\right), \tag{2.69}$$

and so on. Here, $L(\alpha, \beta) = f_u(u; \alpha, \beta)$.

As before, the solvability condition for Eq. (2.68) to have a solution which is bounded in t yields $\beta_1 = 0$ and

$$F_1 = \alpha_1 \frac{\langle \phi_0^\dagger, \partial_\alpha L \phi_0 \rangle}{\langle \phi_0^\dagger, \phi_0 \rangle} A = \alpha_1 \lambda_\alpha A\,. \tag{2.70}$$

The second equality follows from differentiating the eigenvalue equation $[L(\alpha, \beta) - \lambda(\alpha, \beta)]\phi(\alpha, \beta) = 0$ once with respect to α and then setting $\alpha = \beta = 0$:

$$[L_\alpha - \lambda_\alpha]\phi_0 = -[L - i\omega_0]\phi_0'\,, \tag{2.71}$$

where $\phi_0' = \partial_\alpha \phi(0, 0)$. The scalar product of $\phi_0^\dagger$ with the right-hand side of this equality is zero because $L^\dagger \phi_0^\dagger = i\omega_0 \phi_0^\dagger$, and then the left-hand side yields the second equality in Eq. (2.70). Similar arguments yield the equalities

$$\lambda_\beta = \frac{\langle \phi_0^\dagger, L_\beta \phi_0 \rangle}{\langle \phi_0^\dagger, \phi_0 \rangle}\,, \tag{2.72}$$

$$\lambda_{\alpha\alpha} = \frac{\langle \phi_0^\dagger, L_{\alpha\alpha} \phi_0 \rangle}{\langle \phi_0^\dagger, \phi_0 \rangle} + 2\frac{\langle \phi_0^\dagger, [L_\alpha - \lambda_\alpha]\phi_0' \rangle}{\langle \phi_0^\dagger, \phi_0 \rangle}\,. \tag{2.73}$$

The solution of Eq. (2.68) is

$$u^{(2)} = \alpha_1 A \phi_0' e^{i\omega_0 t} + \text{cc} + \frac{A^2}{2} e^{i2\omega_0 t} [2i\omega_0 I - L]^{-1} f^2 \left((\phi_0)^2\right) + \text{cc}$$
$$- |A|^2 L^{-1} f^2 \left(\phi_0, \overline{\phi_0}\right), \tag{2.74}$$

in which we have used Eqs. (2.70) and (2.71). We now insert Eq. (2.74) in Eq. (2.69) and calculate $F_2(A, \overline{A})$ by using its solvability condition. The result is

$$F_2 = \left[\frac{\alpha_1^2}{2} \lambda_{\alpha\alpha} + \alpha_2 \lambda_\alpha + \beta_2 \lambda_\beta\right] A - \mu\, A |A|^2\,, \tag{2.75}$$

where we have used Eq. (2.73) and μ is given by Eq. (2.55).

The amplitude equation is found by substituting Eqs. (2.70) and (2.75) in Eq. (2.45):

$$\frac{dA}{dt} = \left[\left(\epsilon \alpha_1 + \epsilon^2 \alpha_2 \right) i\omega_1 + \epsilon^2 \beta_2 \lambda_\beta + \frac{\epsilon^2 \alpha_1^2}{2} \lambda_{\alpha\alpha} \right] A - \epsilon^2 \mu\, A\, |A|^2 , \quad (2.76)$$

up to terms of order ϵ^3 (recall $\partial_\alpha \lambda(0,0) = i\omega_1$). Note that the coefficient of A in Eq. (2.76) is $\lambda(\alpha,\beta) - \lambda(0,0)$ up to terms of order ϵ^3. If we write $\lambda(\alpha,\beta) - \lambda(0,0)$ instead of $\epsilon^2\alpha_2$ in Eq. (2.56), an expansion of $\lambda(\epsilon\alpha_1 + \epsilon^2\alpha_2, \epsilon^2\beta_2)$ in the resulting expression yields Eq. (2.76).

If we now define $A = e^{i\epsilon(\alpha_1+\epsilon\alpha_2)\omega_1 t}\, B(\tau)$, with $\tau = \epsilon^2 t$, we obtain the following reduced equation for B:

$$\frac{dB}{d\tau} = \left[\beta_2 \lambda_\beta + \frac{\alpha_1^2}{2} \lambda_{\alpha\alpha} \right] B - \mu\, B\, |B|^2 . \quad (2.77)$$

Equation (2.77) could also have been obtained by a modified MS method in which the fast time is defined as $\tilde{t} = (1 + \epsilon\, m_1 + \epsilon^2 m_2)t$, the slow time is $\tau = \epsilon^2 t$, and the m_i are used to eliminate the secular terms in the equation for $u^{(2)}$ and the α_2 term in the amplitude equation ($m_i = \alpha_i \omega_1/\omega_0$, $i = 1, 2$).

To analyze Eq. (2.77), we write the equations for the amplitude and argument of $B = b\, e^{i\theta}$:

$$\frac{db}{d\tau} = \left[\beta_2 \operatorname{Re} \lambda_\beta + \frac{\alpha_1^2}{2} \operatorname{Re} \lambda_{\alpha\alpha} \right] b - \operatorname{Re} \mu\, b^3 , \quad (2.78)$$

$$\frac{d\theta}{d\tau} = \beta_2 \operatorname{Im} \lambda_\beta + \frac{\alpha_1^2}{2} \operatorname{Im} \lambda_{\alpha\alpha} - \operatorname{Im} \mu\, b^2 . \quad (2.79)$$

This system of equations has a periodic solution $b = b_\infty e^{i\Omega\tau}$, where

$$b_\infty = \sqrt{\frac{\beta_2 \operatorname{Re} \lambda_\beta + \alpha_1^2 \operatorname{Re} \lambda_{\alpha\alpha}/2}{\operatorname{Re} \mu}} , \quad (2.80)$$

$$\Omega = \beta_2 \operatorname{Im} \lambda_\beta + \frac{\alpha_1^2}{2} \operatorname{Im} \lambda_\alpha^2 - \frac{\operatorname{Im} \mu}{\operatorname{Re} \mu} \left[\beta_2 \operatorname{Re} \lambda_\beta + \frac{\alpha_1^2}{2} \operatorname{Re} \lambda_{\alpha\alpha} \right] . \quad (2.81)$$

Let us now assume that $\operatorname{Re} \lambda_{\alpha\alpha} < 0$, so that $\operatorname{Re} \lambda$ has a local maximum at $\alpha = \beta = 0$. There are two possibilities:

- $\operatorname{Re} \mu > 0$. Then, if $\beta_2 \operatorname{Re} \lambda_\beta < 0$, there are no bifurcating solutions of the form Eqs. (2.80) and (2.81). On the other hand, if $\beta_2 \operatorname{Re} \lambda_\beta \geq 0$, then there exist a branch of stable periodic solutions that form a semi-ellipse $\alpha_1^2/a^2 + b_\infty^2/z^2 = 1$ with $a^2 = -2\beta_2 \operatorname{Re} \lambda_\beta / \operatorname{Re} \lambda_{\alpha\alpha}$, $z^2 = \beta_2 \operatorname{Re} \lambda_\beta / \operatorname{Re} \mu$. At $\alpha_1 = \pm a$, $b_\infty = 0$ and supercritical Hopf bifurcations issue forth from the trivial state.
- $\operatorname{Re} \mu < 0$. Then, if $\beta_2 \operatorname{Re} \lambda_\beta < 0$, there is a branch of unstable periodic solutions of the form Eqs. (2.80) and (2.81) that form a hyperbola $b_\infty^2/z^2 -$

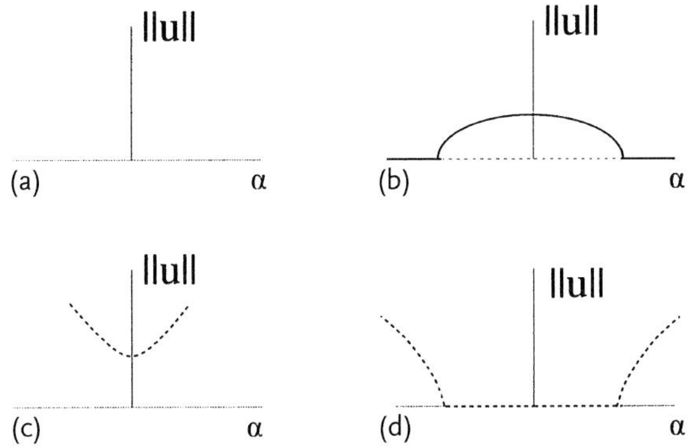

Fig. 2.10 Bifurcation diagram for degenerate simple eigenvalue case showing $\|u\|$ versus α for $\text{Re}\,\lambda_{\alpha\alpha} < 0$. (a) $\text{Re}\,\mu > 0$ and $\beta\,\text{Re}\,\lambda_\beta \leq 0$, (b) $\text{Re}\,\mu > 0$ and $\beta\,\text{Re}\,\lambda_\beta > 0$, (c) $\text{Re}\,\mu < 0$ and $\beta\,\text{Re}\,\lambda_\beta < 0$, (d) $\text{Re}\,\mu < 0$ and $\beta\,\text{Re}\,\lambda_\beta > 0$.

$\alpha_1^2/a^2 = 1$, $b_\infty > 0$, with $a^2 = 2\beta_2\,\text{Re}\,\lambda_\beta/\,\text{Re}\,\lambda_{\alpha\alpha}$, $z^2 = \beta_2\,\text{Re}\,\lambda_\beta/\,\text{Re}\,\mu$. If $\beta_2\,\text{Re}\,\lambda_\beta = 0$, at $\beta = 0$, two branches of unstable periodic solutions, $b_\infty = \alpha_1\sqrt{\text{Re}\,\lambda_{\alpha\alpha}/(2\,\text{Re}\,\mu)}$ (which exist for $\alpha > 0$ and $\alpha < 0$) bifurcate from $u = 0$. On the other hand, if $\beta_2\,\text{Re}\,\lambda_\beta > 0$, then there are two subcritical Hopf bifurcations at $\alpha_1 = \pm a$, with $a = \sqrt{-2\beta_2\,\text{Re}\,\lambda_\beta/\,\text{Re}\,\lambda_{\alpha\alpha}}$, and the bifurcating branches form a semi-hyperbola $\alpha_1^2/a^2 - b_\infty^2/z^2 = 1$ with $z^2 = -\beta_2\,\text{Re}\,\lambda_\beta/\,\text{Re}\,\mu$ and the same a.

These different possibilities are summarized in Figure 2.10. The unfoldings of the Hopf bifurcation for $\text{Re}\,\lambda_{\alpha\alpha} > 0$ (corresponding to the case of a trivial state which is unstable outside the parabola $\beta \sim \pm\kappa\alpha^2$ and stable inside) are similar and we omit their enumeration.

In the example of the FHN system (2.57) and (2.58), there is a degenerate simple eigenvalue if $I_a = (2+a)[B^{-1} + 2a - 2(2+a)^2/9]/3$ and $B = c[-2a + (2+a)^2/3]$ for which $u_* = (2+a)/3$ and $g''(u_*) = 0$. Then, selecting $\alpha = I_a - I_{ac}$ and $\beta = B - B_c$, $\lambda_\alpha = 0$, $\lambda_{\alpha\alpha} = -3c^4 B_c^2[1 - i B_c/(c\omega_0)]/\omega_0^2$ and $\lambda_\beta = -(1 + i B_c/\omega_0)/2$. According to Eq. (2.65), $\mu = 3c(1 - i B_c/\omega_0)/2$. Thus, the bifurcation diagram is that of Figure 2.10a for $B < B_c$ and that of Figure 2.10b for $B > B_c$.

2.3.3.2 **Degenerate Hopf Bifurcation**

Let us assume that in the usual case of the Hopf bifurcation, $\text{Re}\,\mu(\beta) = m\beta$, with $m \neq 0$ in a certain interval $|\beta| < \beta_0$. Then, at $\beta = 0$, $\text{Re}\,\mu = 0$ and we cannot decide whether the bifurcating periodic solutions are sub or supercritical. This problem was also considered by [17]. It is clear that we have to determine more terms in the amplitude equation given by the CEM.

The solution of Eq. (2.47) given Eq. (2.54) is

$$u^{(3)} = (\alpha_2 A\phi'_{0\alpha} + \beta_2 A\phi'_{0\beta} + M_1 A|A|^2)\, e^{i\omega_0 t} + \text{cc} + M_3 A^3 e^{i3\omega_0 t} + \text{cc}\,, \quad (2.82)$$

$$\begin{aligned} M_1 = {} & \frac{1}{2}[i\omega_0 I - L]^{-1}\{f^2(\overline{\phi_0}, [2i\omega_0 I - L(0,0)]^{-1} f^2((\phi_0)^2)) \\ & - 2f^2(\phi_0, L^{-1} f^2(\phi_0, \overline{\phi_0})) + f^3(\overline{\phi_0}, (\phi_0)^2) + 2\mu\phi_0\}\,, \\ & \text{with} \quad \langle \phi_0^\dagger, M_1\rangle = 0\,, \end{aligned} \quad (2.83)$$

$$M_3 = \frac{1}{2}[3i\omega_0 I - L]^{-1}\left\{f^2(\phi_0, [2i\omega_0 I - L]^{-1} f^2((\phi_0)^2)) + \frac{1}{3} f^3((\phi_0)^3)\right\}\,, \quad (2.84)$$

where $\phi'_{0\alpha} = \partial_\alpha \phi(0,0)$, $\phi'_{0\beta} = \partial_\beta \phi(0,0)$ and μ is given by Eq. (2.55) calculated for $\beta = 0$.

The equations for $u^{(4)}$ and $u^{(5)}$ are

$$\begin{aligned} \left[\frac{\partial}{\partial t} - L\right] u^{(4)} = {} & (\alpha_2 L_\alpha + \beta_2 L_\beta) u^{(2)} - F_2 \partial_A u^{(2)} + \text{cc} \\ & + \frac{1}{2} f^2((u^{(2)})^2) + f^2(A\phi_0 e^{i\omega_0 t} + \text{cc}, u^{(3)}) \\ & + \left(\frac{\alpha_2}{2}\partial_\alpha + \frac{\beta_2}{2}\partial_\beta\right) f^2((A\phi_0 e^{i\omega_0 t} + \text{cc})^2) \\ & + \frac{1}{2} f^3(u^{(2)}, (A\phi_0 e^{i\omega_0 t} + \text{cc})^2) + \frac{1}{4!} f^4((A\phi_0 e^{i\omega_0 t} + \text{cc})^4)\,, \end{aligned} \quad (2.85)$$

$$\begin{aligned} \left[\frac{\partial}{\partial t} - L\right] u^{(5)} = {} & (\alpha_2 L_\alpha + \beta_2 L_\beta) u^{(3)} + (\alpha_4 L_\alpha + \beta_4 L_\beta)(A\phi_0 e^{i\omega_0 t} + \text{cc}) \\ & + \left(\frac{\alpha_2^2}{2} L_{\alpha\alpha} + \alpha_2\beta_2 L_{\alpha\beta} + \frac{\beta_2^2}{2} L_{\beta\beta}\right)(A\phi_0 e^{i\omega_0 t} + \text{cc}) \\ & - \left(F_2 \partial_A u^{(3)} + F_4 A\phi_0 e^{i\omega_0 t}\right) + \text{cc} \\ & + f^2(u^{(2)}, u^{(3)}) + f^2\left(A\phi_0 e^{i\omega_0 t} + \text{cc}, u^{(4)}\right) \\ & + \left(\alpha_2\partial_\alpha + \beta_2\partial_\beta\right) f^2(u^{(2)}, A\phi_0 e^{i\omega_0 t} + \text{cc}) \\ & + \frac{1}{2} f^3\left((u^{(2)})^2, A\phi_0 e^{i\omega_0 t} + \text{cc}\right) \\ & + \frac{1}{2} f^3\left(u^{(3)}, \left(A\phi_0 e^{i\omega_0 t} + \text{cc}\right)^2\right) \\ & + \frac{1}{6} f^4\left(u^{(2)}, (A\phi_0 e^{i\omega_0 t} + \text{cc})^3\right) \\ & + \frac{1}{5!} f^5\left(\left(A\phi_0 e^{i\omega_0 t} + \text{cc}\right)^5\right)\,, \end{aligned} \quad (2.86)$$

given that $\alpha = \alpha_2\epsilon^2 + \alpha_4\epsilon^4 + o(\epsilon^4)$ and $\beta = \beta_2\epsilon^2 + \beta_4\epsilon^4 + o(\epsilon^4)$. We have not included a term F_3 in Eq. (2.85) because there are no secular terms in that equation once we set $\alpha_3 = \beta_3 = 0$.

The solution of Eq. (2.85) is

$$\begin{aligned} u^{(4)} &= N_0|A|^4 + (N_2 A^2|A|^2 e^{i2\omega_0 t} + N_4 A^4 e^{i4\omega_0 t}) \\ &\quad + \mathrm{cc} + P_0|A|^2 + P_2 A^2 e^{i2\omega_0 t} + \mathrm{cc} \quad , \end{aligned} \tag{2.87}$$

$$\begin{aligned} N_0 &= L^{-1}\Big\{2\,\mathrm{Re}\,\mu L^{-1} f^2(\phi_0, \overline{\phi_0}) - f^2(\phi_0, \overline{M_1}) + f^2(\overline{\phi}_0, M_1) \\ &\quad - \frac{1}{2} f^2([L(0,0)^{-1} f_{uu}(0;0,0) : \phi_0\overline{\phi_0}]^2) \\ &\quad + \frac{1}{4} f^2((i2\omega_0 - L)^{-1} f^2\left((\phi_0)^2\right), (i2\omega_0 + L)^{-1} f^2\left((\overline{\phi_0})^2\right) \\ &\quad + \frac{1}{4} f^3((\phi_0)^2, (i2\omega_0 + L)^{-1} f^2((\overline{\phi_0})^2)) + \mathrm{cc} \\ &\quad + f^3(\phi_0, \overline{\phi_0}, L^{-1} f^2(\phi_0, \overline{\phi_0})) + \frac{1}{4} f^4((\phi_0)^2, (\overline{\phi_0})^2)\Big\} , \end{aligned} \tag{2.88}$$

$$\begin{aligned} N_2 &= (i2\omega_0 - L)^{-1}\Big\{\mu(i2\omega_0 - L)^{-1} f^2\left((\phi_0)^2\right) + f^2(\phi_0 M_1) + f^2(\overline{\phi_0}, M_3) \\ &\quad - \frac{1}{2} f^2\left((2i\omega_0 I - L)^{-1} f^2\left((\phi_0)^2\right), L^{-1} f^2(\phi_0, \overline{\phi_0})\right) \\ &\quad + \frac{1}{2} f^3\left(\overline{\phi_0}, \phi_0, (i2\omega_0 I - L)^{-1} f^2\left((\phi_0)^2\right)\right) \\ &\quad - \frac{1}{2} f^3\left((\phi_0)^2, L^{-1} f^2(\phi_0, \overline{\phi_0})\right) + \frac{1}{6} f^4\left((\phi_0)^3, \overline{\phi_0}\right)\Big\} , \end{aligned} \tag{2.89}$$

$$\begin{aligned} N_4 &= [i4\omega_0 I - L]^{-1}\Big\{\frac{1}{8} f^2\left(\left[(2i\omega_0 I - L)^{-1} f^2\left((\phi_0)^2\right)\right]^2\right) \\ &\quad + \frac{1}{4} f^3\left((\phi_0)^2, (2i\omega_0 I - L)^{-1} f^2((\phi_0)^2)\right) + f^2(\phi_0, M_3) \\ &\quad + \frac{1}{4!} f^4\left((\phi_0)^4\right)\Big\} , \end{aligned} \tag{2.90}$$

$$P_0 = L^{-1}(\alpha_2 p_{0\alpha} + \beta_2 p_{0\beta}) , \tag{2.91}$$

$$\begin{aligned} p_{0\alpha} &= (L_\alpha - 2\,\mathrm{Re}\,\lambda_\alpha) L^{-1} f^2(\phi_0, \overline{\phi_0}) + f^{2,1,0}(\phi_0, \overline{\phi_0}) + f^2(\phi'_{0\alpha}, \overline{\phi_0}) \\ &\quad + f^2(\phi_0, \overline{\phi'_{0\alpha}}) , \end{aligned} \tag{2.92}$$

$$P_2 = (i2\omega_0 I - L)^{-1}(\alpha_2 p_{2\alpha} + \beta_2 p_{2\beta}) , \tag{2.93}$$

$$p_{2\alpha} = f^2(\phi_0, \phi'_{0\alpha}) + \frac{f^{2,1,0}(\phi_0, \phi_0)}{2} + \frac{1}{2}(L_\alpha - \lambda_\alpha)(i2\omega_0 I - L)^{-1} f^2(\phi_0, \phi_0) . \tag{2.94}$$

The expressions for $p_{0\beta}$ and $p_{2\beta}$ follow by replacing β instead of α in Eqs. (2.92) and (2.94), respectively.

We now find F_4 by using the solvability condition for Eq. (2.86), which yields

$$F_4 = A\left(\alpha_4\lambda_\alpha + \beta_4\lambda_\beta + \frac{\alpha_2^2}{2}\lambda_{\alpha\alpha} + \alpha_2\beta_2\lambda_{\alpha\beta} + \frac{\beta_2^2}{2}\lambda_{\beta\beta}\right) - (\alpha_2\psi_\alpha + \beta_2\psi_\beta)\,A\,|A|^2 - \mu_2 A\,|A|^4\,, \tag{2.95}$$

$$\begin{aligned}\psi_\alpha = \langle\phi_0^\dagger, \phi_0\rangle^{-1}\Big\langle\phi_0^\dagger, (\lambda_\alpha - L_\alpha)M_1 &- \frac{1}{2}f^{2,1,0}\left(\overline{\phi_0}, (i2\omega_0 I - L)^{-1} f^2(\phi_0,\phi_0)\right)\\ &+ f^{2,1,0}\left(\phi_0, L^{-1} f^2(\phi_0,\overline{\phi_0})\right)\\ &- \frac{1}{2}f^2(\overline{\phi'_{0\alpha}}, (i2\omega_0 I - L)^{-1} f^2(\phi_0,\phi_0))\\ &+ f^2\left(\phi'_{0\alpha}, L^{-1} f^2(\phi_0,\overline{\phi_0})\right) - f^2\left(\overline{\phi_0}, (i2\omega_0 I - L)^{-1} p_{2\alpha}\right)\\ &- f^2(\phi_0, L^{-1}p_{0\alpha}) - \frac{1}{2}f^3(\phi_0,\phi_0,\overline{\phi'_{0\alpha}}) + f^3(\phi_0,\overline{\phi_0},\phi'_{0\alpha}) - \mu\phi'_{0\alpha}\\ &+ 2\,\mathrm{Re}\,\lambda_\alpha M_1\Big\rangle\,,\end{aligned} \tag{2.96}$$

$$\begin{aligned}\mu_2 = \langle\phi_0^\dagger, \phi_0\rangle^{-1}\Big\langle\phi_0^\dagger, \frac{1}{2}f^2\left((i2\omega_0 I + L)^{-1} f^2(\overline{\phi_0},\overline{\phi_0}), M_3\right) &+ f^2\left(L^{-1}f^2(\phi_0,\overline{\phi_0}), M_1\right) - f^2(N_0,\phi_0) - f^2(N_2,\overline{\phi_0})\\ &+ \frac{1}{2}f^3\left(\overline{\phi_0}, L^{-1}f^2(\overline{\phi_0},\phi_0), (i2\omega_0 I - L)^{-1} f^2(\phi_0,\phi_0)\right)\\ &+ \frac{1}{4}f^3\left(\phi_0, (i2\omega_0 I + L)^{-1} f^2(\overline{\phi_0},\overline{\phi_0}), (i2\omega_0 I - L)^{-1} f^2(\phi_0,\phi_0)\right)\\ &- \frac{1}{2}f^3(\overline{\phi_0},\overline{\phi_0},M_3)\\ &- \frac{1}{2}f^3\left(\phi_0, [L^{-1}f^2(\overline{\phi_0},\phi_0)]^2\right) - f^3(\phi_0,\overline{\phi_0},M_1) - \frac{1}{2}f^3(\phi_0,\phi_0,M_1)\\ &+ \frac{1}{2}f^4(L^{-1}f^2(\overline{\phi_0},\phi_0),\phi_0,\phi_0,\overline{\phi_0})\\ &- \frac{1}{4}f^4\left((i2\omega_0 I - L)^{-1} f^2(\phi_0,\phi_0),\phi_0,\overline{\phi_0},\overline{\phi_0}\right)\\ &+ \frac{1}{12}f^4\left((i2\omega_0 I + L)^{-1} f^2(\overline{\phi_0},\overline{\phi_0}),(\phi_0)^3\right)\\ &- \frac{1}{12}f^5\left((\phi_0)^3,(\overline{\phi_0})^2\right) - (\mu + 2\,\mathrm{Re}\,\mu)M_1\Big\rangle\,.\end{aligned} \tag{2.97}$$

The expression for ψ_β is obtained from Eq. (2.96) by replacing β instead of α. Note that the last term in Eq. (2.97) is zero because $\langle\phi_0^\dagger, M_1\rangle = 0$ according to the condition $\int_{-\pi/\omega_0}^{\pi/\omega_0} e^{-i\omega_0 t}\langle \phi_0^\dagger, u^{(n)}\rangle\, dt = 0$ which ensures cancellation of terms proportional to $\phi_0 e^{i\omega_0 t}$ and to its complex conjugate in $u^{(n)}$. The amplitude equation is therefore

$$\begin{aligned}\frac{dA}{dt} &= \Big[\epsilon^2\alpha_2\lambda_\alpha + \epsilon^2\beta_2\lambda_\beta \\ &\quad + \epsilon^4\left(\alpha_4\lambda_\alpha + \beta_4\lambda_\beta + \frac{\alpha_2^2}{2}\lambda_{\alpha\alpha} + \alpha_2\beta_2\lambda_{\alpha\beta} + \frac{\beta_2^2}{2}\lambda_{\beta\beta}\right)\Big]A \\ &\quad - \left(\epsilon^2 i\,\mathrm{Im}\,\mu + \epsilon^4\alpha_2\psi_\alpha + \epsilon^4\psi_\beta\beta_2\right) A\,|A|^2 - \epsilon^4\mu_2 A\,|A|^4\,, \end{aligned} \tag{2.98}$$

up to higher order terms. As before, the coefficient of A is simply $\lambda(\alpha,\beta) - \lambda(0,0)$. If $\beta_2 = 0$, this equation has the same form as the one-parameter bifurcation equation, all whose terms depend on β:

$$\begin{aligned}\frac{dA}{dt} &= \left[(\epsilon^2\alpha_2 + \epsilon^4\alpha_4)\lambda_\alpha(\beta) + \frac{\epsilon^4\alpha_2^2}{2}\lambda_{\alpha\alpha}(\beta)\right]A \\ &\quad - \left[\epsilon^2\mu(\beta) + \epsilon^4\alpha_2\psi_\alpha(\beta)\right] A\,|A|^2 - \epsilon^4\mu_2(\beta)A\,|A|^4\,. \end{aligned} \tag{2.99}$$

If we now set $\beta = \epsilon^2\beta_2$ and keep up to terms of order ϵ^4 in Eq. (2.99), we obtain

$$\frac{dA}{dt} = \epsilon^2\varphi\, A - (i\,\mathrm{Im}\,\mu(0) + \epsilon^2\eta)\, A\,|A|^2 - \epsilon^2\mu_2(0)\, A\,|A|^4\,, \tag{2.100}$$

$$\varphi = \alpha_2 + \epsilon^2\left[\alpha_4\lambda_\alpha(0) + \alpha_2\beta_2\lambda_{\alpha\beta}(0) + \frac{\alpha_2^2}{2}\lambda_{\alpha\alpha}(0)\right], \tag{2.101}$$

$$\eta = \beta_2\mu'(0) + \alpha_2\psi_\alpha(0)\,. \tag{2.102}$$

The bifurcation diagram corresponding to Eq. (2.100) results from the equation for $\rho = |A|$:

$$\frac{d\rho}{dt} = \epsilon^2\left(\mathrm{Re}\,\varphi - \epsilon^2\,\mathrm{Re}\,\eta\,\rho^2 - \epsilon^2\,\mathrm{Re}\,\mu_2(0)\,\rho^4\right)\rho\,. \tag{2.103}$$

All the terms are of the same order in this equation if $\mathrm{Re}\,\varphi = O(\epsilon^2)$, which implies $\alpha_2 = 0$. Then, $\varphi = \epsilon^2\alpha_4$ and $\eta = \beta_2\mu'(0) = \beta_2\psi_\beta$ and we have

$$\frac{d\rho}{dT} = (\alpha_4\,\mathrm{Re}\,\lambda_\alpha - \mathrm{Re}\,\mu'\beta_2\rho^2 - \mathrm{Re}\,\mu_2\rho^4)\,\rho\,, \tag{2.104}$$

where $T = \epsilon^4 t$. The nonzero stationary solutions of this equation correspond to periodic solutions of the original problem. They solve

$$x^2 + \frac{\mathrm{Re}\,\mu'}{\mathrm{Re}\,\mu_2}\beta_2 x - \alpha_4\frac{\mathrm{Re}\,\lambda_\alpha}{\mathrm{Re}\,\mu_2} = 0\,, \tag{2.105}$$

with $x = \rho^2 \ge 0$. Clearly, there is only one positive solution if $\alpha_4\,\mathrm{Re}\,\lambda_\alpha/\,\mathrm{Re}\,\mu_2 > 0$ (regardless of the sign of the coefficient of x), whereas if $\alpha_4\,\mathrm{Re}\,\lambda_\alpha/\,\mathrm{Re}\,\mu_2 < 0$, there are two positive solutions for $\beta_2\,\mathrm{Re}\,\mu'/\,\mathrm{Re}\,\mu_2 < 0$ and no positive solution for $\beta_2\,\mathrm{Re}\,\mu'/\,\mathrm{Re}\,\mu_2 > 0$. The corresponding bifurcation diagrams are depicted in Figure 2.11 for the case of $\mathrm{Re}\,\lambda_\alpha > 0$. If $\mathrm{Re}\,\lambda_\alpha < 0$, then the diagrams (a) and (b) correspond to $\mathrm{Re}\,\mu_2 < 0$ and the diagrams (c) and (d) correspond to $\mathrm{Re}\,\mu_2 > 0$ provided we interpret the horizontal axis as the product $\alpha_4\,\mathrm{Re}\,\lambda_\alpha$.

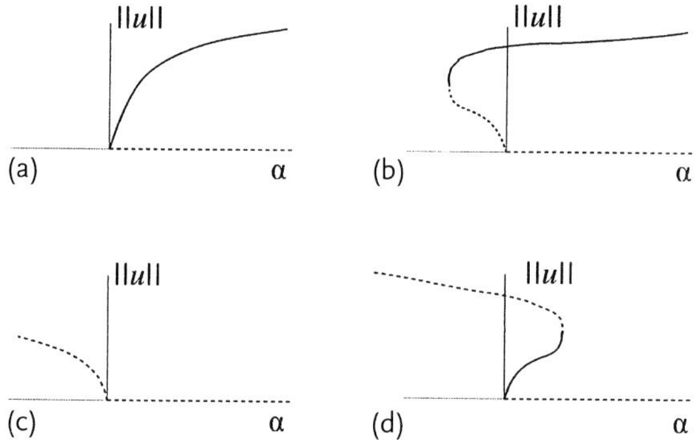

Fig. 2.11 Bifurcation diagrams for the degenerate Hopf bifurcation showing $\|u\|$ versus α for $\mathrm{Re}\,\lambda_\alpha > 0$. (a) $\mathrm{Re}\,\mu_2 > 0$ and $\beta\,\mathrm{Re}\,\mu' < 0$, (b) $\mathrm{Re}\,\mu_2 > 0$ and $\beta\,\mathrm{Re}\,\mu' \geq 0$, (c) $\mathrm{Re}\,\mu_2 < 0$ and $\beta\,\mathrm{Re}\,\mu' < 0$, (d) $\mathrm{Re}\,\mu_2 < 0$ and $\beta\,\mathrm{Re}\,\mu' \geq 0$.

References

1 Kuznetsov, Y.A. (2004) *Elements of Applied Bifurcation Theory*, Springer, New York.

2 Strogatz, S.H. (1994) *Nonlinear Dynamics and Chaos*, Westview Press, Cambridge.

3 Ott, E. (2002) *Chaos in Dynamical Systems*, Cambridge University Press, Cambridge.

4 Guckenheimer, J. and Holmes, P. (1983) *Nonlinear Oscillations, Dynamical Systems, and Bifurcations of Vector Fields*, Springer, New York.

5 Mandelbrot, B.B. (1982) *The Fractal Geometry of Nature*, Freeman, San Francisco.

6 Grassberger, P. and Procaccia, I. (1983) Measuring the strangeness of strange attractors. *Physica D*, **9**, 189.

7 Halsey, T.C., Jensen, M.H., Kadanoff, L.P., Procaccia, I. and Shraiman, B.I. (1986) Fractal measures and their singularities: the characterization of strange sets. *Phys. Rev. A*, **33**, 1141–1151.

8 Teitsworth, S.W. and Westervelt, R.M. (1984) Chaos and broadband noise in extrinsic semiconductors. *Phys. Rev. Lett.*, **53**, 2587–2590.

9 Gwinn, E.G. and Westervelt, R.M. (1986) Frequency locking, quasiperiodicity and chaos in extrinsic Ge. *Phys. Rev. Lett.*, **57**, 1060–1063.

10 Kevorkian, J. and Cole, J. (1996) *Multiple Scale and Singular Perturbation Methods*, Springer, New York.

11 Wallis, C.R. and Teitsworth, S.W. (1994) Hopf bifurcations and hysteresis in resonant tunneling diode circuits. *J. Appl. Phys.*, **76**, 4443–4445.

12 Bonilla, L.L. (2000) Chapman–Enskog method and synchronization of globally coupled oscillators. *Phys. Rev. E*, **62**, 4862–4868.

13 Chapman, S. and Cowling, T.G. (1970) *The Mathematical Theory of Non-Uniform Gases*, 3rd edn, Cambridge University Press, Cambridge.

14 Kogelman, S. and Keller, J.B. (1971) Transient behavior of unstable nonlinear systems with applications to the Bénard and Taylor problems. *SIAM J. Appl. Math.*, **20**, 619–637.

15 Poore, A.B. (1976) On the theory and application of the Hopf-Friedrichs bifurcation theory. *Arch. Ration. Mech. Anal.*, **60**, 371–393.

16 Cross, M.C. and Hohenberg, P.C. (1993) Pattern formation outside of equilibrium. *Rev. Mod. Phys.*, **65**, 851–1112.

17 Golubitsky, M. and Langford, W.F. (1981) Classification of unfoldings of degenerate Hopf bifurcations. *J. Diff. Eqn.*, **41**, 375–415.

3
Excitable Media I: Continuum Systems

3.1
Introduction

In this chapter, we begin by introducing the basic properties that distinguish excitable systems from other nonlinear dynamical systems. This is first done for the space-independent FitzHugh–Nagumo system which serves as a paradigm example for excitable behavior. We then introduce spatial diffusivity into the system which takes us to the study of a class of reaction-diffusion partial differential equations that also exhibit bistability and the possibility of nonlinear traveling waves. A central point of this chapter is the demonstration that localized nonlinear wave excitations, such as fronts and pulses, exist for only specific values of wavespeed. We show how these selected wavespeeds and the underlying wavefunctions are determined using phase plane methods. In addition, the dynamical stability of these nonlinear waves are assessable by studying the appropriate Schrödinger-type equations. The methods developed here find wide application in determining the speeds of nonlinear charge waves in electronic transport problems. The FitzHugh–Nagumo system provides an ideal model for introducing these nonlinear wave techniques due to its relative simplicity, and a number of the key results are analytical or well-approximated by analytical expressions.

3.2
Basic Excitability – the FitzHugh–Nagumo System

Many dynamical systems have a globally stable stationary solution such that an initial disturbance of this solution returns exponentially fast to it, no matter how large the disturbance may be. A classic example is provided by the gradient system,

$$\frac{du}{dt} = -\nabla_u V(u) ,$$

where $u \in \mathbb{R}^n$, and the potential $V(u)$ has a single global minimum at $u = 0$. Furthermore, it is assumed the determinant of the Hessian matrix $\nabla_u \nabla_u V(0)$ is

Nonlinear Wave Methods for Charge Transport. Luis L. Bonilla and Stephen W. Teitsworth

ISBN: 978-3-527-40695-1

not zero, and also that the ratio between the largest and smallest eigenvalues of the Hessian matrix is not very large.

However, there are other types of dynamical systems in which the system returns to its stable stationary solution only after a large excursion, provided that the initial disturbance surpasses a certain threshold. An important example of this behavior is provided by the FitzHugh–Nagumo (FHN) system of equations, which can be written in nondimensional form as

$$\epsilon \frac{du}{dt} = Au(2-u)(u-a) - v + I_a \,, \tag{3.1}$$

$$\frac{dv}{dt} = u - Bv \,, \tag{3.2}$$

with $0 < a < 2$. One example of the applicability of the FHN system (3.1) and (3.2) that should be mentioned is that it provides a good description of the dynamics of nerve conduction in a space-clamped membrane; a specific and famously studied neuron in the natural world for which FHN applies is the outer membrane of a squid giant axon threaded with a metallic conductor [1, 2]. In this case, the excitation variable u denotes the membrane potential and the *recovery* variable v is an outward-flowing ion current. The spatial dependence of u and v may be neglected because the axon is threaded with a metallic conductor. The cubic source term in Eq. (3.1) describes the nonlinear dependence of ionic current on the membrane potential. The constants A and B are selected so that for $I_a = 0$, the source terms in the FHN system are $O(1)$ for u and v of order 1, and the only stationary solution is $u = 0 = v$. A typical choice is $A = 1$ and $B = 0.5$ which we assume for much of our discussion. The constant $\epsilon > 0$ is the ratio between the characteristic time scales of the two variables. We take $\epsilon = 0.01 \ll 1$, that is, fast excitation and slow recovery.

The FitzHugh–Nagumo system and others having similar qualitative behavior are called *excitable systems*. We observe in Figure 3.1 that the *excitation variable* $u(t)$ performs a large excursion before it comes to the unique stationary solution $u = 0$, $v = 0$ provided that the applied current I_a is zero. In this case, we have used as initial conditions $(u_0, v_0) = (0.2, 0)$, and values $a = 0.1$ and $I_a = 0$. The same system also exhibits a sustained oscillatory response corresponding to an underlying stable limit cycle in the dynamical system. This behavior is shown in Figure 3.2 for the case $I_a = 1$. Note that the excitatory nature of the system is still clearly evident in the oscillatory time trace of $u(t)$.

In order to understand the general behavior that the FHN system can exhibit, it is helpful to plot the nullclines of the vector field for different values of the applied current I_a. This is shown in the following set of figures. Figure 3.3 corresponds to the basic excitatory response shown in Figure 3.1, Figure 3.4 corresponds to the oscillatory response shown in Figure 3.2, Figure 3.5 corresponds to behavior in which the response is again of simple excitatory character, and Figure 3.6 shows an example in which the system possesses bistability.

We now generalize to an excitable *medium*, that is, a spatially extended system, by examining two straightforward realizations based on the FHN system. Firstly,

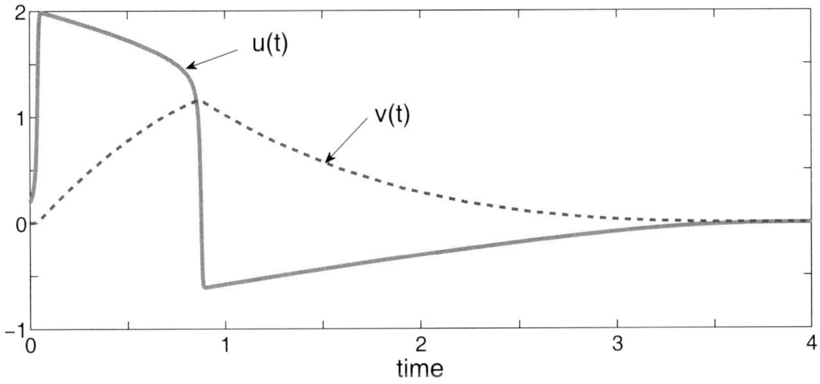

Fig. 3.1 Time evolution of the excitation variable $u(t)$ and recovery variable $v(t)$ showing excitable behavior in response to a sufficiently large disturbance from the stationary state ($I_a = 0$).

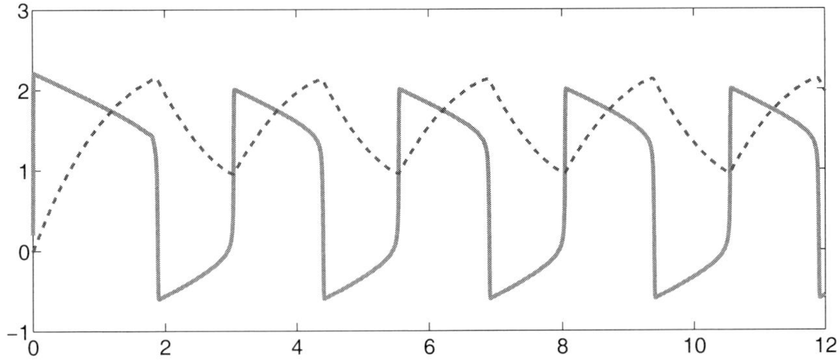

Fig. 3.2 Time evolution of the excitation variable $u(t)$ and recovery variable $v(t)$ showing oscillatory behavior for $I_a = 1$.

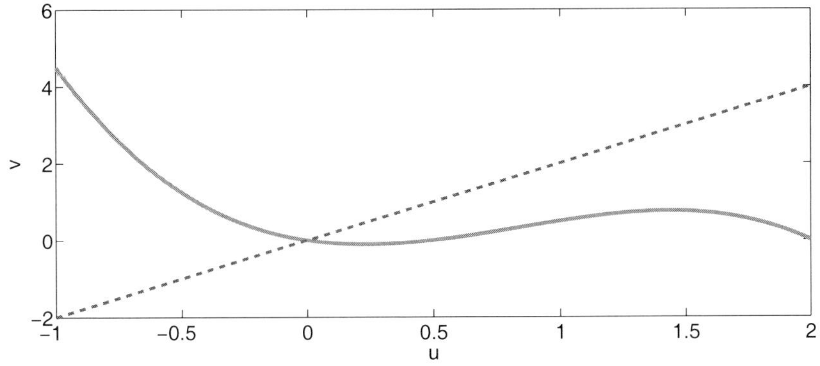

Fig. 3.3 Nullclines for the space-independent FHN model with $a = 0.1$ and $I_a = 0$ corresponding to an excitable system.

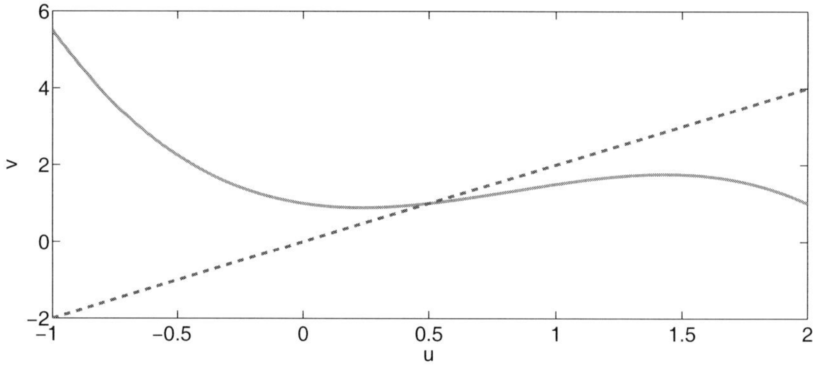

Fig. 3.4 Nullclines for $I_1 < I_a < I_2$ corresponding to an oscillatory system, $I_j = (B^{-1} + 4aA/3 - (2+a)Au_j/3)u_j$, $u_j = [2 + a + (-1)^j \sqrt{4 + a^2 - 2a}]/3$, $j = 1, 2$.

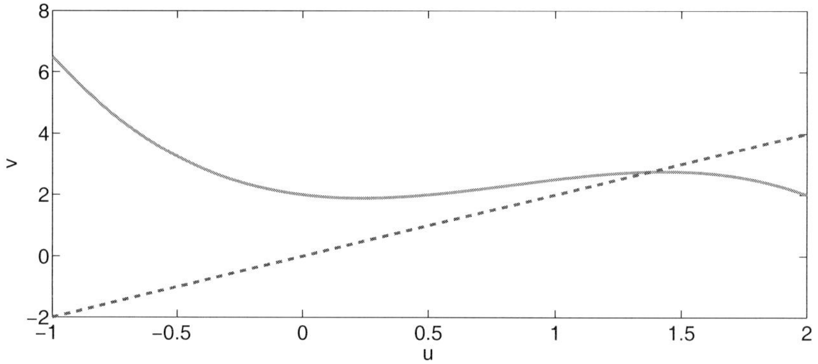

Fig. 3.5 Excitable system for $I_a > I_2$.

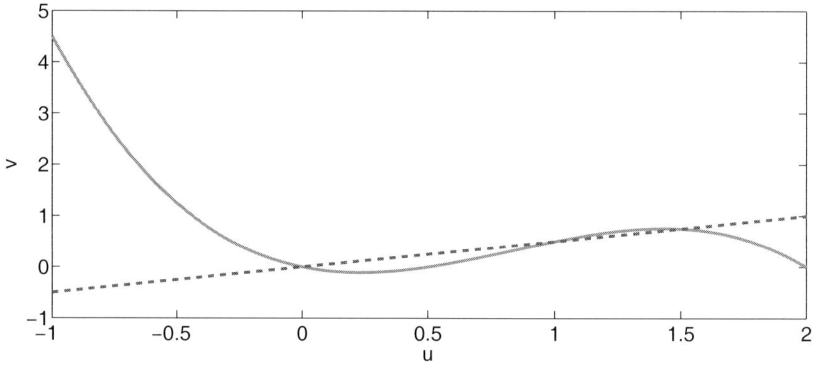

Fig. 3.6 Nullclines for the FHN system with $I_a = 0$, $B > B_1 = 4A/(2-a)^2$ which has three stationary solutions.

let us consider the *continuum* FHN system,

$$\epsilon \frac{\partial u}{\partial t} = \epsilon^2 \frac{\partial^2 u}{\partial x^2} + Au\,(2-u)(u-a) - v\,, \tag{3.3}$$

$$\frac{\partial v}{\partial t} = u - Bv\,, \tag{3.4}$$

where $-\infty < x < \infty$. Note that by rescaling x, we can set the diffusion coefficient in Eq. (3.3) equal to any arbitrarily chosen positive value; in the above rendering, the diffusion constant is effectively equal to ϵ. Equations (3.3) and (3.4) constitute a reaction-diffusion system and describe, for example, propagation of nerve impulses along non-myelinated fibers such as that of the squid giant axon.

However, in some instances, nerve fibers are coated with a lipid material called myelin with periodic gaps of exposure that are termed the nodes of Ranvier. This leads us to a second realization of spatially extended system, namely, the *discrete* FHN system

$$\epsilon \frac{du_n}{dt} = d\,(u_{n+1} - 2u_n + u_{n-1}) + Au_n(2-u_n)(u_n - a) - v_n\,, \tag{3.5}$$

$$\frac{dv_n}{dt} = u_n - Bv_n\,, \tag{3.6}$$

with $n = 0, \pm 1, \ldots$ Conduction in myelinated nerve fibers is more conveniently described by spatially discrete equations such as Eqs. (3.5) and (3.6). In them, u_n and v_n are the membrane potential and the recovery variable at the nth excitable membrane site (i. e., node of Ranvier). The discrete diffusive term is proportional to the difference in internodal currents through a given site. A dimensional version of Eqs. (3.5) and (3.6) was derived in the appendix of [3] from an equivalent-circuit model of myelinated nerves. For background on similar models, see, for example, [1, 4–6].

In the next section, we shall use matched asymptotic expansions to describe excitability and oscillations in the space-clamped system, Eqs. (3.1) and (3.2), for different values of the applied current I_a in the limit as $\epsilon \to 0$. In Section 3.4, we study wave front solutions of the scalar bistable reaction-diffusion equation, cf. Eqs. (3.3) and (3.4), and their stability. Wave fronts of the scalar bistable equation are the basis of a singular perturbation study of wave fronts, pulses and wave trains of the continuum FHN system considered in Section 3.5 below. We return to the study of the discrete FHN model, Eqs. (3.5) and (3.6), in the next chapter.

3.3
Matched Asymptotics: Excitability and Oscillations

In the limit $\epsilon \to 0+$, it is possible to approximate the evolution of a disturbance from the stable stationary solution in an excitable system by matched asymptotic expansions [7]. This is intuitively plausible when one looks at the time trace behavior in Figures 3.1 and 3.2. The regions in which the excitatory variable rapidly changes correspond to fast time dynamics, while the regions of gradual change correspond to the slow time scale. As $\epsilon \to 0$, this separation becomes precise. Below, we carry out the procedure in quantitative detail. Let us assume that $I_a = 0$ in Eq. (3.1), and that $u = A > a$ and $v = 0$ at $t = 0$. Let us denote by $u^{(1)}(v) < u^{(2)}(v) < u^{(3)}(v)$, $v_m < v < v_M$, the three solutions of the equation

$$v = Au(2-u)(u-a)\,. \tag{3.7}$$

The first stage of the time evolution occurs in the fast time scale $\tau = t/\epsilon$: the initial condition rapidly evolves exponentially to the value $u = u^{(3)}(0) = 2$ according to the approximate equations in the limit as $\epsilon \to 0$:

$$\frac{du}{d\tau} = Au(2-u)(u-a) - v\,, \tag{3.8}$$

$$\frac{dv}{d\tau} = 0\,. \tag{3.9}$$

We have $v = 0$ due to the initial condition and the slow nature of the v variable. The solution during this first stage is easily determined by expanding Eq. (3.8) about $u = 2$ and straightforward integration:

$$u^{\mathrm{I}}(\tau) = 2 + O(e^{-2(2-a)A\tau})\,, \quad v^{\mathrm{I}}(\tau) = 0\,,$$

where the order symbol $O(..)$ means that the Stage 1 solution is bounded by some positive constant multiplied by the enclosed expression in the limit $\tau \to \infty$.

After this first stage, the system evolves in the slow time scale t according to Eqs. (3.7) and (3.2). The matching conditions between Stages I and II are

$$|u^{\mathrm{I}}(t/\epsilon) - u^{\mathrm{II}}(t)| \ll 1\,, \quad |v^{\mathrm{I}}(t/\epsilon) - v^{\mathrm{II}}(t)| \ll 1\,,$$

which give

$$2 + O(e^{-2(2-a)At/\epsilon}) - u^{\mathrm{II}}(0) + O(t) \ll 1\,, \quad v^{\mathrm{II}}(0) + O(t) \ll 1\,.$$

The overlap time domain during which the approximate solutions of the two stages *match* is $\epsilon \ll t \ll 1$, provided $u^{\mathrm{II}}(0) = 2$ and $v^{\mathrm{II}}(0) = 0$. The solution of Eq. (3.7) compatible with $u(0) = 2$ is $u = u^{(3)}(v)$, which is inserted into Eq. (3.2) to yield

$$\frac{dv}{dt} = u^{(3)}(v) - Bv\,. \tag{3.10}$$

Since $u^{(3)}(v) > Bv$, in this second stage, $v(t)$ increases until it reaches the maximum of the nullcline, Eq. (3.7), at a time $t_e = \int_0^{v_M} dv/[u^{(3)}(v) - Bv]$. The minimum and maximum of the nullcline in Eq. (3.7) may be expressed analytically and are given by

$$(u_m, v_m) = \left(\frac{2 + a - \sqrt{4 + a^2 - 2a}}{3}, A(2-u_m)(u_m - a)u_m\right)\,,$$

and

$$(u_M, v_M) = \left(\frac{2 + a + \sqrt{4 + a^2 - 2a}}{3}, A(2 - u_M)(u_M - a)u_M \right),$$

respectively. For t close to t_e, we have $u = u^{(3)}(v(t)) = u_M + O(t - t_e)$ and $v(t) = v_M + (u_M - B v_M)(t - t_e) + O((t - t_e)^2)$, which we obtain from Eq. (3.10). At this point, Stage III commences and we need to change to a fast variable $(t - t_e)/\epsilon$. During this stage, the solution jumps to the first branch of the nullcline (3.7) according to Eq. (3.8) with $\tau = (t - t_e)/\epsilon$. The matching conditions for the overlap domain $\epsilon \ll (t_e - t) \ll 1$ common to Stages II and III are:

$$u^{\text{II}}(t) - u^{\text{III}}((t - t_e)/\epsilon) = u_M + O(t - t_e) - u^{\text{III}}(-\infty) + O\left(\frac{\epsilon}{t_e - t}\right) \ll 1,$$
$$v^{\text{II}}(t) - v^{\text{III}}((t - t_e)/\epsilon) = v_M + O(t - t_e) - v^{\text{III}} \ll 1.$$

These conditions imply that the constant value of v during Stage III is v_M and that $u(-\infty) = u_M$. To see how $u^{\text{III}}(\tau)$ departs from $\tau = (t - t_e)/\epsilon \to -\infty$, we rewrite the cubic nonlinear source as $f(u) - v_M = -A(u - u_M)^2(u - u_M + \sqrt{4 + a^2 - 2a})$ and integrate $\tau = \int du/[f(u) - v_M] \sim A^{-1}(4 + a^2 - 2a)^{-1/2}/(u - u_M)$, as $u \to u_M$. This gives $u - u_M \sim A^{-1}(4 + a^2 - 2a)^{-1/2}/\tau$, which provides the order term written above.

At the end of Stage III, as $\tau \to +\infty$, u tends to $u_M - \sqrt{4 + a^2 - 2a} = u^{(1)}(v_M)$ as $u = u^{(1)}(v_M) + O(e^{-A(4+a^2-2a)\tau})$ while $v = v_M$. The equation describing the fourth (recovering) stage is

$$\frac{dv}{dt} = u^{(1)}(v) - Bv, \tag{3.11}$$

with initial condition $v(t_e) = v_M$. In fact, the matching conditions between Stages III and IV are

$$u^{\text{III}}((t - t_e/\epsilon)) - u^{\text{IV}}(t) = u^{(1)}(v_M) + O(e^{-A(4+a^2-2a)(t-t_e)/\epsilon})$$
$$- u^{\text{IV}}(t_e) + O(t - t_e) \ll 1,$$
$$v_M - v^{\text{IV}}(t_e) + O(t - t_e) \ll 1.$$

Thus, we have $u^{\text{IV}}(t_e) = u^{(1)}(v_M)$, which selects $u = u^{(1)}(v)$ as the relevant solution of Eq. (3.7), $v^{\text{IV}}(t_e) = v_M$ and the overlap time domain is $\epsilon \ll (t - t_e) \ll 1$. During Stage IV, v decreases to zero according to Eq. (3.11). The excitation period in Figure 3.1 is thus approximately described.

We can write a uniform approximation to the solution using the results we have from all stages as

$$u = \theta(t_e - t)\,[u^{\text{I}}(t/\epsilon) + u^{\text{II}}(t) - 2 + u^{\text{III}}((t - t_e)/\epsilon) - u_M]$$
$$+ \theta(t - t_e)\,[u^{\text{III}}((t - t_e)/\epsilon) - u^{(1)}(v_M) + u^{\text{IV}}(t)],$$
$$v = v^{\text{II}}(t)\,\theta(t_e - t) + v^{\text{IV}}(t)\,\theta(t - t_e). \tag{3.12}$$

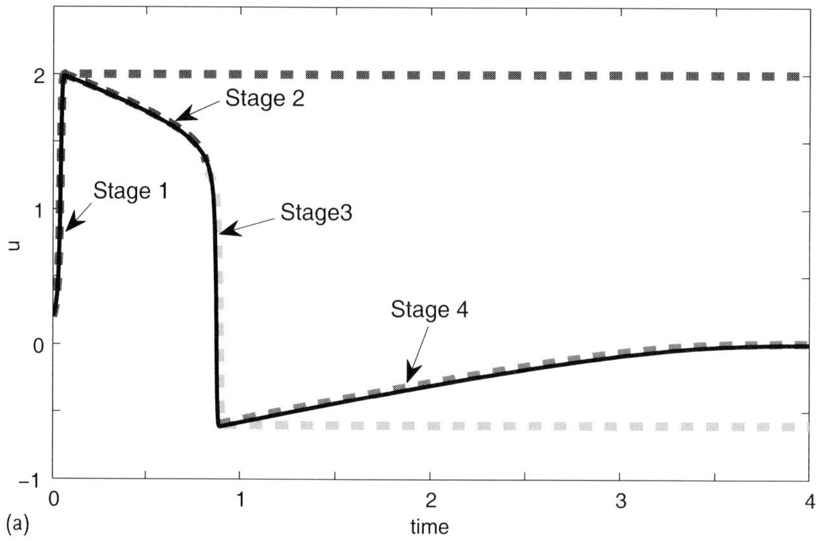

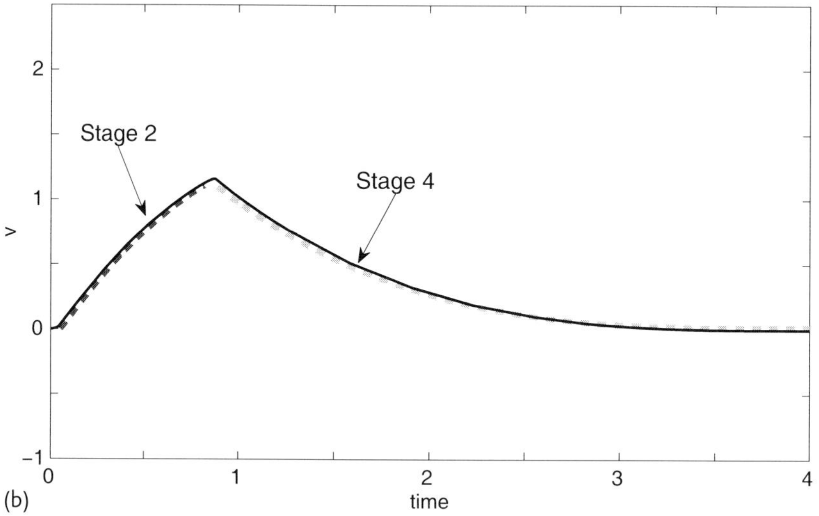

Fig. 3.7 Comparison of the numerical solution of the FHN system to a uniform approximation using match asymptotics. Inner and outer solutions corresponding to the different stages are also shown. (a) Excitatory variable $u(t)$, and (b) recovery variable $v(t)$.

Here $\theta(u)$ is the Heaviside unit step function and we should understand $u^{\mathrm{II}}(t) = u^{(3)}(v^{\mathrm{II}}(t))\,\theta(u - u_M)$, so that $u^{\mathrm{II}} = 0$ for values of u below u_M. Figure 3.7 shows the inner and outer approximations at each stage, the uniform approximation Eq. (3.12) and compares them to the numerical solution of the FHN system.

If we have $I_1 < I_a < I_2$, the stationary solution is on the second branch $u^{(2)}(v)$ and the system is oscillatory, and the solutions approach a limit cycle described by two slow stages. The excitatory stage is given by Eq. (3.10) from $v = v_m$ to $v = v_M$ and has a duration

$$T_e = \int_{v_m}^{v_M} \frac{dv}{u^{(3)}(v) - Bv}, \tag{3.13}$$

while the recovery stage is given by Eq. (3.11) from $v = v_M$ to $v = v_m$ and has a duration

$$T_r = \int_{v_m}^{v_M} \frac{dv}{Bv - u^{(1)}(v)}. \tag{3.14}$$

The two fast stages are described by Eqs. (3.8) and (3.9) with $v = v_m$ or $v = v_M$. The period of the oscillation is simply $T = T_e + T_r$, except for terms of order ϵ which include the duration of the fast stages.

3.4
The Scalar Bistable Equation; Wave Pulses as Heteroclinic Connections

The discussion in the previous section suggests that well-separated time scales can also be used to treat stable nonlinear wave solutions in the related spatially extended system. We begin by discussing a one-component spatially extended system that is closely related to the continuum FHN system, namely, the scalar bistable reaction-diffusion equation

$$\frac{\partial u}{\partial \tau} = \frac{\partial^2 u}{\partial x^2} + f(u) - w, \quad -\infty < x < +\infty,\ t > 0. \tag{3.15}$$

Here, w is a real parameter and $f(u) = w$ has three solutions $u^{(1)}(w) < u^{(2)}(w) < u^{(3)}(w)$ with $sgn(f'(u^{(j)})) = (-1)^j$ for w in a certain interval $w_m < w < w_M$ (here $f'(u) = df/du$).

By using comparison principles, it is possible to show the following statements [8]:

- For fixed w, an initial condition which tends to $u^{(1)}(w)$ (respectively, $u^{(3)}(w)$), as $x \to \pm\infty$ and such that $u(x,0) < u^{(2)}(w)$ (resp., $u(x,0) > u^{(2)}(w)$) for all x, evolves to the *constant solution* $u^{(1)}(w)$ (resp., $u^{(3)}(w)$), as $t \to +\infty$.
- Let the initial condition $u(x,0)$ be monotone increasing (resp., decreasing) and let it tend to $u^{(1)}(w)$ (resp., $u^{(3)}(w)$) as $x \to -\infty$, and to $u^{(3)}(w)$ (resp., $u^{(1)}(w)$), as $x \to +\infty$. Then, the solution of Eq. (3.15) evolves to a *wave front solution* of the form $u(x,t) = U(x - ct)$ with $U(\pm\infty) = u^{(2\pm1)}(w)$ (resp., $U(\pm\infty) = u^{(2\mp1)}(w)$).

In other words, initial conditions that approach one of the stable uniform stationary solutions at $|x| \to \infty$ tend to that stable uniform stationary solution, whereas initial conditions that connect the different stable solutions tend to a wave *front* joining the two stable uniform stationary solutions. Below we find the form of the wavefront solutions and present a phase plane argument for finding the wavespeed.
In the case of the infinite spatial interval, fronts are easy to construct by transforming to a *co-moving reference frame* and then using phase plane arguments. Provided that there exists an increasing wave front with $U(-\infty) = u^{(1)}(w)$ and $U(+\infty) = u^{(3)}(w)$, Eq. (3.15) becomes

$$-cU' = U'' + f(U) - w\,, \quad \xi = x - ct\,, \tag{3.16}$$

where the prime denotes differentiation with respect to the co-moving variable ξ, and c is the assumed wavespeed. By multiplying Eq. (3.16) by U' and integrating over ξ, we obtain

$$-c\int_{-\infty}^{\infty} U'^2 d\xi = \int_{u^{(1)}(w)}^{u^{(3)}(w)} [f(u) - w]\,du\,. \tag{3.17}$$

Then, the velocity of an *increasing* wave front has opposite sign to the integral of the source term in the right-hand side of Eq. (3.17). Note that $c = 0$ for the unique value of $w = w_0$ that satisfies an *equal area* rule of the form

$$\int_{u^{(1)}(w_0)}^{u^{(3)}(w_0)} [f(u) - w_0]\,du = 0\,. \tag{3.18}$$

For $w < w_0$, increasing wave fronts move to the left ($c < 0$), whereas they move to the right ($c > 0$) for $w > w_0$. The same calculation for a *decreasing* wave front indicates that its velocity has the same sign as the integral in the right-hand side of Eq. (3.17). Thus, decreasing wave fronts move from right to left ($c < 0$) for $w > w_0$ and from left to right ($c > 0$) for $w < w_0$, that is, in the opposite direction as for increasing wave fronts.

It can be established that a wave front solution exists for $w \neq w_0$ by the following argument. Suppose $w < w_0$ for definiteness (so that $\int_{u^{(1)}(w)}^{u^{(3)}(w)} [f(u) - w]\,du > 0$ and $c < 0$ for an increasing wave front), and let us define $V(\xi) = U'(\xi)$. From Eq. (3.16) and provided that $c = 0$, we obtain

$$\frac{V^2}{2} + \int_{u^{(1)}(w)}^{U} [f(u) - w]\,du = 0\,. \tag{3.19}$$

If this trajectory were to reach $u^{(3)}(w)$ for a non-zero value $V > 0$, we would have

$$\frac{V^2}{2} + \int_{u^{(1)}(w)}^{u^{(3)}(w)} [f(u) - w]\,du = 0\,, \tag{3.20}$$

which would imply $\int_{u^{(1)}(w)}^{u^{(3)}(w)} [f(u) - w]\, du < 0$ in contradiction with our assumption $w < w_0$. Therefore, U cannot reach $u^{(3)}(w)$ in a finite interval of w. Additionally, it cannot stay in the first quadrant $U > 0$, $V > 0$, because $V > 0$ implies that U is always increasing there. Therefore, the trajectory must intersect $V = 0$ at some value $U < u^{(3)}(w)$ and it cannot be the desired heteroclinic connection between the saddle points $(u^{(1)}(w), 0)$ and $(u^{(3)}(w), 0)$.

Let us now examine the conditions for a heteroclinic connection to exist. In the process, we prove the existence of this connection type and the associated wave pulse. Suppose now that $-c > 0$ is sufficiently large. In the (U, V) phase plane, the slope of the unstable trajectory leaving the saddle point $(u^{(1)}(w), 0)$ is the positive root of $\lambda^2 + c\lambda + f_1' = 0$ (with $f_1' = f'(u^{(1)}(w))$), $\lambda = [|c| + \sqrt{c^2 - 4f_1'}]/2 > |c|$. Let K be the smallest positive number for which $[f(u) - w]/[u - u^{(1)}(w)] \leq K$ for $u^{(1)}(w) < u < u^{(3)}(w)$ and let σ be any positive number. K is the slope of the tangent to $f(u) - w$ (at some u between $u^{(2)}$ and $u^{(3)}$) that passes through $(u^{(1)}(w), 0)$. On the line $V = \sigma[U - u^{(1)}(w)]$, the slope of trajectories crossing it satisfies

$$\frac{dV}{dU} = \frac{w - f(U)}{V} - c = \frac{w - f(U)}{\sigma[U - u^{(1)}(w)]} + |c| \geq |c| - \frac{K}{\sigma}. \qquad (3.21)$$

By selecting a large enough $|c|$, we make sure that $|c| - K/\sigma > \sigma$ so that trajectories that are above the line $V = \sigma[U - u^{(1)}(w)]$ remain above it. Thus, the trajectory leaving the saddle point $(u^{(1)}(w), 0)$ starts above a given straight line $V = \sigma[U - u^{(1)}(w)]$ for large enough $|c|$. Therefore, this trajectory cannot intersect the axis $V = 0$. Hence, we have found that for $c = 0$, the separatrix leaving the saddle point $(u^{(1)}(w), 0)$ with positive slope intersects the $V = 0$ axis at some $u < u^{(3)}(w)$, and that for large enough $|c|$, this separatrix does not intersect the $V = 0$ axis for $u > u^{(1)}(w)$. Since trajectories continuously depend on the parameters of the problem, there must be a negative value of c for which there is a heteroclinic connection between the two saddles. That is, the sought wave front which exists for a unique value of c. A sketch is provided in Figure 3.8.

Let $c_+(w)$ (resp. $c_-(w)$) be the velocity of the increasing (resp. decreasing) wave front corresponding to a given w. Clearly, Eq. (3.16) shows that a decreasing front

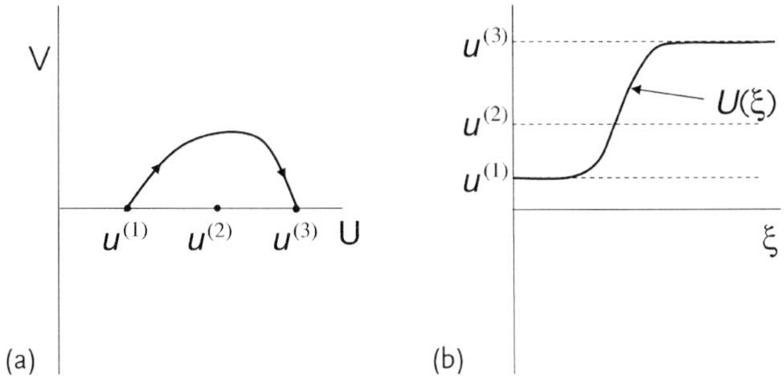

Fig. 3.8 Sketch showing the heteroclinic connection in the effective phase plane (a) and the associated traveling wave pulse in space (b).

is obtained by replacing $-\xi \to \xi$ and $c \to -c$ in the profile of an increasing front. Then,

$$c_+(w) = -c_-(w)\,. \tag{3.22}$$

3.4.1
Wave Fronts Near $w = w_0$ and a Formula for dc/dw

It is possible to show that $c_+(w)$ is an increasing function of w and therefore, $c_-(w)$ is a decreasing function for $w_m < w < w_M$. For a general source $f(u)$, we can construct the wave fronts and determine their velocity near a given value $w = W_0$ by using a standard regular (non-singular) perturbation theory. Suppose now that $w = W_0 + \epsilon w_1$, with $\epsilon \ll 1$. The perturbed wave front will have a nonzero velocity which we assume to be $c = c_0 + c_1\epsilon + O(\epsilon^2)$, and its profile will be

$$U(\xi;\epsilon) = U_0(\xi) + \epsilon\, u_1(\xi) + O(\epsilon^2)\,. \tag{3.23}$$

Here, $\xi = x - c(\epsilon)\tau$ and $U_0(x - c_0\tau)$ is the unperturbed front (which may be increasing or decreasing) corresponding to $w = W_0$. The equation for u_1 is obtained by equating terms of order ϵ in Eq. (3.16)

$$u_1'' + c_0 u_1' + f'(U_0)u_1 = w_1 - c_1 U_0'\,. \tag{3.24}$$

Since $U_0(-\infty) = u^{(1)}(W_0)$ and $U(+\infty) = u^{(3)}(W_0)$ for an increasing front (and similar expressions for a decreasing front), an expansion of the boundary conditions for the perturbed front yields

$$U(-\infty) = u^{(1)}(w) = u^{(1)}(W_0) + \frac{\epsilon\, w_1}{f'(u^{(1)}(W_0))} + O(\epsilon^2)\,,$$
$$U(\infty) = u^{(3)}(w) = u^{(3)}(W_0) + \frac{\epsilon\, w_1}{f'(u^{(3)}(W_0))} + O(\epsilon^2)\,,$$

with similar expressions for a decreasing front. Then, $u_1(\xi)$ satisfies the boundary conditions

$$u_1(-\infty) = \frac{w_1}{f'(u^{(1)}(W_0))}\,, \qquad u_1(\infty) = \frac{w_1}{f'(u^{(3)}(W_0))}\,. \tag{3.25}$$

Due to translational invariance of the unperturbed wave fronts, $u = U_0'(\xi)$ satisfies the homogeneous linear equation $u'' + c_0 u' + f'(U_0)u = 0$ with homogeneous boundary conditions $u(\pm\infty) = 0$. Note that Eq. (3.24) can be written in the self-adjoint form

$$\left(e^{c_0\xi/2}u_1\right)'' + \left[f'(U_0) - \frac{c_0^2}{4}\right]\left(e^{c_0\xi/2}u_1\right) = \left(w_1 - c_1 U_0'\right)e^{c_0\xi/2}\,, \tag{3.26}$$

and that $(e^{c_0\xi/2}U_0')$ is a solution of the homogeneous self-adjoint problem Eq. (3.26) with zero boundary conditions at $\xi = \pm\infty$. Then, if we multiply Eq. (3.26) by

$(e^{c_0\xi/2}U_0')$ and integrate by parts (note that the contributions from the endpoints of the integrals are zero), we find the following solvability condition which yields $dc/dw = c_1/w_1$:

$$0 = w_1 \int_{-\infty}^{\infty} e^{c_0\xi}\, U_0' d\xi - c_1 \int_{-\infty}^{\infty} e^{c_0\xi}\, U_0'^2 d\xi \,.$$

Hence,

$$\frac{dc}{dw} = \frac{\int_{-\infty}^{\infty} e^{c_0\xi}\, U_0'(\xi)\, d\xi}{\int_{-\infty}^{\infty} e^{c_0\xi} [U_0'(\xi)]^2 d\xi} \,. \tag{3.27}$$

It is straightforward to check that the integrals in the right-hand side of Eq. (3.27) are convergent. Equation (3.27) shows that the sign of dc/dw is positive for an increasing wave front ($U_0' > 0$) and negative for a decreasing wave front ($U_0' < 0$).

If $W_0 = w_0$, $c_0 = 0$, then the integrals in Eq. (3.27) become

$$\frac{dc}{dw}(w_0) = \frac{u^{(3)}(w_0) - u^{(1)}(w_0)}{\int_{u^{(1)}(w_0)}^{u^{(3)}(w_0)} V(u)\, du} \,, \tag{3.28}$$

for an increasing wave front and minus this value for a decreasing wave front. We have used $V(U) = U_0'(\xi)$ to change variable in the denominator of Eq. (3.27), which is clearly half the area enclosed by the two heteroclinic orbits that connect the two saddles $(u^{(1)}(W_0), 0)$ and $(u^{(3)}(W_0), 0)$ in the phase plane. In fact, note that at $w = w_0$, $c = 0$, we find an energy integral of Eq. (3.16) by multiplying it by U' and integrating the result:

$$\frac{1}{2}\, U'^2 + \int_{u^{(1)}(w_0)}^{U} [f(u) - w_0]\, du = E \,.$$

Clearly, $E = 0$ for the constant solutions $U' = 0$, $U = u^{(1)}(w_0)$ and $U = u^{(3)}(w_0)$. These two solutions are saddle points in the phase plane (U, U') having the same energy. Then, the separatrices joining these two saddles also have zero energy and are the desired wave fronts. They can be found by solving the previous equation with $E = 0$:

$$\xi - \xi_0 = \int_{u^{(1)}(w_0)}^{U} \frac{du}{\sqrt{2\, F(u, w_0)}} \,, \qquad F(u, w) = \int_{u^{(1)}(w)}^{u} [w - f(u)]\, du \,,$$

for the increasing front with $U' > 0$. The decreasing front is given by replacing $-(\xi - \xi_0)$ instead of $(\xi - \xi_0)$ in the previous formula. In the phase plane (U, V), the decreasing front is obtained from the increasing front by the transformation: $\xi \to -\xi$ and $V \to -V$. Thus, the area enclosed between the two separatrices is twice the area comprised between one separatrix and the U axis in the (U, V) phase plane.

3.4.2
Wave Fronts for a Cubic Source

In the case of a cubic source, $f(u) = Au(2-u)(u-a)$, with $0 < a < 2$, the fronts and their velocity can be explicitly found. In fact, $f(u) - w = -A(u-u^{(1)})(u-u^{(2)})(u-u^{(3)})$ in the cubic case. Suppose that the wave front has the expression

$$V = -B(U-u^{(1)})(U-u^{(3)})\,, \tag{3.29}$$

which certainly connects the saddles $(u^{(1)}(w), 0)$ and $(u^{(3)}(w), 0)$ with $V > 0$. By inserting this assumed form in the phase plane equation,

$$\frac{dV}{dU} = \frac{w - f(U)}{V} - c\,, \tag{3.30}$$

we find

$$-B(2U - u^{(1)} - u^{(3)}) = -\frac{A}{B}(U - u^{(2)}) - c\,.$$

The terms containing U are canceled provided that $B = \sqrt{A/2}$, which then yields

$$c_+(w) = -\sqrt{\frac{A}{2}}\,[u^{(1)}(w) + u^{(3)}(w) - 2u^{(2)}(w)]\,, \tag{3.31}$$

where we have written $c = c_+(w)$ to emphasize that we have considered an increasing front. Note that $c'_+(w) > 0$ because $du^{(j)}/dw = 1/f'(u^{(j)}(w))$ and $f'(u^{(j)}(w)) < 0$ for $j = 1, 3$, whereas $f'(u^{(2)}(w)) > 0$. For a decreasing front of a cubic nonlinearity,

$$c_-(w) = \sqrt{\frac{A}{2}}\,[u^{(1)}(w) + u^{(3)}(w) - 2u^{(2)}(w)]\,. \tag{3.32}$$

With Eqs. (3.31) and (3.32), it is immediately obvious that $c_+(w)$ increases with w and that $c_-(w)$ decreases. Integration of Eq. (3.29) provides the exact profile of the increasing front,

$$U(\xi) = \frac{u^{(1)} + u^{(3)}}{2} + \frac{u^{(3)} - u^{(1)}}{2}\,\tanh\left[\sqrt{\frac{A}{2}}\,\frac{u^{(3)} - u^{(1)}}{2}\,(\xi - \xi_0)\right], \tag{3.33}$$

after some algebra. Figure 3.9 shows the functions $c_+(w)$ and $c_-(w)$ for $0 < a < 1$ and for $1 < a < 2$.

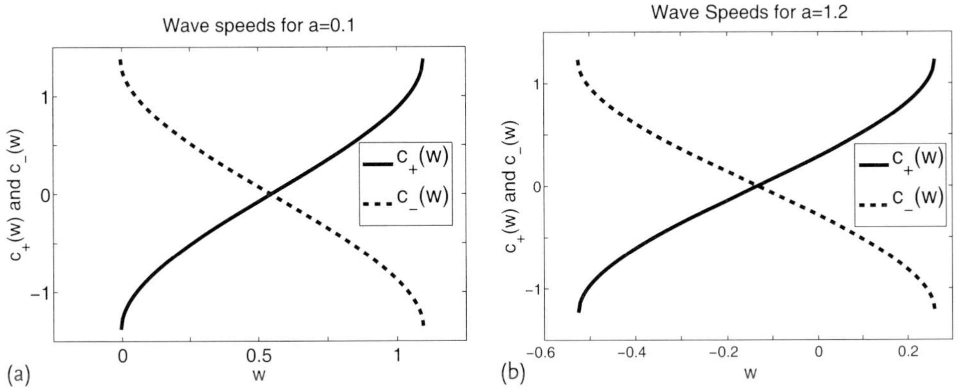

Fig. 3.9 Velocities $c_+(w)$ and $c_-(w)$ of the increasing and decreasing wave fronts of Eq. (3.15) for a cubic source $f(u) = Au(2-u)(u-a)$. (a) $0 < a < 1$, (b) $1 < a < 2$.

3.4.3 Linear Stability of the Wave Fronts

There is an elegant and relatively simple argument to show that wave fronts constructed in the previous sections are linearly stable. Let us consider a small disturbance about a wave front solution:

$$u(\xi, \tau) = U(\xi) + \epsilon\, e^{\lambda\tau}\, W(\xi)\,. \tag{3.34}$$

By inserting Eq. (3.34) in Eq. (3.15), we obtain the following eigenvalue problem for W:

$$W'' + c\,W' - [f'(U) + \lambda]\, W = 0\,, \quad W(\pm\infty) = 0\,. \tag{3.35}$$

We may convert this problem into the form of a stationary Schrödinger equation through the use of the transformation $W = e^{-c\xi/2}\psi$ as follows

$$-\psi'' + \left[\frac{c^2}{4} - f'(U) + \lambda\right]\psi = 0\,, \quad \psi(\pm\infty) = 0\,. \tag{3.36}$$

Here, $W = U'$ is a solution of the eigenvalue problem Eq. (3.35) with eigenvalue $\lambda = 0$. However, $-\lambda$ is the energy of the Schrödinger equation, Eq. (3.36), and we know that there is a smaller eigenvalue $-\lambda_1$ that the others are ordered so that $-\lambda_n < -\lambda_{n+1}$, and that the eigenfunction corresponding to $-\lambda_n$ has $n-1$ real zeros. Since U' does not vanish, the corresponding energy is $-\lambda_1 = 0$. Then, all of the other energies are positive which means all other eigenvalues λ_n are negative. Thus, the wave fronts are linearly stable. Nonlinear stability of the wave fronts is provided by the comparison principle discussed in the next chapter [8].

3.5 Traveling Waves of the FitzHugh–Nagumo System

3.5.1 Wave Fronts

If the FHN system (3.1) and (3.2) has three stationary solutions ($B > B_1$), then the diffusive system (3.3) and (3.4) may have increasing and decreasing wave fronts joining the two saddles $(0, 0)$ and (u_*, v_*). We can show that a step-like initial condition that approaches (u_*, v_*) as $x \to -\infty$ and $(0, 0)$ as $x \to +\infty$ evolves toward a stable wave front. In fact, for most values of x, Eq. (3.3) can be approximated by $f(u) - v = 0$ and therefore, Eq. (3.4) becomes

$$\frac{\partial v}{\partial t} = u^{(j)}(v) - Bv\,, \quad j = 1, 3\,. \tag{3.37}$$

Most of the points for $x < 0$ will follow Eq. (3.37) with $j = 3$ and therefore, will evolve toward v_* with $u^{(3)}(v_*) = u_*$. Similarly, most of the points with $x > 0$ will follow Eq. (3.37) with $j = 1$ and therefore, will evolve toward $v = 0 = u$.

Suppose that the initial condition $u(x,0)$ is monotonically decreasing from $u(-\infty) = u_*$ to $u(+\infty) = 0$ and that $v(x,0) = 0$. Let $X_-(0)$ be the point at which $u(x,0) = a$. According to the previous discussion, we should solve Eq. (3.37) with $j = 3$ for $x < X_-(0)$ and with $j = 1$ for $x > X_-(0)$. As time elapses, v and X_- change, but the $x = X_-(t)$ will be such that $u(x,t) \to u^{(3)}(v(X_-, t))$ as $x \to X_-(t)-$ and $u(x,t) \to u^{(1)}(v(X_-, t))$ as $x \to X_-(t)+$. Let $v_-(t) = v(X_-(t), t)$ be the instantaneous value of v corresponding to the point $X_-(t)$. To find out how u jumps down from $u^{(3)}(v_-)$ to $u^{(1)}(v_-)$, we define the fast coordinate $\xi = [x - X_-(t)]/\epsilon$ moving with the front and insert in Eq. (3.3). The leading order term of the result yields the following eigenvalue problem for the velocity $c_-(v_-) = dX_-/dt$:

$$\frac{\partial^2 u}{\partial \xi^2} + c_-(v_-)\frac{\partial u}{\partial \xi} + f(u) - v_- = 0\,, \quad \xi = \frac{x - X_-(t)}{\epsilon}\,, \tag{3.38}$$

$$u(-\infty) = u^{(3)}(v_-)\,, \quad u(+\infty) = u^{(1)}(v_-)\,. \tag{3.39}$$

For $x > X_-(t)$ and our initial conditions, we have $v(x,t) = 0$. However, even initially if $v(x,0) \neq 0$, we still find that $v(x,t) \to 0$ as $t \to +\infty$, so that $v_-(t) \to 0$ and therefore, $c_-(v_-) \to c_-(0)$, and $X(t) \sim c_-(0)t$ as $t \to +\infty$. For $x < X_-(t)$, v will become a function of $y = x - X_-(t)$ satisfying

$$-c_-(0)\,\frac{\partial v}{\partial y} = u^{(3)}(v) - Bv\,, \quad v(0) = u^{(3)}(0)\,, \tag{3.40}$$

so that

$$y = x - X_-(t) = c_-(0) \int_{u^{(3)}(0)}^{v} \frac{dv}{Bv - u^{(3)}(v)}\,. \tag{3.41}$$

For a cubic source, the explicit formula (3.32) yields $c_-(0) = (2A)^{1/2}(1-a)$, and the decreasing front moves to the right if $0 < a < 1$. Note that in this case, $\partial v/\partial y < 0$ according to Eq. (3.40) so that v increases as $|y|$ increases to the left of $X_-(t)$. Then, u decreases as $x < X_-(t)$ becomes more negative until $v = v_*$ and $u^{(3)}(v_*) = u_*$ is reached. An increasing front can be similarly constructed.

If $a > 1$, the previous construction does not work. In this case, we have $c_-(0) < 0$ and the decreasing front should move to the left. However then, $\partial v/\partial y > 0$ according to Eq. (3.40) so that v decreases as $|y|$ increases to the left of $X_-(t)$. Then, u increases as $x < X_-(t)$ and (u, v) moves in the opposite direction to that where the fixed point $(u, v) = (u_*, v_*)$ is located. This means that we cannot construct a steady decreasing wave front in the case $a > 1$. Similar arguments show that we cannot construct a steady increasing wave front either.

3.5.2
Pulses of the FHN System

We adapt the asymptotic construction in [9]. For $B < B_1$, the FHN system is either excitable or oscillatory depending on the value of the applied current I_a. In the excitable case, we may have either single pulses or periodic wave trains. In the oscillatory case, we may construct periodic wave train solutions. Let us consider the excitable case with $I_a = 0$ with a single uniform stationary solution at $u = 0 = v$. Let us imagine an initial condition with a single maximum of sufficient height in the variable u (located at $x = 0$) such that u and v tend to zero as $x \to \pm\infty$. Such an initial condition will evolve towards a pulse moving either to the left or to the right depending on the shape of the source term in the equation for the excitation variable.

Suppose the pulse moves to the right. The regions sufficiently far from $x = 0$ will obey Eq. (3.37) with $j = 1$, whereas there will be an excited region (initially surrounding $x = 0$) where Eq. (3.37) with $j = 3$ will be satisfied. The excited region will be separated from the recovering region by a leading decreasing front at $x = X_-(t)$ and by a trailing increasing front at $x = X_+(t) < X_-(t)$. While v is continuous at the fronts (with values $v = v_\pm(t)$ at $x = X_\pm(t)$), in the outer approximation $u(x, t)$ jumps from $u^{(1)}(v_-)$ to $u^{(3)}(v_-)$ at the leading front and from $u^{(3)}(v_+)$ to $u^{(1)}(v_+)$ at the trailing front. The leading front is again a solution of the eigenvalue problem Eqs. (3.38) and (3.39), whereas the trailing front solves the following eigenvalue problem:

$$\frac{\partial^2 u}{\partial \xi^2} + c_+(v_+)\frac{\partial u}{\partial \xi} + f(u) - v_+ = 0\,, \quad \xi = \frac{x - X_+(t)}{\epsilon}\,, \tag{3.42}$$

$$u(-\infty) = u^{(1)}(v_+)\,, \quad u(+\infty) = u^{(3)}(v_+)\,, \tag{3.43}$$

with $dX_+/dt = c_+(v_+)$.

We still need to relate v_-, v_+ and determine the size of the pulse and its speed. Let τ be the time delay between leading and trailing fronts. According to Eq. (3.37), it is given by

$$\tau = \int_{v_-(t-\tau)}^{v_+(t)} \frac{dv}{u^{(3)}(v) - Bv} . \tag{3.44}$$

The length of the excited region of the pulse is

$$l = \frac{1}{\epsilon} \int_{t-\tau}^{t} c_-(v_-(s))\, ds , \tag{3.45}$$

and clearly it obeys the equation

$$\frac{dl}{dt} = \frac{c_-(v_-) - c_+(v_+)}{\epsilon} . \tag{3.46}$$

The function $v_-(t)$ is determined by solving Eq. (3.37) with $j = 1$ at the region $x > X_-(t)$ ahead of the leading front. Then, the three unknowns $v_+(t)$, τ and l are found by solving Eqs. (3.44)–(3.46).

After some transient, $v \to 0$ ahead of the leading front. Then $v_-(t) \to 0$, Eq. (3.45) provides $l = c_-(0)\tau/\epsilon$ and Eq. (3.44) yields

$$l = \frac{c_-(0)}{\epsilon} \int_{0}^{v_+} \frac{dv}{u^{(3)}(v) - Bv} . \tag{3.47}$$

The solution of this equation is $v_+ = V_+(l)$. By inserting it into Eq. (3.46), we obtain

$$\frac{dl}{dt} = \frac{c_-(0) - c_+(V_+(l))}{\epsilon} , \tag{3.48}$$

which is an autonomous equation for l with a fixed point l_* such that

$$c_+(V_+) = c_-(0) , \tag{3.49}$$

and $V_+ = V_+(l_*)$. Once we know the value V_+ from Eq. (3.49), the length l_* of the pulse which moves rigidly with speed $c_-(0)$ is given setting $v_+ = V_+$ in Eq. (3.47). From Eq. (3.47) with $v_+ = V_+(l)$, we can obtain dV_+/dl as

$$\frac{dV_+}{dl} = c_-(0)\, \frac{u^{(3)}(V_+) - BV_+}{\epsilon} , \tag{3.50}$$

by using the chain rule. Since $u^{(3)}(v) > Bv$, the sign of dV_+/dl is the same as $c_-(0)$ which is positive because we assumed that the pulse moves from left to right. Then, the fixed point of Eq. (3.48) is stable.

In the long time limit, $u = 0 = v$ ahead of the pulse, $v(x,t)$ is given by (3.41) in the excited region $X_+(t) < x < X_-(t)$, and $v(x,t)$ is given by

$$x - X_+(t) = c_-(0) \int_{u^{(1)}(V_+)}^{v} \frac{dv}{Bv - u^{(1)}(v)} , \tag{3.51}$$

in the tail region $x < X_+(t)$. Clearly, $v \to 0$ as $(x - X_+) \to -\infty$.

For cubic nonlinearity, we have $c_-(0) = (2A)^{1/2}(1 - a)$ and the pulse moves with this positive velocity if $0 < a < 1$. For the same interval $0 < a < 1$, a pulse moving with negative velocity (towards the left) can be found by repeating the same construction from left to right. Now, the leading front will be an increasing one, moving with velocity $c_+(v_+) < 0$. Instead of Eqs. (3.44)–(3.46), we obtain:

$$\tau = \int_{v_+(t-\tau)}^{v_-(t)} \frac{dv}{u^{(3)}(v) - Bv}, \tag{3.52}$$

$$l = \frac{1}{\epsilon} \int_{t-\tau}^{t} c_+(v_+(s))\, ds, \tag{3.53}$$

$$\frac{dl}{dt} = \frac{c_+(v_+) - c_-(v_-)}{\epsilon}. \tag{3.54}$$

As $t \to +\infty$, $v_+ \to 0$ and these equations become

$$l = \frac{c_+(0)}{\epsilon} \int_0^{v_-} \frac{dv}{u^{(3)}(v) - Bv}, \tag{3.55}$$

$$\frac{dl}{dt} = \frac{c_+(0) - c_-(v_-)}{\epsilon}. \tag{3.56}$$

Again, these equations have a stable fixed point

$$v_- = V_-(l), \quad c_+(0) = c_-(V_-). \tag{3.57}$$

The region ahead of the leading front, $x < X_+(t)$, has $u = 0 = v$. In the excited region, $X_+(t) < x < X_-(t)$, we have

$$x - X_+(t) = -c_+(0) \int_{u^{(3)}(0)}^{v} \frac{dv}{u^{(3)}(v) - Bv}. \tag{3.58}$$

Finally, in the tail region behind the trailing front, $x > X_-(t)$, we have

$$x - X_-(t) = c_+(0) \int_{u^{(1)}(V_-)}^{v} \frac{dv}{Bv - u^{(1)}(v)}. \tag{3.59}$$

Again, for the cubic nonlinearity, $c_+(0) = -c_-(0) = -(2A)^{1/2}(1 - a)$, which is negative for $0 < a < 1$.

For $1 < a < 2$, it is not possible to construct steady wave fronts for $B > B_1$ (the case with three fixed points). In the excitable case, we can show that we cannot construct pulses either. In fact, we get $c_-(0) < 0 < c_+(0)$ so that the decreasing wave front of a pulse moves to the left while the increasing wave front moves to the right. Thus, the region between these two wave fronts shrinks with time until it disappears when both wave fronts come together.

3.5.3
Wave Trains

Suppose we have $0 < a < 1$ in the case of a cubic source and let us look for a solution which is a periodic succession of alternating decreasing and increasing fronts moving to the right. Let the increasing fronts occur at $v = v_+$ and the decreasing ones at $v = v_- < v_+$, with $v_- > 0$. The increasing fronts solve the problems of Eqs. (3.42) and (3.43) while the decreasing fronts solve Eqs. (3.38) and (3.39). Both fronts should move at the same speed, which means

$$c_-(v_-) = c_+(v_+) \,. \tag{3.60}$$

The time that the variable v spends at the excited region where $u = u^{(3)}(v)$ is

$$T_e = \int_{v_-}^{v_+} \frac{dv}{u^{(3)}(v) - Bv} \,, \tag{3.61}$$

whereas v spends a time

$$T_r = \int_{v_-}^{v_+} \frac{dv}{Bv - u^{(1)}(v)} \,, \tag{3.62}$$

at the recovering region. Note that both T_r and T_e are positive and the expressions are identical to those given previously for the related ordinary differential equation system, Eqs. (3.1) and (3.2). Given $v_- > 0$, Eq. (3.60) determines v_+ and the period $T = T_e + T_r$ is a function of v_-. Therefore, we have found a one-parameter family of wave trains. A similar construction yields a one-parameter family of wave trains in an oscillatory medium with an applied current $I_a \in (I_1, I_2)$. This implies that there is a unique (unstable) fixed point for the spatially uniform solutions, $v = v_*$, $u = u^{(2)}(v_*) = u_*$.

More general considerations about stability and evolution toward wave trains can be found in [9].

References

1 Keener, J.P. and Sneyd, J. (1998) *Mathematical Physiology*, Springer, New York, Chapt. 9.

2 Murray, J.D. (1993) *Mathematical Biology*, 2nd edn., Springer, Berlin.

3 Bell, J. and Cosner, C. (1984) Threshold behavior and propagation for nonlinear differential-difference systems motivated by modeling myelinated axons. *Q. Appl. Math.*, **42**, 1–13.

4 Scott, A.C. (1975) The electrophysics of a nerve fiber. *Rev. Mod. Phys.*, **47**, 487–533.

5 Struijk, J.J. (1997) The extracellular potential of a myelinated nerve fiber in an unbounded medium and in nerve cuff models. *Biophys. J.*, **72**, 2457–2469.

6 McIntyre, C.C. and Grill, W.M. (1999) Excitation of central nervous system neurons by nonuniform electric fields. *Biophys. J.*, **76**, 878–888.
7 Bender, C.M. and Orszag, S.A. (1978) *Advanced Mathematical Methods for Scientists and Engineers*. McGraw Hill, N.Y.
8 Fife, P.C. (1979) Mathematical aspects of reacting and diffusing systems. *Lect. Notes Biomath.*, vol. 28, Springer, New York.
9 Keener, J.P. (1980) Waves in excitable media. *SIAM J. Appl. Math.*, **39**, 528–548.

4
Excitable Media II: Discrete Systems

4.1
Introduction

Effects of spatial discreteness are important in many physical and biological systems comprised of smaller interacting components such as, for example, atoms, quantum wells and cells. Examples include the motion of dislocations [1], crystal growth and interface motion in crystalline materials [2], the motion of domain walls in semiconductor superlattices [3, 4], sliding of charge density waves [5], and pulse propagation through myelinated nerves [6, 7].

In spatially extended continuum media, traveling waves are functions of a moving coordinate and therefore, they can be found by phase plane studies, as shown in the previous chapter. In contrast with these relatively simple investigations, the mathematical study of spatially discrete models involves differential difference equations for the profiles of traveling waves, which brings forth new and challenging phenomena that are absent if the continuum limit of these models is taken. Paramount among these phenomena, is the pinning or propagation failure of wave fronts in spatially discrete equations. For values of a control parameter in a certain interval, wave fronts joining two different constant states fail to propagate [8]. When the control parameter surpasses a threshold, the wave front depins and starts moving [5, 9–11]. Physically, the pinning of wave fronts is related to the existence of Peierls stresses in continuum mechanics [12], relocation of electric field domains [13] and self sustained oscillations of the current in semiconductor superlattices [3, 14], electric current due to the sliding of charge density waves [5], saltatory propagation of impulses in myelinated fibers and its failure [7], and so on.

In this chapter, we focus our attention on the spatially discrete FitzHugh–Nagumo system:

$$\epsilon \frac{du_n}{dt} = d\,(u_{n+1} - 2u_n + u_{n-1}) + u_n(2 - u_n)(u_n - a) - v_n\,, \tag{4.1}$$

$$\frac{dv_n}{dt} = u_n - Bv_n\,, \tag{4.2}$$

$n = 0, \pm 1, \ldots$, where $\epsilon \ll 1$. As in the case of the continuum FHN system, solutions of Eqs. (4.1) and (4.2) can be approximately constructed in the limit $\epsilon \to 0+$

Nonlinear Wave Methods for Charge Transport. Luis L. Bonilla and Stephen W. Teitsworth

ISBN: 978-3-527-40695-1

by exploiting the solutions of the spatially discrete scalar *Nagumo* equation:

$$\frac{du_n}{d\tau} = d\,(u_{n+1} - 2u_n + u_{n-1}) + u_n(2 - u_n)(u_n - a) - w\,, \tag{4.3}$$

which will be analyzed in Section 4.2. Section 4.3 contains the asymptotic construction of pulses and wave trains for the discrete FHN system. These ideas and our results are tested by numerically solving the FHN system with appropriate boundary conditions in Section 4.4. Comments on propagation failure of pulses in the FHN system are made in Section 4.5. Section 4.6 briefly discusses how a pulse may be generated by applying a temporary stimulus at one end of a fiber with finitely many nodes.

4.2
The Spatially Discrete Nagumo Equation

We consider the equation

$$\frac{du_n}{d\tau} = d\,(u_{n+1} - 2u_n + u_{n-1}) + h(u_n, w, a)\,. \tag{4.4}$$

This model is the spatially discrete Nagumo equation if $h(u, w, a) = u(2 - u)(u - a) - w$ (w is constant), and is the overdamped Frenkel–Kontorova (FK) model if $h(u, w, a) = w - a \sin u$, and so on. As long as $\min h(u, 0, a) < w < \max h(u, w, a)$, this is a "cubic" source having three zeros $u^{(i)}(w, a)$, $i = 1, 2, 3$, $u^{(1)} < u^{(2)} < u^{(3)}$. Wave front solutions joining $u^{(1)}$ and $u^{(3)}$ (the two stable zeros) exist. This can be simply numerically shown by drawing the trajectories of $u_n = u(n - c\tau)$ at adjacent sites, as done in Figure 4.1c. Note that these trajectories correspond to exactly the same profile shifted by one or more sites. In addition to the empirical demonstration provided by the numerical solution of Figure 4.1c, we would like to know why solutions of the Nagumo equation are wave front profiles that move rigidly with constant velocity. Proving the existence of wave fronts requires more complicated arguments than the phase plane arguments of Chapter 3, [15, 16]. We will simply rephrase a chain of arguments given by Hankerson and Zinner in their existence proof, [16].

1. Consider an initial profile, $u_n(0)$, which is monotonically increasing in n and such that $u_n \to 0$ as $n \to -\infty$ and $u_n \to 2$ as $n \to \infty$. Prove by using a comparison principle that the solution of the Nagumo equation is an increasing profile with $0 < u_n(\tau) < 2$. Then, there is a sub-succession of times τ_k such that $u_{n_k}(\tau_k) \to b \in (0, 2)$ as $k \to \infty$.
2. Prove that the limit of $du_{n_k}/d\tau$ is not zero in order to exclude the possibility of a pinned wave front (see below). Let us assume that this limit is positive (a similar argument holds if it is negative).
3. For any n, show that there is another sub-succession of times τ_j such that $u_{-n_j}(\tau_{n_j}) = b$.

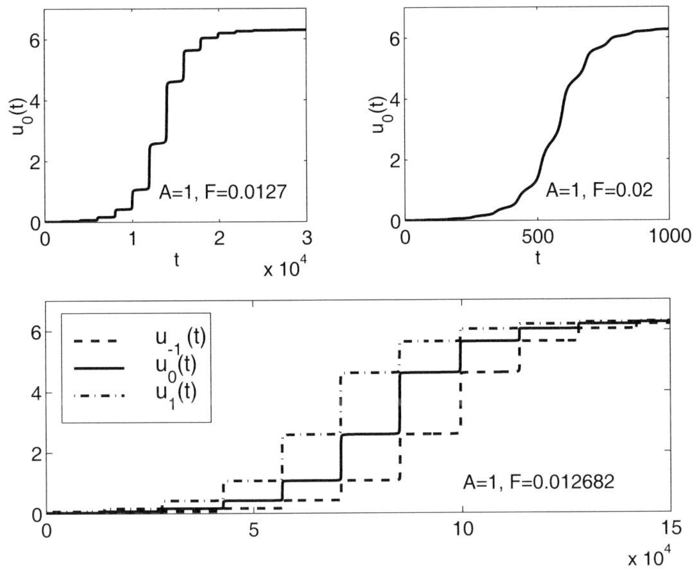

Fig. 4.1 Wave front profiles for the overdamped FK model $h(u, F, A) = F - A \sin u$ when $A = 1$ near $F = F_c$ (from [18]).

4. Prove that the numbers $x_l^{n_j} = u_{-n_j+l}(\tau_{n_j})$ converge to some x_l as $j \to \infty$.
5. Consider the time evolution operator $T(s)x_l^{n_j} = u_{-n_j+l}(\tau_{n_j} + s)$. Obviously, $T(0)x_l^{n_j} = x_l^{n_j}$. Let $S f_j = f_{j+1}$ be the spatial shift operator and $\sigma_j = \tau_{n_j+1} - \tau_{n_j}$. We have

$$T(\sigma_j)x_l^{n_j} = u_{-n_j+l}(\tau_{n_j+1}) = S u_{-n_j-1+l}(\tau_{n_j+1}) = S x_l^{n_j+1} . \tag{4.5}$$

6. As $j \to \infty$, prove that $\sigma_j \to \sigma$ and that $x_l^{n_j} \to x_l$. Then, the previous equation shows that $T(\sigma)x_l = S x_l = x_{l+1}$. Thus, a σ time shift of the profile x_l yields a one site spatial shift x_{l+1}. If we denote $x_l = u(j)$, we have shown that

$$T(\sigma)u(l) = u(l+1) = u(l + |c|\,\sigma) , \tag{4.6}$$

where $|c| = 1/\sigma$ is the velocity with which the wave front moves to the left. The wave front profile is $u_n(\tau) = u(n - c\tau)$, with $c = -1/\sigma$.

A wave front solution of Eq. (4.4) solves the following eigenvalue problem:

$$-c\frac{\partial u}{\partial \xi} = d\,[u(\xi+1) - 2u(\xi) + u(\xi-1)] + h(u(\xi), w, a) , \tag{4.7}$$

$$u(\mp\infty) = u^{(1)}(w, a) , \quad u(\pm\infty) = u^{(3)}(w, a) .$$

Here, $\xi = n - c\tau$ and c is the front velocity. By multiplying this equation by $\partial u/\partial\xi$ and integrating, we obtain

$$-c\int_{-\infty}^{\infty}\left(\frac{\partial u}{\partial\xi}\right)^2 d\xi = \int_{u^{(1)}}^{u^{(3)}} h(u,w,a)\,du\,. \tag{4.8}$$

This equation relates the speed of a propagating front to the integral of the source term. Let w_0 be the value of w for which the integral in the right-hand side of Eq. (4.8) is zero. For the cubic nonlinearity, $h = u(2-u)(u-a) - w$, the sign of c is the same as the sign of $(w - w_0)$, whereas for the overdamped FK model, $h = w - a\sin u$, the sign of c is opposite to the sign of w ($w_0 = 0$ in this case).

For most nonlinearities (including our two examples), the wave fronts of the Nagumo equation are pinned at the lattice sites for w in an interval of nonzero width. This means that there exist front-like stationary solutions $u_n(a,w)$ such that $u_n \to u^{(1)}$ as $n \to \mp\infty$ and $u_n \to u^{(3)}$ as $n \to \pm\infty$. For both examples, there exists a critical pinning strength, w_c, such that wave fronts move with nonzero velocity if $|w - w_0| > w_c$ and are pinned with zero velocity if $|w - w_0| < w_c$. For $0 < |w - w_0| < w_c$, the right-hand side of Eq. (4.8) is a finite nonzero number and $c = 0$ in the left-hand side of Eq. (4.8), which implies that the profile of a pinned front is piecewise constant with jump discontinuities [17]. Then, we can visualize the transition from a moving to a pinned front at $w = w_c$ as the loss of continuity of the front profile as $|w - w_0| \to w_c+$. This can be observed in Figure 4.1 corresponding to the wave front profiles of the overdamped FK model near the critical field $w = w_c$ (recall $w_0 = 0$ in this case). Individual points undergo abrupt jumps at particular times, which gives the misleading impression that the motion of the discrete fronts proceeds by successive jumps. Actually, the points remain very close to their stationary values at $w = w_c$, say $u_n(a, w_c)$, during a very long time interval of order $|w - w_c|^{-\frac{1}{2}}$. Then, at a specific time, *all* the points $u_n(t)$ jump to a vicinity of $u_{n+1}(a, w_c)$. The method of matched asymptotic expansions can be used to describe this two-stage motion of the points $u_n(t)$. Then, the wave front profile can be reconstructed by using the definition $u_n(\tau) = u(n - c\tau)$. The slow stage of front motion is described by the normal form of a saddle-node bifurcation and yields an approximation to the wave front velocity that scales with the field as $|w - w_c|^{\frac{1}{2}}$.

There are examples of bistable sources for which Eq. (4.4) has a one-parameter family of stationary solutions with smooth profiles corresponding to increasing or decreasing pinned wave fronts. However, one can then show that the pinning interval in w consists of only one point, $w = w_0$, that the wave fronts move for all other values of w, and that their speed increases linearly with $|w - w_0|$ (cf. Theorem 2.1 in [18]). One example is $h = 2\gamma u(1-u^2)/(1-\gamma u^2) - w$ with $\gamma = \tanh^2(1)$, $d = 1$, which has solutions $u_n = \tanh(n + p)$, where p is any real constant, provided $w = 0$. For nonzero w, increasing or decreasing wave fronts exist and move with nonzero speed. Moving and pinned wave fronts of the discrete Nagumo equation cannot coexist for the same value of w.

4.2.1
Depinning Transition of Wave Fronts

The discrete Nagumo equation with nonlinearity $h(u, w, a)$ has pinned wave front solutions (with discontinuous profile) for w in an interval of finite width except for very particular sources (for which stationary wave fronts of smooth profile exist only for a single value of w). For these usual sources, a theory of the pinning and propagation of fronts for Eq. (4.4) has been developed in [4, 17, 18] and we shall describe its more salient features now.

Let us consider a cubic source: $h(u, w, a) = u(2 - u)(u - a) - w$. First, assume $w = 0$ so that the asymmetry of the cubic source is controlled by the parameter a. For d fixed, there are values $a_{cl}(d)$ and $a_{cr}(d)$ such that the following hold:

- The fronts joining $u = 0$ and $u = 2$ are stationary if $a_{cl}(d) \leq a \leq a_{cr}(d)$. No front propagation is possible.
- Outside this interval, there exist traveling wave fronts $u_n(\tau) = u(n - c\tau)$ joining 0 and 2. For $a > a_{cr}(d)$, increasing fronts move to the right and decreasing fronts move to the left. For $a < a_{cl}(d)$, fronts move in the opposite way: decreasing fronts move to the right and increasing fronts move to the left.

The values $a_{cl}(d)$ and $a_{cr}(d)$ can be approximately calculated as follows. In a large lattice, we decrease or increase a from 1 until we obtain a stationary solution $u_n(a)$ whose linear stability problem has a zero eigenvalue; see Figure 4.2.

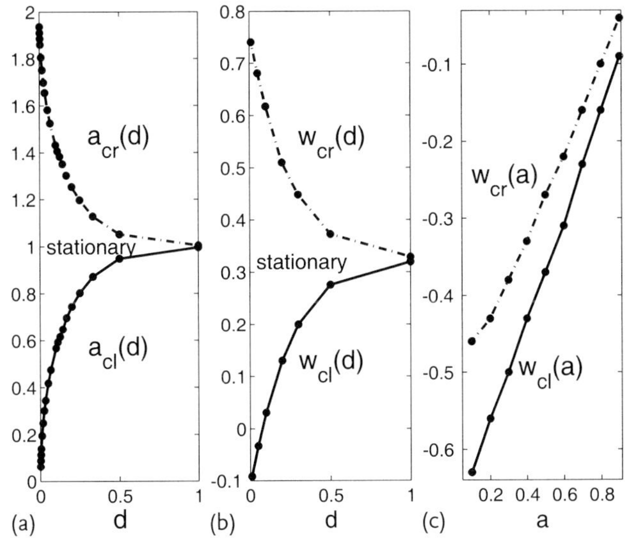

Fig. 4.2 (a) Critical values a_{cl} and a_{cr} as functions of d. (b) Critical $w_{cl}(a, d)$ and $w_{cr}(a, d)$ for $a = 0.5$. (c) Critical $w_{cl}(a, d)$ and $w_{cr}(a, d)$ for $d = 0.5$ (from [19]).

Now, we fix a and vary w. The asymmetry of the source is controlled by a and w. For fixed d and a, critical values $w_{cl}(a, d)$ and $w_{cr}(a, d)$ are found such that the following hold:

- The fronts joining $u^{(1)}(w, a)$ and $u^{(3)}(w, a)$ are stationary if $w_{cl}(a, d) \leq w \leq w_{cr}(a, d)$.
- Outside this interval, there exist traveling wave fronts $u_n(\tau) = u(n - c\tau)$ joining $u^{(1)}(w, a)$ and $u^{(3)}(w, a)$. For $w < w_{cl}$, these fronts move to the left if they increase from $u^{(1)}$ to $u^{(3)}$, and to the right if they decrease from $u^{(3)}$ to $u^{(1)}$. For $w > w_{cr}$, fronts decreasing from $u^{(3)}$ to $u^{(1)}$ move to the left, and increasing fronts move to the right.

To calculate w_{cl} and w_{cr}, we start by fixing a and finding a value $w = w_0$ at which stationary solutions exist for a large lattice. We now decrease or increase w from this value until we obtain a stationary solution $u_n(w)$ whose linear stability problem has a zero eigenvalue; see Figure 4.2.
For w near any of its critical values, we can use the following formula to predict the speed of the fronts for $w > w_{cr}$ and for $w < w_{cl}$:

$$c(a, d, w) \sim \text{sign}(w - w_c)\frac{\sqrt{\alpha\beta(w - w_c)}}{\pi}, \tag{4.9}$$

where w_c is either w_{cl} or w_{cr}. The parameters α and β, given by $\alpha = \sum \phi_n$, $\beta = \frac{1}{2}\sum[-6u_n(w_c) + 2(2 + a)]\phi_n^3$ (see [17, 18]), are functions of a, d, and the critical value of w. In these formulas, ϕ is a positive eigenfunction of the linear stability problem for $u_n(w_c)$ with $\sum \phi_n^2 = 1$, and $u_n(w_c)$ is a stationary solution of Eq. (4.4) with $w = w_c$ [18]. If w is not close to its critical values, the speed $c(a, d, w)$ should be calculated numerically.

A peculiarity of the Nagumo equation is the scenario for front propagation failure. As we approach the critical values for a, w, or any other appropriate pinning control parameter, the front profiles become less smooth and a number of steps appear. In the limit as the control parameter tends to its critical value, the transition regions between steps become infinitely steep, the front profile becomes discontinuous, and its velocity vanishes; cf. Figure 4.1 and [17, 18].

4.2.2
Construction of the Wave Front Profile Near the Depinning Transition

Let us consider the extremely discrete case in which $d \ll 1$ and $a = 1$ so that $h = F - g(u)$ and $g(u)$ is smooth, it has three zeros and it is odd about the middle one so that $\int_{u^{(1)}}^{u^{(3)}} f(u)du = 0$. An example is $g(u) = u(u - 1)(u - 2)$ which is odd about $u = 1$. For sufficiently small d, the stable stationary solution does not contain points u_n in the region where $g'(u) < 0$ [18]. When $F > 0$, this stationary solution is no longer symmetric with respect to $u^{(2)}(F)$. For larger F and generic potentials, numerical simulations show that $g' < 0$ only for one point, labeled $u_0(F, d)$. This property persists until F_c is reached; see Figure 4.3.

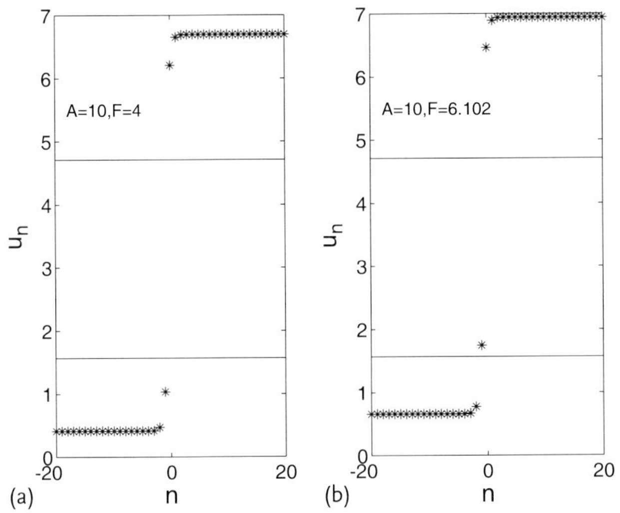

Fig. 4.3 Stationary solutions for the FK model with $A = 10$ ($d = 0.1$): (a) No points are found in the region $g' < 0$ for sufficiently small F; (b) one point enters the region $g' < 0$ for sufficiently large $F < F_c$ (from [18]).

Let us first consider the symmetric stationary profile with $u_n \neq u^{(2)}$ for $F = 0$, $u_n \leq u_{n+1}$ (for any n) and $u_{-\infty} = u^{(1)}(0)$, $u_{\infty} = u^{(3)}(0)$. The front profile consists of two tails with u_n very close to $u^{(1)}(0)$ and to $u^{(3)}(0)$, plus two symmetric points u_0, u_1 in the gap region between $u^{(1)}(0)$ and $u^{(3)}(0)$. As $F > 0$ increases, this profile changes slightly but the two tails are still very close to $u^{(1)}(F)$ and $u^{(3)}(F)$. As for the two middle points, u_0 gets closer and closer to $u^{(3)}$, whereas u_1 moves away from $u^{(1)}$. This structure is preserved by the traveling fronts above the critical field. Most of the time, there is only one active point between $u^{(1)}$ and $u^{(3)}$, which we can adopt as our u_0. Then, the wave front profile Eq. (4.7) can be calculated as $u(-c\tau) = u_0(\tau)$. In Eq. (4.4), we can approximate $u_{-1} \sim u^{(1)}$, $u_1 \sim u^{(3)}$, thereby obtaining

$$\frac{du_0}{d\tau} \approx d\left[u^{(1)}(F) + u^{(3)}(F) - 2u_0\right] - g(u_0) + F\,, \tag{4.10}$$

where $h = F - g(u)$. This equation has three stationary solutions for $F < F_c$, two stable and one unstable, and only one stable stationary solution for $F > F_c$. Let us consider $F < F_c$. Only two out of the three solutions of Eq. (4.10) approximate stationary fronts for the exact system: those having smaller values of u_0. The one having smallest u_0 approximates the stable stationary front while the other one approximates the unstable stationary front. Recall that the unstable front had a value $u_0 = [u^{(1)}(0) + u^{(3)}(0)]/2$ at the middle of the gap for $F = 0$. As $F > 0$ increases, u_0 decreases towards $u^{(1)}(F)$. Thus, one active point will also approximate the profile of the unstable stationary front. The stationary solution of Eq. (4.10), having the largest value of u_0 (slightly below $u^{(3)}(F)$), is not consistent with the assumptions we made to derive Eq. (4.10), and therefore, it does not approximate a physically existing stationary front. If $F > F_c$, the only stationary solution of Eq. (4.10) is the unphysical one. The critical field F_c is such that the expansion of the right-hand side of Eq. (4.10) about the two coalescing stationary solutions has zero linear term, $2d + g'(u_0) = 0$, and

$$2du_0 + g(u_0) \sim d\left[u^{(1)}(F_c) + u^{(3)}(F_c)\right] + F_c\,. \tag{4.11}$$

These equations for F_c and $u_0(F_c)$ have been solved for the FK model, $g(u) = \sin u$, for which $u_0 = \cos^{-1}(-2d)$ and $u^{(1)} + u^{(3)} = 2\sin^{-1}(F_c) + 2\pi$. The results are depicted in Figure 4.4 and show excellent agreement with those of direct numerical simulations for $d < 0.1$. Our approximation performs less well for larger d, and it breaks down at $d = 0.5$ with the wrong prediction $F_c = 0$. Notice that $F_c(d) \sim 1$ as d decreases. In practice, only steady solutions are observed for very small d.

Let us now construct the profile of the traveling wave fronts after depinning for F slightly above F_c. Then, $u_0(\tau) = u_0(F_c, d) + v_0(\tau)$ obeys the following equation:

$$\frac{dv_0}{d\tau} = \alpha\,(F - F_c) + \beta\, v_0^2\,, \tag{4.12}$$

$$\alpha = 1 + \frac{d}{g'(u^{(1)}(F_c))} + \frac{d}{g'(u^{(3)}(F_c))}\,, \tag{4.13}$$

$$\beta = -\frac{1}{2}\, g''(u_0)\,, \tag{4.14}$$

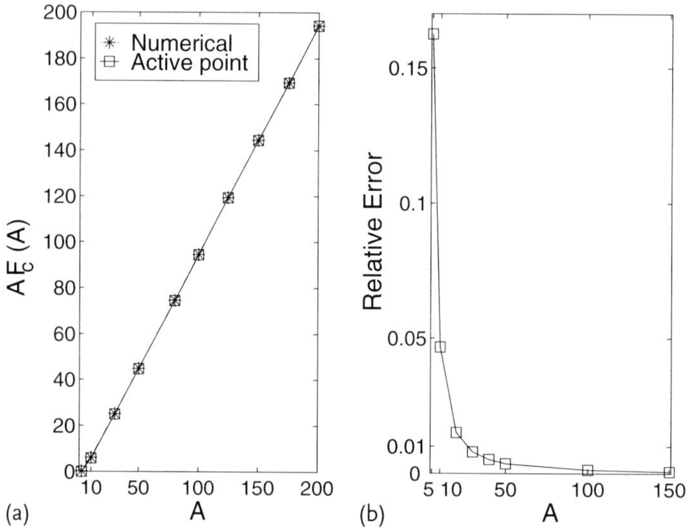

Fig. 4.4 Approximation of Eq. (4.4) by the equation with one active point for the FK potential and $A > 2$ ($d < 0.5$): (a) Critical force versus $A = 1/d$. (b) Error in the approximation of $F_c(A)$ (from [18]).

where we have used $2d + g'(u_0) = 0$, Eq. (4.11) and ignored terms of order $(F - F_c)\, v_0$ and higher. These terms are negligible after rescaling $v_0 = (F - F_c)^{\frac{1}{2}}\varphi$ and $s = (F - F_c)^{1/2}\tau$. The coefficients α and β are positive because $g'(u^{(i)}) > 0$ for $i = 1, 3$ and $g''(u_0) < 0$ since $u_0 \in (u^{(1)}(0), u^{(2)}(0))$. For the FK potential, $\alpha = 1 + 2d/\sqrt{1 - F_c^2}$ and $\beta = \sqrt{1 - 4d^2}/2$. Equation (4.12) has the (outer) solution

$$v_0(\tau) \sim \sqrt{\frac{\alpha(F - F_c)}{\beta}}\, \tan\left(\sqrt{\alpha\beta\,(F - F_c)}\,(\tau - \tau_0)\right)\,, \tag{4.15}$$

which is very small most of the time, though it blows up when the argument of the tangent function approaches $\pm\pi/2$. Thus, the outer approximation holds over a time interval $(\tau - \tau_0) \sim \pi/\sqrt{\alpha\beta\,(F - F_c)}$, which equals $\pi\sqrt{2/\alpha}(1 - 4d^2)^{-\frac{1}{4}}(F - F_c)^{-\frac{1}{2}}$ for the FK potential. The reciprocal of this time interval yields an approximation for the wave front velocity,

$$c(A, F) \sim -\frac{\sqrt{\alpha\beta\,(F - F_c)}}{\pi}\,, \tag{4.16}$$

or $c \sim -(1 - 4d^2)^{\frac{1}{4}}(1 + 2d/\sqrt{1 - F_c^2})^{\frac{1}{2}}(F - F_c)^{\frac{1}{2}}/(\pi\sqrt{2})$ for a FK potential. The minus sign reminds us that wave fronts move towards the left for $F > F_c$. In

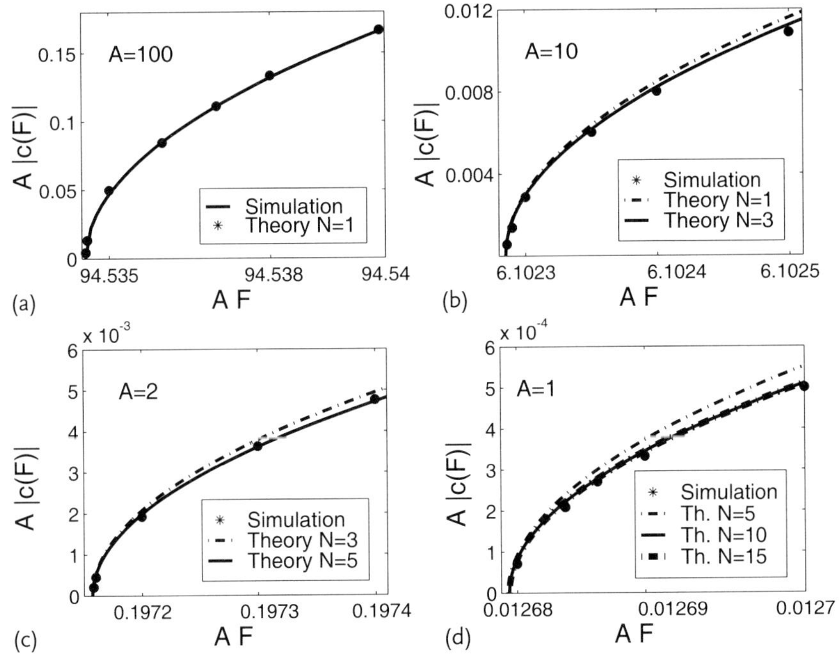

Fig. 4.5 Comparison of theoretically predicted and numerically calculated wave front velocities near F_c for the FK model with N active points and the following values of the parameter A: (a) $A = 1/d = 100$, (b) $A = 10$, (c) $A = 2$, (d) $A = 1$ (from [18]).

Figures 4.5a and b, we compare this approximation with the numerically computed velocity for $A = 100$ ($d = 0.01$) and $A = 10$ ($d = 0.1$).

When the solution begins to blow up, the outer solution Eq. (4.15) is no longer a good approximation, for $u_0(\tau)$ departs from the stationary value $u_0(F_c, d)$. We must go back to Eq. (4.10) and obtain an inner approximation to this equation. As F is close to F_c and $u_0(\tau) - u_0(F_c, d)$ is of order 1, we numerically solve Eq. (4.10) at $F = F_c$ with the matching condition that

$$u_0(\tau) - u_0(F_c, d) \sim \frac{2}{\pi\sqrt{\beta/[\alpha(F - F_c)]} - 2\beta(\tau - \tau_0)},$$

as $(\tau - \tau_0) \to -\infty$ (recall that $\tan x \sim 1/(\pi/2 - x)$ as $x \to \pi/2 - 0$). This inner solution describes the jump of u_0 from $u_0(F_c, d)$ to values on the largest stationary solution of Eq. (4.10), which is close to $u^{(3)}$. During this jump, the motion of u_0 forces the other points to move. Thus, $u_{-1}(\tau)$ can be calculated by using the inner solution in Eq. (4.4) for u_0, with $F = F_c$ and $u_{-2} \approx u^{(1)}$.

We can construct a composite expansion with these inner and outer solutions as follows. First, note that the blow up times are $\tau_m = (2c)^{-1} + m/c$, $m = 0, \pm 1, \pm 2, \ldots$ Let us denote by $u_n^{(m)}(\sigma)$, $\sigma = \tau - \tau_m$, the solution of Eq. (4.4) with $h = F - g(u)$ and the boundary conditions $u_n^{(m)}(\sigma) \to u_{n+m}(F_c, d)$ as $\sigma \to -\infty$, and $u_n^{(m)}(\sigma) \to u_{n+m+1}(F_c, d)$ as $\sigma \to \infty$. $u_n(F_c, d)$ is the profile of the pinned wave front at $F = F_c$. During the time interval (τ_{-n-1}, τ_{-n}) that $u_n(\tau)$ needs to go

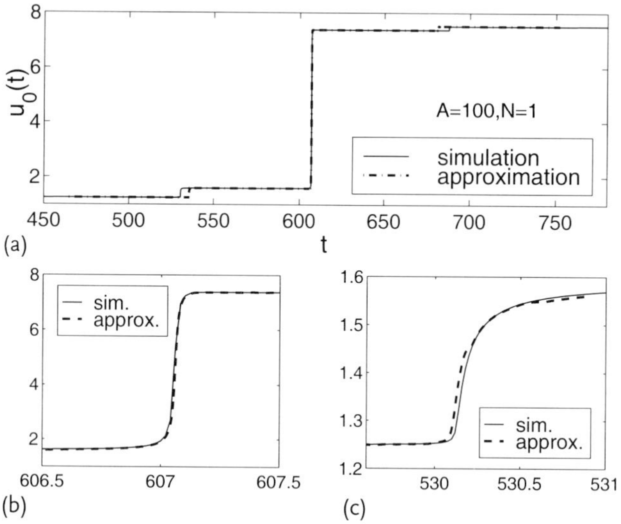

Fig. 4.6 Comparison of asymptotic and numerically calculated wave front profiles near F_c: (a) Complete wave front profile as indicated by the trajectory $u_0(t)$, with $t = \tau/A$. (b) Zoom near the largest jump in the profile. (c) Zoom near the jump preceding the largest one after translating the asymptotic profile. The latter has been calculated by inserting the approximate $u_0(t)$ in the equation for $u_{-1}(t)$ (from [18]).

from $u^{(1)}(F_c)$ to $u^{(3)}(F_c)$, the uniform approximation to the wave front is

$$u_n(\tau) \sim \sum_{m=-n-1}^{-n} \Big[u_n^{(m)}(\tau - \tau_m) + u_n^{(m-1)}(\tau - \tau_{m-1}) - u_{n+m}(F_c, d) + v_0\left(\tau - \frac{m}{c}\right) - \frac{1}{\beta(\tau_m - \tau)} + \frac{1}{\beta(\tau - \tau_{m-1})} \Big] \chi(\tau_{m-1}, \tau_m) \,. \quad (4.17)$$

Then, $u_n(\tau_{-n-1}) \sim u^{(1)}(F_c)$ and $u_n(\tau_{-n}) \sim u^{(3)}(F_c)$. In Eq. (4.17), the indicator function $\chi(\tau_{m-1}, \tau_m)$ is 1 if $\tau_{m-1} < \tau < \tau_m$ and 0 otherwise. Figure 4.6 compares the composite expansion with the numerically evaluated wave front solution of the Nagumo equation.

Notice that Eq. (4.12) is the normal form associated with a saddle-node bifurcation in a one-dimensional phase space. The wave front depinning transition is a *global* bifurcation with generic features: each individual point $u_n(\tau)$ spends a long time, which scales as $|F - F_c|^{-\frac{1}{2}}$, near discrete values $u_n(F_c, d)$, and then jumps to the next discrete value on a time scale of order 1. The traveling wave ceases to exist for $F \le F_c$.

4.2.3
Wave Front Velocity Far from the Depinning Transition

Far from the depinning transition at $w = w_c$, the formula (4.9) ceases to accurately approximate the wave front velocity. For $d > 1$, the length of the pinning intervals is below 0.001 for the Nagumo equation with cubic nonlinearity. Then, the wave front velocities are corrections of those for the spatially continuous Nagumo equation. To evaluate these corrections, we use the variable $\eta = \tau + n/(c\sqrt{d})$, so that the wave front velocity is $c\sqrt{d}$. Then, the discrete diffusion can be approximated by the Taylor expansion:

$$u\left(\eta + \frac{1}{c\sqrt{d}}\right) + u\left(\eta - \frac{1}{c\sqrt{d}}\right) - 2u(\eta) = \frac{1}{c^2 d}\frac{d^2u}{d\eta^2} + \frac{1}{12c^4d^2}\frac{d^4u}{d\xi^4} + \frac{1}{360c^6d^3}\frac{d^6u}{d\xi^6} + \dots,$$

and the Nagumo equation (4.7) becomes

$$\frac{1}{c^2}\frac{d^2u}{d\eta^2} - \frac{du}{d\eta} + f(u) - w = -\frac{1}{12c^4d}\frac{d^4u}{d\eta^4} + O\left(\frac{1}{c^6d^2}\right), \quad (4.18)$$

where we have set $h(u, w, a) = f(u) - w$. By inserting the ansatz $u(\eta, \epsilon) \sim U_0(\eta) + d^{-1}U_1(\eta)$ and $c \sim c_0 + d^{-1}c_1$ in Eq. (4.18), we find the hierarchy of equations

$$\frac{d^2U_0}{d\eta^2} - c_0^2\frac{dU_0}{d\eta} + c_0^2[f(U_0) - w] = 0\,, \quad (4.19)$$

$$\frac{d^2U_1}{d\eta^2} - c_0^2\frac{dU_1}{d\eta} + c_0^2 f'(U_0)\,U_1 = \frac{2c_1}{c_0}\frac{d^2U_0}{d\eta^2} - \frac{1}{12c_0^2}\frac{d^4U_0}{d\eta^4}\,. \quad (4.20)$$

The corresponding boundary conditions are

$$U_0(\mp\infty) = u^{(1)}(w)\,, \quad U_0(\pm\infty) = u^{(3)}(w)\,, \quad U_1(\pm\infty) = 0\,. \tag{4.21}$$

The solution of Eqs. (4.19) and (4.21) is an increasing or a decreasing front of the continuous bistable equation, $U_0 = \phi_\pm(\eta)$, whose velocity $c_\pm(w)$ is known in the case of the cubic source, as indicated in Chapter 3. We have

$$c_0^2 = \pm\frac{c_0}{\sqrt{2}}[u^{(1)}(w) + u^{(3)}(w) - 2u^{(2)}(w)]\,.$$

Then,

$$c_{0\pm}(w) = \pm\sqrt{\frac{1}{2}}\,[u^{(1)}(w) + u^{(3)}(w) - 2u^{(2)}(w)]\,. \tag{4.22}$$

Recall that the + sign (resp., − sign) indicates an increasing (resp., decreasing) front. Equation (4.20) is equivalent to

$$\left(\frac{1}{c_{0\pm}^2}\frac{d^2}{d\eta^2} + f'(\phi_\pm) - \frac{1}{4}\right)\left(e^{-c_{0\pm}^2\eta/2}U_1\right) = \frac{2e^{-c_{0\pm}^2\eta/2}}{c_{0\pm}^3}\left(c_1\frac{d^2\phi_\pm}{d\eta^2} - \frac{1}{24c_{0\pm}}\frac{d^4\phi_\pm}{d\eta^4}\right)\,.$$

The corresponding linear homogeneous problem has a solution $e^{-c_{0\pm}^2\eta/2}d\phi_\pm/d\eta$, and therefore, the non-homogeneous problem has a bounded solution only if its right hand side is orthogonal to this function. This yields

$$c_{1\pm} = \frac{1}{24c_{0\pm}}\frac{\int_{-\infty}^{\infty} e^{-c_{0\pm}^2\eta}\frac{d^4\phi_\pm}{d\eta^4}\frac{d\phi_\pm}{d\eta}\,d\eta}{\int_{-\infty}^{\infty} e^{-c_{0\pm}^2\eta}\frac{d^2\phi_\pm}{d\eta^2}\frac{d\phi_\pm}{d\eta}\,d\eta}\,. \tag{4.23}$$

Once Eq. (4.23) has been evaluated, an approximate expression for the front velocity is $c \approx \sqrt{d}[c_{0\pm}(w) + c_{1\pm}/d]$ which, for the case of the cubic source, becomes

$$c_\pm = \sqrt{d}\,c_{0\pm} + \frac{\int_{-\infty}^{\infty} e^{-c_{0\pm}^2\eta}\frac{d^4\phi_\pm}{d\eta^4}\frac{d\phi_\pm}{d\eta}\,d\eta}{24c_{0\pm}\sqrt{d}\int_{-\infty}^{\infty} e^{-c_{0\pm}^2\eta}\frac{d^2\phi_\pm}{d\eta^2}\frac{d\phi_\pm}{d\eta}\,d\eta}\,. \tag{4.24}$$

Here, $c_{0\pm}$ is given by Eq. (4.22).

4.3
Asymptotic Construction of Pulses

Our construction of pulses as discussed in the remaining sections of this chapter is based on Reference [19]. As we will discuss below, an appropriate initial condition evolves towards a pulse. In particular, we need to fix the parameters $d > 0$, $a <$

$a_{cl}(d)$ (the case $a > a_{cr}(d)$ follows by symmetry), and ϵ smaller than a certain critical value, $\epsilon_c(a, d)$. This last condition also holds for the spatially continuous FHN system, which has two pulse solutions (one stable and one unstable) for $\epsilon < \epsilon_c$. These solutions coalesce at ϵ_c and cease to exist for larger ϵ (see [20, 21]). A pulse consists of regions of smooth variation of u on the time scale t, separated by sharp interfaces in which u varies rapidly on the time scale $T = t/\epsilon$. In the regions where u varies smoothly, we can set $\epsilon = d = 0$, thereby obtaining the reduced problem

$$u_n(2 - u_n)(a - u_n) - v_n = 0\,, \tag{4.25}$$

$$\frac{dv_n}{dt} = u_n - B\,v_n\,. \tag{4.26}$$

These regions are separated by sharp interfaces (moving fronts), at which u_n varies rapidly as $u_n(t) = u(z)$, $v_n(t) = v(z)$, with $z = n - ct/\epsilon$. There, to leading order,

$$-c\frac{du}{dz} = d\,[u(z+1) - 2u(z) + u(z-1)] + u(z)\,(2 - u(z))\,(a - u(z)) - v\,, \tag{4.27}$$

$$-c\,\frac{dv}{dz} = 0\,. \tag{4.28}$$

Thus v is a constant equal to the value $v_n(t)$ at the last point in the region of smooth variation before the front. Equation (4.27) has a wave front solution as discussed in the previous section. We can now discuss different regions in the asymptotic description of a pulse as follows:

1. The region of smooth variation of u in front of the pulse, described by Eqs. (4.25) and (4.26). In this region, $u_n = u^{(1)}(v_n)$, so that

   $$\frac{dv_n}{dt} = u^{(1)}(v_n) - B\,v_n\,,$$

   and initial data evolve exponentially fast towards equilibrium, $u_n = v_n = 0$.
2. The pulse leading edge. Let $v(t)$ be the value of v_n at the last point of the region in front of the pulse. Eventually, $v \to 0$. At the leading edge, $u_n(t) = u(n - ct/\epsilon)$ is a wave front moving towards the right with speed $C = c(a, d, v)/\epsilon$ measured in points per unit time t. We have the boundary conditions $u(-\infty) = u^{(3)}(v)$ and $u(\infty) = u^{(1)}(v)$ for the monotone decreasing profile $u(z)$ which satisfies Eq. (4.27). It is convenient to call $c_-(v) = c(a, d, v)$. Eventually, $C \sim c_-(0)/\epsilon$, and u_n decreases from $u_n = 2$ to $u_n = 0$ across the leading edge of the pulse.
3. The region between fronts: $u_n = u^{(3)}(v_n)$ and

   $$\frac{dv_n}{dt} = u^{(3)}(v_n) - B\,v_n\,.$$

   There is a finite number of points in this region. On its far right, $v_n = v \to 0$. As we move towards the left, v_n increases until it reaches a certain value $V(t)$ corresponding to that in the trailing wave front.

4. The trailing wave front: $v_n(t) = v(z) = V$, and $u_n(t) = u(z)$ obeys Eq. (4.27) with boundary conditions $u(-\infty) = u^{(1)}(V)$ and $u(\infty) = u^{(3)}(V)$. This front increases monotonically with z, and it moves with speed $C = c(a, d, V)/\epsilon$ measured in points per unit time t. It is convenient to denote $c_+(V) = c(a, d, V)$. We shall indicate how to determine V below. Clearly, if the pulse is to move rigidly, we should have $c_+(V) = c_-(0)$ after a sufficiently long transient period.
5. Pulse tail. Again, $u_n = u^{(1)}(v_n)$ and $dv_n/dt = u^{(1)}(v_n) - Bv_n$. Sufficiently far to the left, $v_n = u_n = 0$.

The number of points between wave fronts of the pulse is not arbitrary: it can be calculated following the same ideas explained in Chapter 3 for the spatially continuous case. Let τ be the delay between fronts, that is, the time elapsed from the instant at which the leading front traverses the point $n = N$ to the instant when the trailing front is at $n = N$. Clearly,

$$\tau = \int_{v(t-\tau)}^{V(t)} \frac{dv}{u^{(3)}(v) - Bv} . \tag{4.29}$$

The number of points between fronts, $l(t)$, can be calculated as

$$l = \frac{1}{\epsilon} \int_{t-\tau}^{t} c_-(v(t))\, dt . \tag{4.30}$$

On the other hand, the separation between fronts satisfies the equation

$$\frac{dl}{dt} = \frac{c_-(v(t)) - c_+(V(t))}{\epsilon} . \tag{4.31}$$

The three Eqs. (4.29)–(4.31) can be solved to obtain the three unknowns τ, l, and $V(t)$. The function $v(t)$ is determined by solving Eq. (4.26) with $u_n = u^{(1)}(v_n)$ in the region to the left of the leading front.

After a transient period, $v(t) \to 0$ and $V(t) \to V$ (a constant value), so that we have the simpler expressions

$$\tau = \int_0^V \frac{dv}{u^{(3)}(v) - Bv} , \tag{4.32}$$

$$\frac{dl}{dt} = \frac{c_-(0) - c_+(V)}{\epsilon} , \tag{4.33}$$

instead of Eqs. (4.29) and (4.31), respectively. The number of points at the pulse top is now

$$l = \frac{c_-(0)\tau}{\epsilon} = \frac{c_-(0)}{\epsilon} \int_0^V \frac{dv}{u^{(3)}(v) - Bv} . \tag{4.34}$$

This equation yields V as a function of l. Then, Eq. (4.33) becomes an autonomous differential equation for l that has a stable constant solution at $l = l^*$ such that $c_-(0) = c_+(V(l))$: at $l = l^*$, the right-hand side of Eq. (4.33) has a slope $-[u^{(3)}(V) - B\,V]\,c'_+(V)/c_-(0) < 0$.

By recapitulating for appropriate initial conditions, the leading and trailing fronts of a pulse evolve until l reaches its stable value at which $c_-(0) = c_+(V(l^*))$ and Eq. (4.34) holds. To compute l^*, we first determine $V^* = V(l^*)$ by using $c_-(0) = c_+(V(l^*))$. Then, we calculate $\tau = \tau^*$ (which does not depend on ϵ!) from Eq. (4.32) and $l^* = c_-(0)\tau^*/\epsilon$. Our construction breaks down if the number of points between fronts falls below 1. This yields an upper bound for the critical value of ϵ above which pulse propagation fails: $\epsilon_c \sim c_-(0)\tau^*$.

The asymptotic length of the pulse tail is obtained by first calculating the time needed for v_n to go from 0 to $V(l^*)$ to the left of the trailing front: $T = \int_0^V dv/[u^{(1)}(v) - v]$. The tail length is then $L = c_-(0)\,T/\epsilon$.

4.4 Numerically Calculated Pulses

We shall compare numerical solutions for different representative values of d with the approximate pulses provided by our theory. As initial data, we have adopted our approximate pulses. We have also used hump-like profiles with compact support

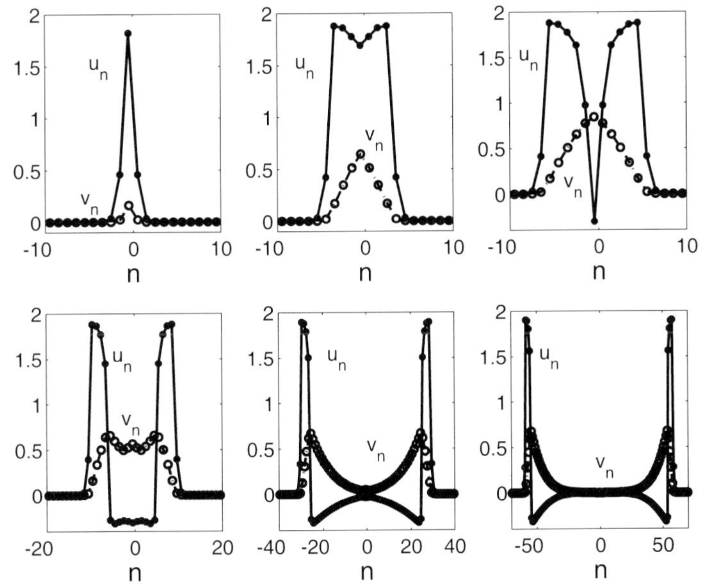

Fig. 4.7 Splitting of an initial profile into two pulses propagating in opposite directions for $d = 0.1$, $a = 0.5$, and $\epsilon = 0.006$ (from [19]).

for $u_n(0)$ and $v_n(0)$. It is important that $v_n(0) = 0$ at the leading edge of $u_n(0)$ and to its right, and that $v_n(0) = V \approx w_{cr}(d)$ at the trailing edge where $v_n(0)$ reaches its maximum. Had we chosen $v_n(0) = 0$ for all n, the u_n profile would have split into two pulses traveling in opposite directions as time elapsed; see Figure 4.7. The region between leading and trailing fronts of the pulse acquires its asymptotic shape quite quickly, but the pulse tail is usually rather long and evolves slowly towards its final form.

- $d = 1$. In this case, $a_{cl} = 0.996$. We choose $a = 0.99 < a_{cl}$ and analyze front propagation for the rescaled Nagumo equation (4.4) first. The front propagation thresholds for these values of d and a are $w_{cl} = 0.0038$ and $w_{cr} = 0.0095$. Figure 4.8a shows the speeds of leading and trailing fronts as functions of w, as predicted by Eq. (4.9). For $w = 0$, the leading front should move at speed $c_-(0) = 0.0093$. The relation $c_+(V) = c_-(0)$ yields the asymptotic value $V^* = 0.0133$ at the trailing front joining $u^{(1)}(V^*) = -0.00665$ to $u^{(3)}(V^*) = 1.9933$. The time elapsed between fronts is $\tau^* = 0.00652$, as calculated from Eq. (4.32). Then, our upper bound for the critical value of ϵ is $\epsilon_c = 0.000064$. Choosing a smaller value, $\epsilon = 0.000005$, we obtain a pulse speed of $C = c_-(0)/\epsilon = 1869$ points

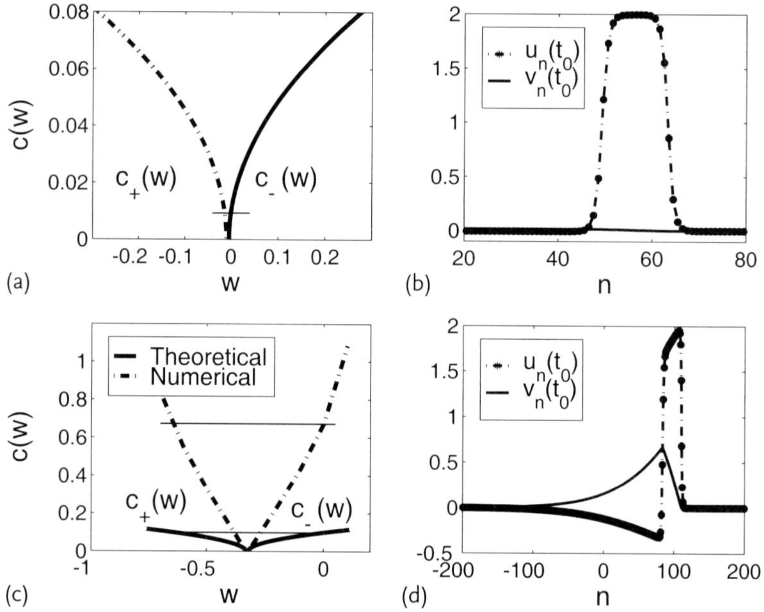

Fig. 4.8 (a) Predicted speeds for wave fronts of the Nagumo equation (4.4) with $d = 1$, $a = 0.99$. The horizontal line marks the condition $c_+(w) = c_-(0)$, thereby graphically yielding $w = V^*$. (b) FHN pulse for $\epsilon = 0.000005$. (c) Predicted and numerical speeds for wave fronts of Eq. (4.4) with $d = 1$, $a = 0.5$. The horizontal lines mark $c_+(w) = c_-(0)$. (d) FHN pulse for $\epsilon = 0.01$ (from [19]).

per unit time and a pulse width of $l^* = C\tau^* \sim 13$ points. Our numerical solution of the full FHN system (4.1) and (4.2) yields a pulse speed $C = 2000$ and a width of 13 points for $\epsilon = 0.000005$. The trailing front joins -0.006647 and 1.993 with $V^* = -0.0133$; see Figure 4.8b. Note that the relative error in the predicted speed C is 0.0655. Obviously, rescaling the speed to $C = c_-(0)/\epsilon$ amplifies the error in our predictions. We have not been able to observe pulses for $\epsilon \geq 0.0000076$, which is smaller, but not far from our estimation $\epsilon_c = 0.000064$. Let us now choose $a = 0.5$, which is far from a_{cl}. Then, $w_{cl} = 0.3194$ and $w_{cr} = 0.3287$. Equation (4.9) predicts $c_-(0) = 0.09983$, whereas the trailing front joins -0.316 to 1.71 at $V^* = 0.6$. If $\epsilon = 0.01$, the speed and width of the pulse are $C = 9.983$ and $l^* = 0.351C \sim 4$, according to our theory. Numerically, we observe $C = 64.7$ and $l^* = 25$. The source of these large errors is the value $c_-(0) = 0.09983$ predicted with formula (4.9). If we replace this value by the numerical front speed calculated directly, $c_-(0) = 0.673$, we obtain $C = 67.3$ and $l^* \sim 24$ points, which better fit the numerically observed values.

- $d = 0.1$. In this case, $a_{cl} = 0.567$, and we shall choose $a = 0.5 < a_{cl}$. Let us first analyze front propagation for the rescaled Nagumo equation (4.4). For these values of d and a, we obtain $w_{cl} = 0.0307$ and $w_{cr} = 0.6175$. Figure 4.9a shows the predicted speeds of the leading and trailing fronts as functions of v, as given by formula (4.9). For $w = 0$, the leading front should move with speed $c_-(0) = 0.075662$. At the trailing front, $c_+(V) = c_-(0)$ yields $V^* = 0.648$, $u^{(1)}(V^*) = -0.33328$, and $u^{(3)}(V^*) = 1.666$. The time elapsed between fronts is $\tau^* = 0.39266$, which gives $l^* = 0.0297/\epsilon$. Our bound for the critical value of ϵ is $\epsilon_c = c_-(0)\tau^* = 0.029$. Selecting $\epsilon = 0.003$, we predict $C = 25.22$ and $l^* \sim 10$ points. Direct numerical calculations yield a pulse speed $C = 26.38$ and a pulse width of about 10 points. The trailing front joins -0.3269 to 1.675 with $V^* = 0.6578$; see Figure 4.9b. We have not been able to obtain pulses for $\epsilon \geq 0.007$, which is four times smaller than our upper bound of 0.029.
- $d = 0.01$. In this case, $a_{cl} = 0.195$ and we shall choose $a = 0.1 < a_{cl}$. Let us first analyze front propagation for the rescaled Nagumo equation (4.4). The front propagation thresholds for these values of d and a are $w_{cl} = 0.0136$ and $w_{cr} = 1.0784$. Figure 4.10a shows the predicted speeds of leading and trailing fronts as functions of w according to Eq. (4.9). For $w = 0$, the leading front should move with speed $c_-(0) = 0.052$. Then, the trailing front has $V^* = 1.092$ corresponding to $c_+(V) = c_-(0)$, and it joins $u^{(1)}(V^*) = -0.6$ to $u^{(3)}(V^*) = 1.4$. The time elapsed between fronts is $\tau^* = 0.748$, and the pulse width, $l^* = 0.297/\epsilon$. Our bound for the critical value of ϵ is $\epsilon_c = c_-(0)\tau^* = 0.058$. Selecting $\epsilon = 0.001$, we predict $C = 52$ and $l^* = 39$ points. Numerical observations yield $C = 77.7$ (a relative error of 0.3) and a pulse width of 59 points. Furthermore, the trailing front joins -0.59 to 1.4 with $V^* = 1.095$; see Figure 4.10b. Again, the observed errors in the

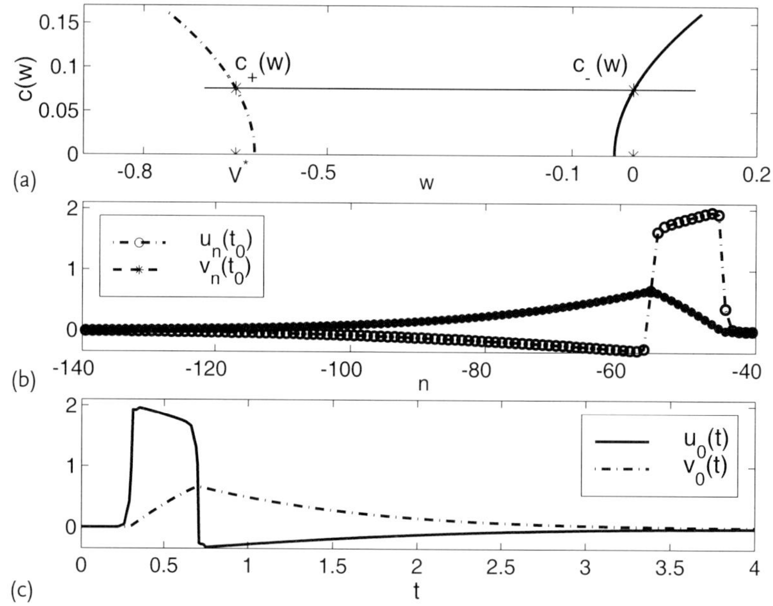

Fig. 4.9 (a) Predicted speeds for the Nagumo equation (4.4) with $d = 0.1$ and $a = 0.5$. The horizontal line graphically yields V^* such that $c_+(V^*) = c_-(0)$. (b) Profiles of the FHN pulse for $\epsilon = 0.003$ (c) Trajectories of one point, $u_0(t)$, $v_0(t)$, as the FHN pulse propagates through it (from [19]).

pulse speed and width are due to errors in the prediction of $c_-(0)$ given by formula (4.9). By replacing this value by the numerically computed front speed $c_+(0) = 0.078$, we obtain $C = 78$ and $l^* \sim 58$, which is a better fit to the real values.

Let us now describe the situation for other values of d. Our asymptotic theory agrees with the numerical results provided that ϵ is sufficiently small, but the velocity of the Nagumo wave fronts should be either approximated by Eq. (4.9) or calculated numerically depending on how close to zero w_{cl} happens to be. For $d < 0.01$, the length of the intervals in which fronts of the Nagumo equation propagate is very small. Then, the front speeds are always very small and given by Eq. (4.9) with great accuracy. Our asymptotic description of the pulse agrees very well with numerical solutions of the FHN system. If $d > 1$, the spatially discrete FHN system can be approximated by its continuum limit. The length of the pinning intervals for the Nagumo equation is below 0.001, and the wave front velocities are essentially a correction of the wave front velocities for the spatially continuous Nagumo equation given by Eq. (4.24) for the appropriate front solution of Eqs. (4.19) and (4.21).

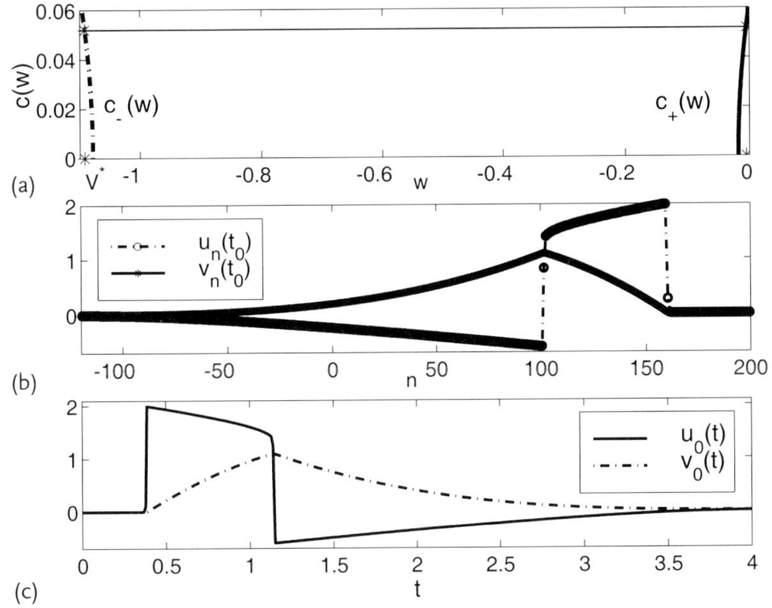

Fig. 4.10 (a) Predicted speeds for the Nagumo equation (4.4) with $d = 0.01$ and $a = 0.1$. (b) FHN pulse for $\epsilon = 0.001$. (c) Trajectory of one point, $u_0(t)$, as the FHN pulse propagates through it (from [19]).

4.5 Propagation Failure

Two facts may lead to propagation failure: a value of ϵ that is too large or $a \in (a_{cl}(d), a_{cr}(d))$.

Let us consider the first cause of propagation failure. If ϵ surpasses a certain critical value ϵ_c, recovery is too fast and a stable pulse cannot be sustained. This situation also occurs in spatially continuous FHN systems. In these systems, there exist two pulses (one pulse is stable, the other unstable) for $\epsilon < \epsilon_c$; they coalesce at ϵ_c and cease to exist for larger ϵ. In the discrete FHN system, the phenomenon of wave front propagation failure implies that pulses may propagate only if $a < a_{cl}(d)$ or $a > a_{cr}(d)$. As indicated by Eq. (4.34), the number of points l between the two fronts of the stable pulse decreases as ϵ increases towards $\epsilon_c(a, d)$. Eventually, the two fronts coalesce and it is no longer possible to propagate a stable pulse for $\epsilon > \epsilon_c(a, d)$. If we start with an appropriate pulse-like initial condition, we find the scenario of propagation failure depicted in Figures 4.11 and 4.12. For small d ($d = 0.1$), the variable v_n ceases to be almost constant at the leading edge of the pulse, and the distance between the two fronts diminishes. While $v_n \sim 0$ at the rightmost point of the leading front, $v_n \sim w > 0$ at the leftmost point. Thus, u_n in this front decreases from $u^{(3)}(w)$ to zero as n increases. The value w increases with ϵ, and $u^{(3)}(w)$ decreases. At the same time, the leading front speed diminishes

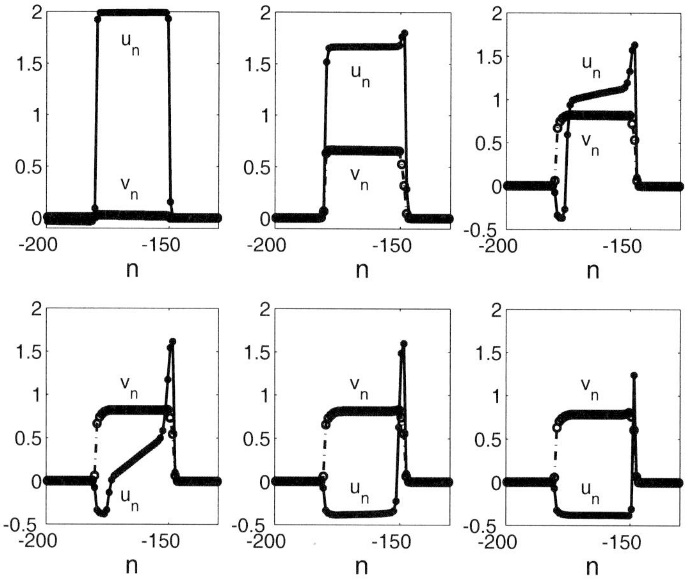

Fig. 4.11 Snapshots of the excitation and recovery variables for $d = 0.1$, $a = 0.5$, and $\epsilon = 0.007$, illustrating propagation failure of the pulse for $\epsilon > \epsilon_c$ (from [19]).

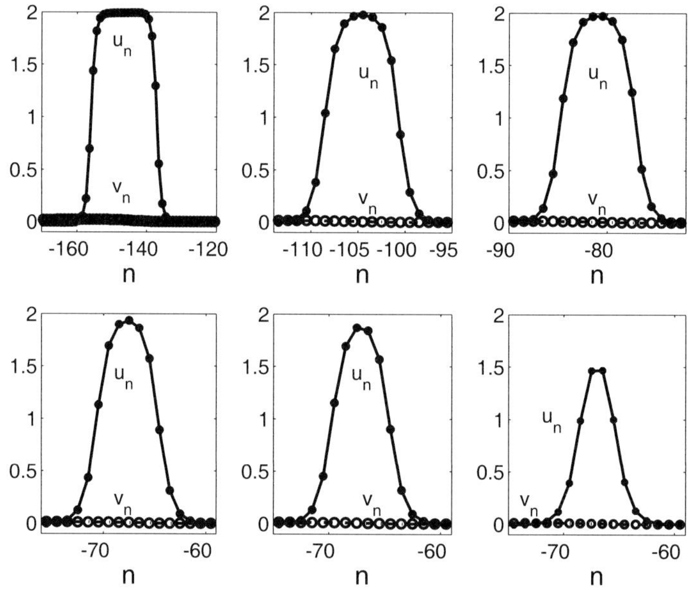

Fig. 4.12 Same as Figure 4.11 for $d = 1$, $a = 0.99$, and $\epsilon = 0.0000076$ (from [19]).

as w increases until w surpasses the propagation threshold and the leading front stops. Since the back front continues moving, the pulse vanishes; see Figure 4.11. For large d ($d = 1$), a decremental pulse is formed. Its width and height decrease as it moves until it disappears; see Figure 4.12. Numerical simulations of the FHN system show that $\epsilon_c \to 0$ as a tends to either $a_{cl}(d)$ or $a_{cr}(d)$.

Let us now assume that $a \in (a_{cl}(d), a_{cr}(d))$. Then, the leading front cannot propagate with $v_n = v = 0$. We need $v_n \sim v < w_{cl}(a, d) < 0$. However, in the region in front of the leading edge, v_n and u_n evolve towards 0, whereas we have $u_n > 0$ at the leading front. Thus, $dv_n/dt = u_n - Bv_n \geq 0$ there, and v_n will increase until $v_n > 0$, which contradicts our previous assumption. Thus, we cannot have stable propagating pulses. Furthermore, there are no stationary pulses of the type we have discussed for this range of a: if $v_n = u_n/B$, the source $u_n(2 - u_n)(a - u_n) - v_n = u_n(2 - u_n)(u_n - a) - u_n/B$ has only one zero, not three as in our construction. This does not preclude the existence of other pulses, such as those corresponding to the homoclinic orbit in the phase space of the spatially continuous FHN system. However, we have not observed stable stationary pulses of this type in the spatially discrete FHN system.

4.6 Pulse Generation at a Boundary

So far, we have considered the motion of a pulse (or its failure) in a sufficiently large myelinated nerve fiber. We have not discussed how such a pulse might be created in a more realistic situation. Clearly, nerve fibers have finitely many nodes of Ranvier, and pulses are typically generated at the fiber boundary. Thus, we are led to consider how a pulse might be generated by an excitation at a boundary and how the pulse propagates or fails in a finite fiber. This problem was tackled by Booth and Erneux [22] using parameter values for which the FHN pulse fails to propagate. We shall now discuss different parameter ranges.

Nerve fibers may have either a few nodes of Ranvier (e. g., 20 for neurons of the central nervous system [23]) or several hundred nodes (in the peripheral nervous system [24]). Thus, we shall consider a finite FHN system with N nodes and a Neumann boundary condition at the right end, $u_{N+1} = u_N$. At the left end, we impose $u_0(t) = 2$ for $0 \leq t \leq 0.05$, and $u_0(t) = 0$ for $t > 0.05$. The results corresponding to parameter values $d = 0.1$ and $a = 0.5$ are depicted in Figures 4.13 (for which $\epsilon = 0.006$) and 4.14 (for which $\epsilon = 0.003$). The asymptotic theory predicts that fully developed FHN pulses (corresponding to $N = \infty$) would have widths of $l^* \approx 5$ and $l^* \approx 10$, respectively. The left boundary condition ensures that the membrane potential u_n is excited during sufficient time, so that a wave is generated at the left end of the fiber.

The excitation at the left boundary induces a wave front that propagates with a velocity given approximately by $C = c_-(0)/\epsilon$ along the finite fiber for the parameter values we consider. For example, $C \approx 12.6$ for $\epsilon = 0.006$, which is close to the numerically observed value of 10 in Figure 4.13. Similarly, $C \approx 25.22$ for $\epsilon = 0.003$,

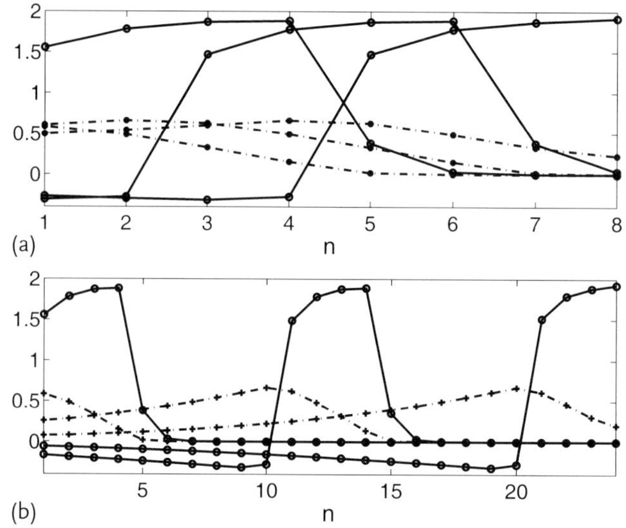

Fig. 4.13 Snapshots of the excitation (solid line) and recovery (dotted line) variables for an FNH system with N nodes and $d = 0.1$, $a = 0.5$, and $\epsilon = 0.006$. (a) Profiles at times 0.4, 0.6, and 0.8 for $N = 8$. (b) Profiles at times 0.4, 1.4, and 2.4 for $N = 24$ (from [19]).

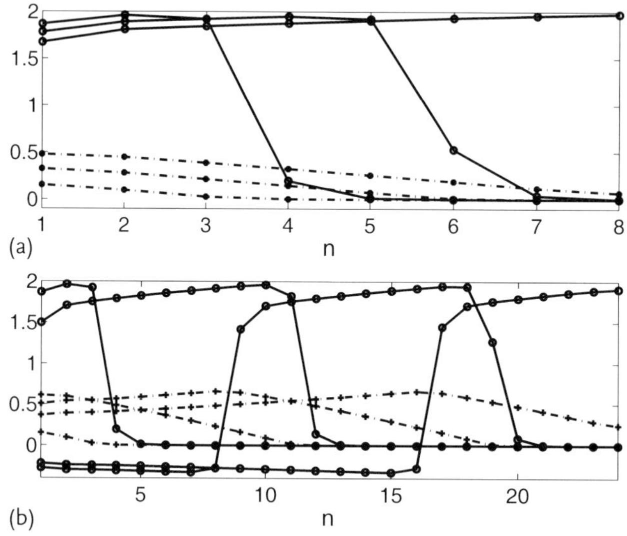

Fig. 4.14 Snapshots of the excitation (solid line) and recovery (dotted line) variables for an FNH system with N nodes and $d = 0.1$, $a = 0.5$, and $\epsilon = 0.003$. (a) Profiles at times 0.1, 0.2, and 0.3 for $N = 8$. (b) Profiles at times 0.1, 0.4, 0.7, and 1.0 for $N = 24$ (from [19]).

which is close to the numerically observed value of 26 in Figure 4.14. If the fiber is long enough, a second wave front follows the first one and their mutual distance rapidly approaches the asymptotic value l^*. (The number of nodes between fronts is four in Figure 4.13, while the asymptotic theory predicts $l^* \approx 5$; in Figure 4.14, numerical observation confirms the asymptotic value $l^* \approx 10$.) The numerical solution of the finite FHN system shows that an eventually truncated FHN pulse comprising the two wave fronts and the region between them is formed, provided that N is at least twice l^*. Otherwise, at best, only the first wave front is shed at the boundary, as shown in Figure 4.14a. Pulses fail to propagate in fibers whose parameters fall in the propagation failure region, as discussed in Section 4.5.

4.7 Concluding Remarks

In this chapter, we have constructed stable pulses of the spatially discrete FHN system by asymptotic methods. In a pulse, there are regions where the excitation variable varies smoothly, separated by sharp fronts. These fronts are solutions of the discrete Nagumo equation with a constant value of the recovery variable. Their shape and speed can be approximately calculated near parameter values corresponding to front propagation failure or near the continuum limit. For long times, their width is given by the only stable solution of a one-dimensional autonomous system. We have compared the asymptotic results with numerical solutions of the FHN system and analyzed different scenarios for failure of pulse propagation. Besides the classical scenario of small separation between the time scales of excitation and recovery (large ϵ as in the spatially continuous FHN system), propagation failure of fronts for the spatially discrete Nagumo equation provides a different mechanism of propagation failure of pulses for the discrete FHN system. Wave fronts and pulses can be generated at a boundary and propagate or fail to propagate along a finite FHN system. If the number of nodes is sufficiently large, the two wave fronts comprising an FHN pulse can be shed at the boundary, and their separation rapidly reaches the value given by the asymptotic theory. This is true even if the fiber is too short to accommodate the slowly varying regions at the back of the second wave front of the pulse. In long fibers, a fully developed FNH pulse may be generated by an over-threshold stimulus applied during a short time at one end of the fiber.

In a similar fashion, Carpio [25] has constructed wave trains for discrete excitable systems, among which is the FHN system. In the same paper, propagation failure and generation of wave trains by periodic excitation at a boundary are studied. If the system has oscillatory dynamics, wave trains cannot be excited from an equilibrium state by periodic excitation at a boundary. Instead, synchronization phenomena may occur [25]. The original model of nerve excitation by Hodgkin and Huxley consisted of one equation for the electric potential (the excitatory variable) and three equations for different ion concentrations (recovery). Wave trains and pulses for these more complicated systems have been asymptotically constructed and analyzed by Carpio [26].

References

1 Frenkel, J. and Kontorova, T. (1938) On the theory of plastic deformation and twinning. *J. Phys. USSR*, **13**, 1–10.
2 Cahn, J.W. (1960) Theory of crystal growth and interface motion in crystalline materials. *Acta Metall.*, **8**, 554–562.
3 Bonilla, L.L. (2002) Theory of nonlinear charge transport, wave propagation and self-oscillations in semiconductor superlattices. *J. Phys. Condens. Matter*, **14**, R341–R381.
4 Carpio, A., Bonilla, L.L., and Dell'Acqua, G. (2001) Wave front motion in semiconductor superlattices. *Phys. Rev. E*, **64**, 036204.
5 Grüner, G. (1988) The dynamics of charge-density waves. *Rev. Mod. Phys.*, **60**, 1129–1181.
6 Scott, A.C. (1975) The electrophysics of a nerve fiber. *Rev. Mod. Phys.*, **47**, 487–533.
7 Anderson, A.R.A. and Sleeman, B.D. (1995) Wave front propagation and its failure in coupled systems of discrete bistable cells modelled by FitzHugh–Nagumo dynamics. *Int. J. Bifurc. Chaos Appl. Sci. Eng.*, **5**, 63–74.
8 Keener, J.P. and Sneyd, J. (1998) *Mathematical Physiology*, Springer, New York, Chap. 9.
9 Keener, J.P. (1987) Propagation and its failure in coupled systems of discrete excitable cells. *SIAM J. Appl. Math.*, **47**, 556–572.
10 Nabarro, F.R.N. (1987) *Theory of Crystal Dislocations*, Dover, New York.
11 Carpio, A., Bonilla, L.L., Schöll, E., and Wacker, A. (2000) Wavefronts may move upstream in doped semiconductor superlattices. *Phys. Rev. E*, **61**, 4866–4876.
12 Hobart, R. (1965) Peierls-barrier minima. *J. Appl. Phys.*, **36**, 1948–1952.
13 Amann, A., Wacker, A., Bonilla, L.L., and Schöll, E. (2001) Dynamic scenarios of multistable switching in semiconductor superlattices. *Phys. Rev. E*, **63**, 066207.
14 Kastrup, J., Hey, R., Ploog, K., Grahn, H.T., Bonilla, L.L., Kindelan, M., Moscoso, M., Wacker, A., and Galán, J. (1997) Electrically tunable GHz oscillations in doped GaAs-AlAs superlattices. *Phys. Rev. B*, **55**, 2476–2488.
15 Zinner, B. (1992) Existence of traveling wavefront solutions for the discrete Nagumo equation. *J. Differ. Equ.*, **96**, 1–27.
16 Hankerson, D. and Zinner, B. (1993) Wave fronts for a cooperative tridiagonal system of differential equations. *J. Dyn. Differ. Equ.*, **5**, 359–373.
17 Carpio, A. and Bonilla, L.L. (2001) Wave front depinning transitions in discrete one-dimensional reaction diffusion equations. *Phys. Rev. Lett.*, **86**, 6034–6037.
18 Carpio, A. and Bonilla, L.L. (2003) Depinning transitions in discrete reaction-diffusion equations. *SIAM J. Appl. Math.*, **63**, 1056–1082.
19 Carpio, A. and Bonilla, L.L. (2003) Pulse propagation in discrete systems of coupled excitable cells. *SIAM J. Appl. Math.*, **63**, 619–635. Copyright (2003) by the Society for industrial and Applied Mathematics.
20 Nagumo, J., Arimoto, S. and Yoshizawa, S. (1962) An active pulse transmission line simulating nerve axon. *Proc. Inst. Radio Eng.*, **50**, 2061–2070.
21 Rinzel, J. and Keller, J.B. (1973) Traveling wave solutions of a nerve conduction equation. *Biophys. J.*, **13**, 1313–1337.
22 Booth, V. and Erneux, T. (1995) Understanding propagation failure as a slow capture near a limit point. *SIAM. J. Appl. Math.*, **55**, 1372–1389.
23 McIntyre, C.C. and Grill, W.M. (1999) Excitation of central nervous system neurons by nonuniform electric fields. *Biophys. J.*, **76**, 878–888.
24 Struijk, J.J. (1997) The extracellular potential of a myelinated nerve fiber in an unbounded medium and in nerve cuff models. *Biophys. J.*, **72**, 2457–2469.
25 Carpio, A. (2005) Wave trains, self-oscillations and synchronization in discrete media. *Physica D*, **207**, 117–136.
26 Carpio, A. (2005) Asymptotic construction of pulses in the Hodgkin Huxley model for myelinated nerves. *Phys. Rev. E*, **72**, 011905.

5
Electronic Transport in Condensed Matter: From Quantum Kinetics to Drift-diffusion Models

5.1
Introduction

In this chapter, we study electron transport using quantum and semiclassical kinetic theory. The first part of the chapter is devoted to general considerations on kinetic equations and drift-diffusion equations that can be derived from them in the diffusive limit. In the second part of the chapter, we provide an in depth study of the same problems particularized to a semiconductor superlattice. We introduce the Wigner function for non-interacting spinless particles in an external potential and find the Vlasov equation as its semiclassical limit. We then consider collisions between particles. Electron-electron collisions are treated in the Hartree approximation and therefore, the electric potential seen by the electrons solves the Poisson equation with the charge density produced by the electrons themselves. As a simple collision model, we consider the Bhatnagar–Gross–Krook (BGK) collision model, in which the distribution function tries to reach a local equilibrium distribution having the same electron density. We prove the H theorem for the resulting kinetic equation. Then, we study how to derive drift-diffusion equations in the parabolic limit by using multiple scales and the Chapman–Enskog method. In the parabolic limit, collisions dominate and the distribution function is close to local equilibrium and both multiple scales, and the Chapman–Enskog method produces the same drift-diffusion equation. In it, drift and diffusion are related by the Einstein relation.

In the second part of the chapter, we study nonlinear charge transport in strongly coupled semiconductor superlattices. It is described by Wigner–Poisson kinetic equations involving one or more minibands. Again, electron–electron collisions are treated within the Hartree approximation, whereas other inelastic collisions are described by a modified BGK model. However, we study these equations in the hyperbolic limit such that the collision frequencies are of the same order as the Bloch frequencies due to the electric field and the corresponding terms in the kinetic equation are dominant. In this limit, spatially nonlocal drift-diffusion balance equations for the miniband populations and the electric field are derived by means of the Chapman–Enskog perturbation technique. Numerical solutions of the drift-

Nonlinear Wave Methods for Charge Transport. Luis L. Bonilla and Stephen W. Teitsworth

ISBN: 978-3-527-40695-1

diffusion equation show stable Gunn-type self-sustained oscillations of the current through a voltage biased superlattice.

5.1.1
Wigner Function for Non-interacting Particles in an External Potential

Suppose we have a single spinless quantum particle whose wave function obeys the Schrödinger equation

$$i\hbar \frac{\partial \psi}{\partial t} = \left(-\frac{\hbar^2}{2m}\nabla^2 + V(x)\right)\psi . \tag{5.1}$$

An arbitrary state can be described by the density matrix whose equation is obtained from Eq. (5.1) by differentiating $\psi^*(x)\psi(y) = \rho(x, y)$ (we drop the dependence on time from ρ for the time being):

$$i\hbar \frac{\partial}{\partial t}\rho(x, y) = \left(-\frac{\hbar^2}{2m}\nabla_y^2 + V(y)\right)\rho(x, y) - \left(-\frac{\hbar^2}{2m}\nabla_x^2 + V(x)\right)\rho(x, y) . \tag{5.2}$$

The solutions of Eq. (5.2) are not only pure states as that used to derive it, but also mixed states such as $\rho(x, y) = \sum_n c_n \psi_n^*(x)\psi_n(y)$ which are used in statistical descriptions. Employing Dirac's notation, $\rho(x, y) = \langle x|\hat{\rho}|y\rangle$, this is just the particularization to space coordinates of the Landau–von Neumann operator equation

$$i\hbar \frac{\partial}{\partial t}\hat{\rho} = [\hat{\rho}, \hat{H}] \equiv \hat{\rho}\hat{H} - \hat{H}\hat{\rho} , \tag{5.3}$$

where $\hat{H} = \hat{p}^2/(2m) + V(x)$ ($\hat{p} = -i\hbar\nabla_x$) is simply the Hamiltonian.

By definition, the density matrix is a complex function satisfying the Hermitian property $\rho(x, y)^* = \rho(y, x)$. To make contact with classical statistical physics, it is convenient to use the Wigner transform [1]

$$f(x, k) = \int_{\mathbb{R}^3} \rho\left(x + \frac{\xi}{2}, x - \frac{\xi}{2}\right) e^{ik\cdot\xi} d^3\xi , \tag{5.4}$$

$$\rho(x, y) = \frac{1}{(2\pi)^3}\int_{\mathbb{R}^3} f\left(\frac{x + y}{2}, k\right) e^{ik\cdot(x-y)} d^3k . \tag{5.5}$$

$f(x, k)$ is called the Wigner function and the Hermitian property of the density matrix implies that it has real values: $f^*(x, k) = f(x, k)$. The first two moments of the Wigner function are the number density:

$$n(x) = \rho(x, x) = \frac{1}{(2\pi)^3}\int_{\mathbb{R}^3} f(x, k)\, d^3k , \tag{5.6}$$

and the particle current:

$$J_n(x) = \frac{i\hbar}{2m}(\nabla_x - \nabla_y)\rho(x, y)|_{y=x} = \frac{1}{(2\pi)^3}\int_{\mathbb{R}^3} \frac{\hbar k}{m} f(x, k)\, d^3k . \tag{5.7}$$

Although the number density must be positive, the Wigner function may take on negative values which means we cannot interpret it as a classical distribution function which is always positive.

Starting from Eqs. (5.2) and (5.4), a short exercise in algebra shows that the Wigner function satisfies the following equation

$$\frac{\partial f}{\partial t}+\frac{\hbar k}{m}\cdot\nabla_x f+\frac{1}{i\hbar}\left[V\left(x+\frac{1}{2i}\nabla_k\right)-V\left(x-\frac{1}{2i}\nabla_k\right)\right]f(x,k)=0. \quad (5.8)$$

The third term in the left hand side of this equation means

$$\left[V\left(x+\frac{1}{2i}\nabla_k\right)-V\left(x-\frac{1}{2i}\nabla_k\right)\right]f(x,k)=$$
$$\int_{\mathbb{R}^3} e^{ik\cdot\xi}\left[V\left(x+\frac{\xi}{2}\right)-V\left(x-\frac{\xi}{2}\right)\right]\rho\left(x+\frac{\xi}{2},x-\frac{\xi}{2}\right)d^3\xi\,, \quad (5.9)$$

in terms of the inverse Fourier transform of the Wigner function, which is $\rho(x+\xi/2, x-\xi/2)$.

5.1.2
Classical Limit

Let us denote by x_V the length scale over which the potential $V(x)$ varies appreciably and let x_ρ denote the quantum correlation length, that is, the characteristic distance above which $\rho(x,y)$ decays as $|x-y|$ increases. The Wigner function varies appreciably in k for k-space length scales that are less than order $1/x_\rho$. The classical limit is reached as $x_\rho/x_V \to 0$. In this limit, the integrand in Eq. (5.9) can be approximated by

$$\xi\cdot\nabla_x V(x)\,\rho\left(x+\frac{\xi}{2},x-\frac{\xi}{2}\right),$$

which, when inserted in Eq. (5.9), yields $-i\nabla_x V\cdot\nabla_k f(x,k)$. Then, the Wigner equation (5.8) becomes the Vlasov equation of classical kinetic theory:

$$\frac{\partial f}{\partial t}+\frac{p}{m}\cdot\nabla_x f-\nabla_x V(x)\cdot\nabla_p f(x,k)=0\,, \quad (5.10)$$

in which $p=\hbar k$ is the momentum. Note that by integrating this equation over momentum we obtain the continuity equation:

$$\frac{\partial n}{\partial t}+\nabla_x\cdot J_n(x)=0\,, \quad (5.11)$$

where density and particle current are given by Eqs. (5.6) and (5.7), respectively.

5.1.3
Boltzmann Transport Equation and BGK Collision Model

The Vlasov equation describes an ensemble of non-interacting particles in a force field given by the potential energy $V(x)$ such as the particles comprising a collisionless plasma. Very often we are interested in describing ensembles of interacting particles which undergo scattering processes. In a simple mean-field Hartree description, electron–electron scattering can be approximately described by means of the Poisson equation

$$-\varepsilon \nabla_x^2 V = e^2(n - N_D)\,, \tag{5.12}$$

where $\varepsilon, -e > 0$ and N_D are the permittivity, the electron charge and a background constant density of positively charged impurities, respectively. Note that V is a potential energy, not the electric potential $V/(-e)$. Provided N_D is independent of time, the continuity equation (5.11) and the Poisson equation (5.12) are equivalent to the following form of Ampère's law:

$$\varepsilon \frac{\partial F}{\partial t} + e^2 J_n(x) = e^2 J\,, \quad \nabla_x \cdot J = 0\,. \tag{5.13}$$

Here, $F = -\nabla_x V$ is the force, J is the total particle current density and $-e\,J$ the total current density. We will explain later how to find a relation between J_n and F from the kinetic equation by using perturbation methods.

In classical physics, other scattering processes are modeled by collision terms which are added to the right-hand side of Eq. (5.10). These collision terms usually have the form of an integral over momentum involving the distribution function f and are therefore nonlocal in momentum though local in space and time. For physically meaningful collision models, the resulting Boltzmann equation typically has an associated H theorem, indicating that the entropy increases with time as a result of scattering and that equilibrium is reached as time goes to infinity. Bhatnagar, Gross and Krook (1954) [2] proposed a simple model of atomic collisions for gases, thereafter called the BGK model. They assumed that collisions drive the gas to a state of local equilibrium given by the Maxwell–-Boltzmann distribution with density, average velocity and temperature given by the instantaneous values of the first moments of the time dependent distribution function. These moments correspond to the five collisional invariants of the usual collision term of the Boltzmann equation (mass, 3D velocity and kinetic energy: the momentum integral of the collision operator times 1, p/m or $p^2/(2m)$ is zero) [3]. In the case we are considering, the density is the first moment of the distribution and the continuity equation (5.11) should be satisfied. Thus, the integral of the collision term over momentum should vanish so that 1 is a collisional invariant. These requirements hold for the BGK equation:

$$\frac{\partial f}{\partial t} + \frac{\hbar k}{m} \cdot \nabla_x f - \frac{1}{\hbar} \nabla_x V(x) \cdot \nabla_k f(x,k) = -\nu\,(f - f^B)\,, \tag{5.14}$$

$$f^B(k; n) = n\hbar^3 \left(\frac{2\pi}{m k_B T_0}\right)^{3/2} e^{-\hbar^2 |k|^2/(2 m k_B T_0)}\,, \tag{5.15}$$

where ν, k_B and T_0 are the constant collision frequency, the Boltzmann constant and the equilibrium temperature, because

$$n = \frac{1}{(2\pi)^3} \int_{\mathbb{R}^3} f^B(k; n)\, d^3k\,.$$

We can show that a certain quantity related to the relative entropy

$$\eta(t) = \frac{1}{(2\pi)^3} \int_{\mathbb{R}^3}\int_{\mathbb{R}^3} f \ln\left(\frac{f}{f^B}\right) d^3k\, d^3x \tag{5.16}$$

is a Lyapunov functional, which is the content of the H theorem. In fact, the inequality $x \ln x \geq x - 1$ for $x > 0$ proves that $\eta \geq \int (f - f^B) = 0$, so that the relative entropy is bounded from below. Let us now calculate the time derivative of the relative entropy and figure out which related quantity is a Lyapunov functional:

$$\frac{d\eta}{dt} = \iint \left(1 + \ln \frac{f}{f^B}\right) \frac{\partial f}{\partial t} \frac{d^3k d^3x}{(2\pi)^3} - \iint \frac{f}{n} \frac{\partial n}{\partial t} \frac{d^3k d^3x}{(2\pi)^3}\,.$$

The first and third terms in this expression cancel each other. We now substitute $\partial f/\partial t$ by the expression we get from Eq. (5.14) and obtain

$$\frac{d\eta}{dt} = \iint \ln\left(\frac{f}{f^B}\right)\left[-\nabla_x f \cdot \frac{\hbar k}{m} + \frac{\nabla_x V}{\hbar} \cdot \nabla_k f - \nu\,(f - f^B)\right] \frac{d^3k d^3x}{(2\pi)^3}\,.$$

In this integrand, $\ln f \nabla f = \nabla(f \ln f - f)$ and the terms containing $\ln f \nabla_{x,k} f$ are integrals of gradients of some functions which, therefore, contribute boundary terms that vanish as $|k|$ or $|x|$ go to infinity. Thus, we are left with

$$\begin{aligned}\frac{d\eta}{dt} &= -\nu \iint (f - f^B) \ln\left(\frac{f}{f^B}\right) \frac{d^3k d^3x}{(2\pi)^3} \\ &\quad + \iint \ln f^B \left[\nabla_x f \cdot \frac{\hbar k}{m} - \frac{\nabla_x V}{\hbar} \cdot \nabla_k f\right] \frac{d^3k d^3x}{(2\pi)^3}\,.\end{aligned}$$

By integrating by parts, we obtain

$$\frac{d\eta}{dt} = -\frac{\nu}{(2\pi)^3} \iint (f - f^B) \ln\left(\frac{f}{f^B}\right) d^3k d^3x - \int J_n \cdot \nabla_x \left(\frac{\mu_e}{k_B T_0}\right) d^3x\,, \tag{5.17}$$

$$\mu_e = V + k_B T_0 \ln \frac{n}{n_0}\,. \tag{5.18}$$

Here, $\mu_e(x, t)$ is the electrochemical potential and n_0 is an arbitrary unit of density which we take as the constant density at equilibrium, N_D. The second term in the right-hand side of Eq. (5.17) can be rewritten as the total derivative of a new functional $-v$. In fact, the Ampère's law Eq. (5.13) implies

$$-\int J_n \cdot \nabla_x V = -\int \left(J - \frac{\varepsilon}{e^2} \frac{\partial F}{\partial t} \right) \cdot \nabla_x V \, d^3x$$
$$= -\int \left[\nabla_x \cdot (J V) + \frac{\varepsilon}{e^2} F \cdot \frac{\partial F}{\partial t} \right] d^3x = -\frac{d}{dt} \int \frac{\varepsilon}{2e^2} |F|^2 . \tag{5.19}$$

Similarly, integration by parts and the continuity equation (5.11) imply

$$-\int J_n \cdot \nabla_x \ln \frac{n}{N_D} \, d^3x = \int (\nabla_x \cdot J_n) \ln \frac{n}{N_D} \, d^3x$$
$$= -\int \frac{\partial n}{\partial t} \ln \frac{n}{N_D} \, d^3x$$
$$= -\frac{d}{dt} \int \left(n \ln \frac{n}{N_D} - n + N_D \right) d^3x . \tag{5.20}$$

Using Eqs. (5.19) and (5.20), we can therefore write the second term of the right-hand side of Eq. (5.17) as the time derivative of $-v$, with

$$v \equiv \int \left[\frac{\varepsilon}{2e^2 k_B T_0} |F|^2 + N_D \left(\frac{n}{N_D} \ln \frac{n}{N_D} - \frac{n}{N_D} + 1 \right) \right] d^3x . \tag{5.21}$$

Then, Eq. (5.17) is equivalent to

$$\frac{dH}{dt} = -\frac{\nu}{(2\pi)^3} \iint (f - f^B) \ln \left(\frac{f}{f^B} \right) d^3k \, d^3x \le 0, \tag{5.22}$$

$$H = \eta + v \equiv \frac{1}{(2\pi)^3} \int_{\mathbb{R}^3} \int_{\mathbb{R}^3} f \ln \left(\frac{f}{f^B} \right) d^3k \, d^3x$$
$$+ \int_{\mathbb{R}^3} \left[\frac{\varepsilon |F|^2}{2e^2 k_B T_0} + N_D \left(\frac{n}{N_D} \ln \frac{n}{N_D} - \frac{n}{N_D} + 1 \right) \right] d^3x \ge 0 , \tag{5.23}$$

where we have used the inequality $(x-1) \ln x \ge 0$ for all $x > 0$. Note that H is non-negative and bounded, provided $|F|^2$ is integrable and that $n \sim N_D$ is sufficiently rapidly as $|x| \to \infty$. The inequalities of Eqs. (5.22) and (5.23) imply that $H(t)$ is a Lyapunov functional. Then, any initial condition of the BGK equation evolves towards the equilibrium solution $f = f^B$ with constant density N_D as $H(t)$ decreases to zero. Thus, the Boltzmann distribution with constant density N_D is the globally asymptotically stable solution of the BGK equation.

5.1.4
Parabolic Scaling

The BGK-Poisson system (5.12) and (5.14) contains more information than we typically want. When the collision term in Eq. (5.14) dominates the convective terms and the time derivative of f, the distribution quickly approaches the local equilibrium of Eq. (5.15) while the density and the potential evolve in a slower time scale towards the global equilibrium.

How can these considerations become more precise? Firstly, we need to compare order of magnitude estimates of each term in the BGK equation (5.14). Its right-hand side is of order $\nu\,[f]$, where $[f]$ is a typical scale over which f varies. Equation (5.6) implies that f is dimensionless, however, its variation is a dimensionless number $[f]$ which will be calculated below in terms of combinations of dimensional units that characterize the variation of other magnitudes. A typical unit of velocity is the thermal velocity $\sqrt{k_B T_0/m}$, which can be obtained from the average energy $3k_B T_0/2$ of the Boltzmann distribution (5.15). Then, the second term in the left-hand side of Eq. (5.14) has the order

$$\left[\frac{\hbar k}{m}\cdot\nabla_x f\right] = \frac{[f]\sqrt{k_B T_0}}{[x]\sqrt{m}}\,,$$

where $[x]$ measures the length over which the force and the distribution function vary since it is physically reasonable that these two length scales are similar. If $[F]$ gives the order of magnitude of the force $F = -\nabla_x V$, the third term in the left-hand side of Eq. (5.14) is of the order

$$\left[-\frac{\nabla_x V}{\hbar}\cdot\nabla_k f\right] = \frac{[F][f]}{\sqrt{m k_B T_0}}\,.$$

Assuming that both convective terms in the BGK equation have the same order, we obtain $[F] = k_B T_0/[x]$. Furthermore, the Poisson equation (5.12) indicates that

$$[-\nabla_x\cdot\nabla_x V] = \frac{[F]}{[x]} = \frac{e^2 N_D}{\varepsilon} = \left[\frac{e^2(n-N_D)}{\varepsilon}\right].$$

Inserting $[F] = k_B T_0/[x]$ in this expression, we find

$$[x] = \frac{\sqrt{\varepsilon k_B T_0}}{e\sqrt{N_D}}\,,\qquad [F] = \frac{e\sqrt{N_d k_B T_0}}{\sqrt{\varepsilon}}\,.$$

The convective terms in the BGK equation are therefore of order $\nu_{\text{conv}}[f]$, where

$$\nu_{\text{conv}} = \frac{\sqrt{k_B T_0/m}}{[x]} = e\sqrt{\frac{N_D}{\varepsilon m}}\,. \tag{5.24}$$

For a typical doping density, $N_D = 10^{24}\,\text{m}^{-3}$, $\nu_{\text{conv}} = 10^7$ Hz, which is much smaller than typical collision frequencies of 10^{13} Hz. Then, the dimensionless ratio

$$\lambda = \frac{\nu_{\text{conv}}}{\nu} = \frac{e}{\nu}\sqrt{\frac{N_D}{\varepsilon m}} \ll 1 \tag{5.25}$$

is very small and the collision term dominates both convective terms. In the previous considerations, $[f]$ drops from all formulas. However, using $[n] = N_D$ and $[k] = \sqrt{m k_B T_0}/\hbar$ in Eq. (5.6) yields

$$[f] = \frac{N_D\hbar^3}{(m k_B T_0)^{3/2}}\,.$$

Table 5.1 Parabolic scaling.

f	$F = -\nabla_x V$	n	J_n	x	k	t
$\frac{N_D \hbar^3}{(m k_B T_0)^{3/2}}$	$\frac{e\sqrt{N_d k_B T_0}}{\sqrt{\varepsilon}}$	N_D	$N_D\sqrt{\frac{k_B T_0}{m}}$	$\frac{\sqrt{m k_B T_0}}{\hbar}$	$\frac{\sqrt{\varepsilon k_B T_0}}{e\sqrt{N_D}}$	$\frac{\varepsilon m v}{e^2 N_D}$

The scales of f, F, x and k thus far obtained are neatly presented in Table 5.1. This table also includes a unit of time to be discussed shortly.

Equations (5.12), (5.14), and so on. contain many physical constants which can be swept away if we measure all variables in natural units such as those in Table 5.1. To do so, we shall replace each variable in the equations by its scale times the symbol of the variable with a tilde. Thus, we shall write $n = N_D \tilde{n}$, and so on. Variables and parameters in the resulting equations will be dimensionless and the resulting procedure is called *nondimensionalization*. We obtain the uncluttered nondimensional equations:

$$\lambda^2 \frac{\partial \tilde{f}}{\partial \tilde{t}} + \lambda \left(\tilde{k} \cdot \nabla_{\tilde{x}} \tilde{f} + \tilde{F} \cdot \nabla_{\tilde{k}} \tilde{f}\right) = -(\tilde{f} - \tilde{f}^B), \tag{5.26}$$

$$\tilde{f}^B = (2\pi)^{3/2} \tilde{n}\, e^{-|\tilde{k}|^2/2}, \tag{5.27}$$

$$\tilde{n} = \frac{1}{(2\pi)^3} \int_{\mathbb{R}^3} \tilde{f}\, d^3\tilde{k} = \frac{1}{(2\pi)^3} \int_{\mathbb{R}^3} \tilde{f}^B d^3\tilde{k}, \tag{5.28}$$

$$\nabla_{\tilde{x}} \cdot \tilde{F} = \tilde{n} - 1, \tag{5.29}$$

$$\tilde{J}_n = \frac{1}{(2\pi)^3} \int_{\mathbb{R}^3} \tilde{k} \tilde{f}\, d^3\tilde{k}. \tag{5.30}$$

The continuity and Ampère equation are now:

$$\lambda \frac{\partial \tilde{n}}{\partial \tilde{t}} + \nabla_{\tilde{x}} \cdot \tilde{J}_n = 0, \tag{5.31}$$

$$\lambda \frac{\partial \tilde{F}}{\partial \tilde{t}} + \tilde{J}_n = \tilde{J}. \tag{5.32}$$

We have chosen the scale of time in Table 5.1 in such a way that the first term in the left-hand side of Eq. (5.26) is of order λ^2. The reason for this choice is technical, and it concerns the perturbation method we shall use to derive reduced equations for the electron density and the force. We have chosen *parabolic scaling*, using λ and λ^2, respectively, in order to rewrite the BGK equation such that collision terms, convective terms and time derivatives of f are of orders 1. This scaling is so named because the resulting leading order equation for the force F is a *parabolic* partial differential equation. This scaling is traditionally used when deriving the Navier-Stokes equations from the Boltzmann equation for gases.

5.1.5
Derivation of a Drift-Diffusion Equation

To derive reduced equations for n and F, we shall use two different methods: multiple scales and the Chapman–Enskog method. It is useful to see these two methods side-by-side because this brings out certain advantages of the Chapman–Enskog method, which of the two, is less intuitive: the method of multiple scales yields reduced equations all whose terms are of the same order, whereas the Chapman–Enskog method may give equations containing terms of different order. With parabolic scaling, both methods yield the same result. In the hyperbolic scaling discussed later in this chapter, the method of multiple scales yields a hyperbolic equation but cannot correct it, whereas the Chapman–Enskog method gives small diffusive corrections to the hyperbolic equation. We shall omit the tildes on nondimensional variables to simplify our notation.

5.1.5.1 Method of Multiple Scales

There are two different time scales in Eq. (5.26), the slow time t and a fast time $\tau = t/\lambda^2$, such that $\partial f/\partial \tau$ is of order 1, the same as the collision part of the equation. We assume that all unknowns are functions of both times. Any function $G(t, t/\lambda^2)$ satisfies

$$\frac{d}{dt} G = \frac{\partial}{\partial t} G + \frac{1}{\lambda^2} \frac{\partial}{\partial \tau} G \,. \tag{5.33}$$

The multiscale ansatz consists of the following expansions:

$$f(x, k, t; \lambda) \sim f^{(0)}(x, k, t, \tau) + \lambda\, f^{(1)}(x, k, t, \tau) + \lambda^2 f^{(2)}(x, k, t, \tau) \,, \tag{5.34}$$

$$n(x, t; \lambda) \sim \sum_{j=0}^{2} \lambda^j n^{(j)}(x, t, \tau) \,, \quad F(x, t; \lambda) \sim \sum_{j=0}^{2} \lambda^j F^{(j)}(x, t, \tau) \,. \tag{5.35}$$

By inserting Eqs. (5.33)–(5.35) into Eqs. (5.26)–(5.29) and equating like powers of λ in the resulting equations, we obtain the following hierarchy:

$$\nabla_x \cdot F^{(j)} = n^{(j)} - \delta_{j0} \,, \quad j = 0, 1, 2 \,, \tag{5.36}$$

$$n^{(j)} = \frac{1}{(2\pi)^3} \int_{\mathbb{R}^3} f^{(j)} d^3k \,, \quad j = 0, 1, 2 \,, \tag{5.37}$$

$$\mathcal{L} f^{(0)} \equiv \left(\frac{\partial}{\partial \tau} + 1 \right) f^{(0)} - n^{(0)} (2\pi)^{3/2} e^{-|k|^2/2} = 0 \,, \tag{5.38}$$

$$\mathcal{L} f^{(1)} = -\left(k \cdot \nabla_x + F^{(0)} \cdot \nabla_k \right) f^{(0)} \,, \tag{5.39}$$

$$\mathcal{L} f^{(2)} = -\left(k \cdot \nabla_x + F^{(0)} \cdot \nabla_k \right) f^{(1)} - F^{(1)} \cdot \nabla_k f^{(0)} - \frac{\partial f^{(0)}}{\partial t} \,, \tag{5.40}$$

and so on. Equation (5.36) are consequences of the Poisson equation, whereas Eq. (5.37) are the compatibility conditions obtained from Eq. (5.28).

The solution of Eq. (5.38) is

$$f^{(0)}(k,x,t,\tau) = n^{(0)}(2\pi)^{3/2} e^{-|k|^2/2}(1-e^{-\tau}) + f_0 e^{-\tau} ,$$
$$n^{(0)}(x,t,\tau) = n^{(0)}(x,t)\,(1-e^{-\tau}) + n_0 e^{-\tau} , \tag{5.41}$$

where $f_0(x,k)$ is the initial distribution function with density $n_0(x)$, and $n^{(0)}(x,t)$ has yet to be determined. Clearly, $f^{(0)}$ tends exponentially fast to the local equilibrium distribution Eq. (5.27) with $n = n^{(0)}(x,t)$.

If we ignore exponentially small terms proportional to $e^{-\tau}$, the solution of Eq. (5.39) is

$$f^{(1)} = \left[n^{(1)} - k\cdot(\nabla_x - F^{(0)})n^{(0)}\right]\phi , \tag{5.42}$$

$$\phi(k) = (2\pi)^{3/2} e^{-|k|^2/2} , \qquad \int_{\mathbb{R}^3} \phi\, d^3k = 1 , \tag{5.43}$$

which satisfies the compatibility condition (5.37). The solution of Eq. (5.40) is

$$f^{(2)} = \left[n^{(2)} - k\cdot(\nabla_x - F^{(0)})n^{(1)}\right]\phi + (k\cdot F^{(1)})\phi\, n^{(0)}$$
$$+ \left(k\cdot\nabla_x + F^{(0)}\cdot\nabla_k\right)\left[k\cdot\left(\nabla_x - F^{(0)}\right)(n^{(0)}\phi)\right] - \phi\,\frac{\partial n^{(0)}}{\partial t} . \tag{5.44}$$

The compatibility condition (5.37) yields

$$n^{(2)} = n^{(2)} + \nabla_x\cdot\int k\,k\cdot(\nabla_x - F^{(0)})(n^{(0)}\phi)\,d^3k - \frac{\partial n^{(0)}}{\partial t} .$$

Since $\int k\,k\,\phi(k)\,d^3k = I/2$, we get

$$\frac{\partial n^{(0)}}{\partial t} + \frac{1}{2}\nabla_x\cdot(n^{(0)}F^{(0)} - \nabla_x n^{(0)}) = 0 . \tag{5.45}$$

Together with the Poisson equation (5.36), Eq. (5.45) forms the reduced system of equations that describes the evolution of density and field on the slow time scale t. The particle current J_n can be obtained by inserting Eqs. (5.41), (5.42) and (5.44) in Eqs. (5.34) and (5.30):

$$J_n = \frac{\lambda}{2}\, n^{(0)}F^{(0)} - \frac{\lambda}{2}\nabla_x n^{(0)} . \tag{5.46}$$

When Eq. (5.46) is inserted in the continuity equation (5.11), the result is again Eq. (5.45). The current J_n comprises a drift term proportional to the force and a diffusion term.

5.1.5.2 Chapman–Enskog Method

The Chapman–Enskog method has a different philosophy from the method of multiple scales, as we discussed in Chapter 2. From the solution of Eq. (5.26) with $\lambda = 0$, we obtain $f = f^B(k; n) = n\,\phi(k)$, which is a function of the density n. As we have seen before, an initial distribution decays to f^B exponentially fast as $\tau = t/\lambda^2 \to \infty$. Since $\lambda \to 0$, for any finite time in the slow scale t, $f \sim f^B$ to order λ^0 and we can consider that the distribution function is the local equilibrium f^B plus terms that vanish as $\lambda \to 0$. We assume that f^B depends on the slow scales x and t only through its dependence on n or the force F. The equations of the latter contain certain functionals of n and F to be determined. In particular, we assume

$$f(x,k,t;\lambda) \sim n(x,t;\lambda)\,\phi(k) + \lambda\, f^{(1)}(k;n,F) + \lambda^2 f^{(2)}(k;n,F)\,, \tag{5.47}$$

$$\frac{\partial n}{\partial t} \sim N^{(0)}(n,F) + \lambda\, N^{(1)}(n,F)\,, \tag{5.48}$$

$$\frac{\partial F}{\partial t} + J_n^{(0)}(n,F) + \lambda\, J_n^{(1)}(n,F) \sim J \quad \text{with} \quad \nabla_x \cdot J = 0\,. \tag{5.49}$$

For the zeroth order term in Eq. (5.47), we have used the same functional form as the limit of the expression given by the zeroth order term of the method of multiple scales because the fast time variable goes to infinity. Note that, in contradiction to the method of multiple scales, n and F are not functions of the small parameter λ: In the Chapman–Enskog method, we expand the equations for the slowly varying functions n and F, not the slow functions themselves. The only unknown to be expanded in powers of λ is the distribution function itself. The density and the force are related through the Poisson equation (5.29): $\nabla_x \cdot F = n - 1$. If we take the divergence of Eq. (5.49) and use the Poisson equation, we find an equation for $\partial n/\partial t$ that should be identical to Eq. (5.48). We obtain

$$N^{(j)} = -\nabla_x \cdot J_n^{(j)}\,. \tag{5.50}$$

Inserting Eq. (5.47) into Eq. (5.28), we find the compatibility conditions for $f^{(j)}$:

$$\frac{1}{(2\pi)^3} \int_{\mathbb{R}^3} f^{(j)}\, d^3k = 0\,, \quad j = 1, 2\,. \tag{5.51}$$

The functionals $J_n^{(j)}$ are found by imposing Eq. (5.51).

Inserting Eqs. (5.47)–(5.50) in Eq. (5.26), we obtain the hierarchy

$$f^{(1)} = -\,(k \cdot \nabla_x + F \cdot \nabla_k)\,(n\phi)\,, \tag{5.52}$$

$$f^{(2)} = (k \cdot \nabla_x + F \cdot \nabla_k)^2\,(n\phi) - N^{(0)}\phi\,. \tag{5.53}$$

$f^{(1)}$ in Eq. (5.52) satisfies the compatibility condition (5.51). However, this condition for Eq. (5.53) yields

$$N^{(0)} = \int_{\mathbb{R}^3} (k \cdot \nabla_x + F \cdot \nabla_k)^2\, \frac{n\, e^{-|k|^2/2}}{(2\pi)^{3/2}}\, d^3k = \frac{1}{2} \nabla_x \cdot (\nabla_x n - nF)\,. \tag{5.54}$$

Then, Eqs. (5.48) and (5.50) give the drift-diffusion expression

$$\frac{\partial n}{\partial t} + \nabla_x \cdot J_n^{(0)} = 0\,, \quad J_n^{(0)} = \frac{1}{2}\,(n\,F - \nabla_x n)\,, \tag{5.55}$$

up to terms of order λ. We see that Eq. (5.46) obtained by the method of multiple scales is λ times $J_n^{(0)}$ given by Eq. (5.55). The reduced equations for n and F are then the continuity equation (5.55) and the Poisson equation (5.29). The same equations are also obtained by using the method of multiple scales when Eq. (5.46) is inserted in the continuity equation (5.31) because the common factor λ drops out from the resulting equation.

5.1.5.3 Einstein Relation

To return the dimensional units in the continuity equation (5.45),

$$\frac{\partial \tilde{n}}{\partial \tilde{t}} + \frac{1}{2}\nabla_{\tilde{x}} \cdot (\tilde{n}\tilde{F} - \nabla_{\tilde{x}}\tilde{n}) = 0 \tag{5.56}$$

(note that the tildes are back), we use Table 5.1: We multiply Eq. (5.56) by $-e\,N_D/[t]$, use $\tilde{n} = n/N_D$, $\tilde{x} = x/[x]$, $\tilde{F} = F/[F]$ and write the result in the following way:

$$-e\,\frac{\partial(n - N_D)}{\partial t} + \nabla_x \cdot \left(-e\,n\,\frac{-e}{2m\nu}\,\frac{F}{-e} - (-e)\frac{-e}{2m\nu}\,\frac{k_B T_0}{-e}\,\nabla_x n\right) = 0\,. \tag{5.57}$$

The first term in left-hand side of this equation is the time derivative of the electric charge density, $e\,(N_D - n)$, and the second term is the divergence of the electric current density. The drift part of the current density is $-en\mu_n E$, where $E = F/(-e)$ is the electric field, $\mu_n = -e/(2m\nu)$ is the electron mobility and $\mu_n E$ is the electron velocity. These expressions are familiar as the phenomenological drift-diffusion expressions for the electrical current flowing in a bulk three-dimensional semiconductor [4]. The diffusion part of the current density is $e\,D_n \nabla_x n$, with a diffusion coefficient D_n satisfying the Einstein relation:

$$D_n = \frac{\mu_n k_B T_0}{-e}\,, \quad \mu_n = \frac{-e}{2m\nu}\,. \tag{5.58}$$

5.2
Superlattices

In this section, we shall use the previous ideas to analyze a more realistic kinetic description of electron transport in useful materials. Semiconductor superlattices are essential ingredients in fast nanoscale oscillators, quantum cascade lasers and infrared detectors. Quantum cascade lasers are used to monitor environmental pollution in gas emissions, to analyze breath in hospitals and in many other industrial applications [5]. A superlattice (SL) is a convenient approximation to a quasi-one-dimensional crystal that was originally proposed by Esaki and Tsu to observe Bloch oscillations, that is, the periodic coherent motion of electrons in a miniband in the

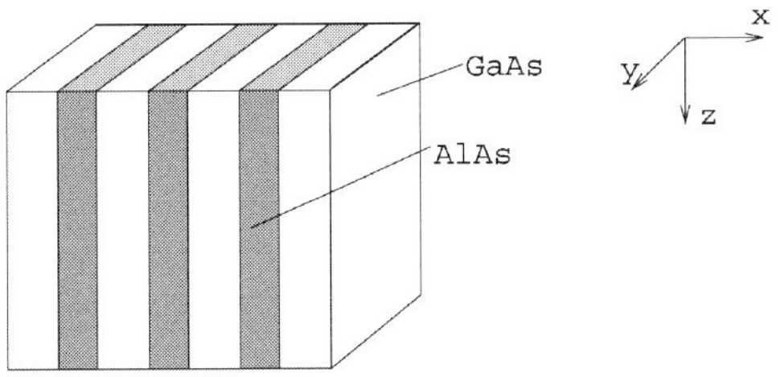

Fig. 5.1 Schematic drawing of a superlattice.

presence of an applied electric field. Figure 5.1a shows a simple realization of a N-period SL. Each period of length l consists of two layers of semiconductors with different energy gaps, but with similar lattice constants. The SL lengths in the lateral directions L_y and L_z are much larger than l, typically tens of microns compared to about ten nanometers. The energy profile of the conduction band of this SL can be modeled as a succession of square quantum wells and barriers along the x direction (Kronig–Penney model) and, for a n-doped SL, we do not have to consider the valence band. The wave functions of a single electron in the conduction band of a SL can be expanded in terms of plane waves in the long directions $\boldsymbol{x}_\perp = (y, z)$ times 1D Bloch wave functions:

$$\varphi_\nu(\boldsymbol{x}, \boldsymbol{k}) = \frac{1}{\sqrt{S}} e^{i\boldsymbol{k}_\perp \cdot \boldsymbol{x}_\perp} \varphi_\nu(x, k)\,, \quad \varphi_\nu = e^{ikx} u_\nu(x, k)\,, \tag{5.59}$$

where ν is the miniband index and $\boldsymbol{k}_\perp = (k_y, k_z)$. We have used the notation $\boldsymbol{x} = (x, \boldsymbol{x}_\perp) = (x, y, z)$, $\boldsymbol{k} = (k, k_y, k_z)$. The function $u_\nu(x, k)$ is l-periodic in x and $2\pi/l$-periodic in k. S is the area of the lateral cross section, equal to $L_y L_z$ for a rectangular cross section. Strictly speaking, the above form of wavefunction is only true for a structure subject to a zero applied electric field. We examine methods for generalizing these expressions to non-vanishing applied fields below.

Many interesting nonlinear phenomena have been observed in voltage biased SL comprising finitely many periods, including self-oscillations of the current through the SL due to motion of electric field pulses, multistability of stationary charge and field profiles, and so on [5]. It is important to distinguish between strongly and weakly coupled SLs depending on the coupling between their component QWs. Roughly speaking, if barriers are narrow, QWs are strongly coupled and we can use the electronic states Eq. (5.59) as a convenient basis in a quantum kinetic description. The resulting reduced balance equations for electron density and electric field are partial differential equations. On the other hand, for SLs having wide barriers, their QWs are weakly coupled and the electronic states of a single well provide a good basis in a quantum kinetic description, replacing the Bloch functions $e^{ikx} u_\nu(x, k)$ in Eq. (5.59). In this case, the balance equations are spatially discrete and phenomena such as multistability of stationary field profiles, formation and pinning of electric field domains, and so on, are theoretically predicted and observed in experiments. See the review [5].

5.2.1
Kinetic Theory Description of a Superlattice with a Single Populated Miniband[1)]

The Hamiltonian describing electron transport in a SL is

$$\hat{H} = \hat{H}_0 + \hat{H}_{e-e} + \hat{H}_{sc} . \tag{5.60}$$

We have separated the electron–electron interaction $\hat{H}_{e-e}$ and other scattering processes (impurity, phonon, ...) $\hat{H}_{sc}$ from the one-electron Hamiltonian $\hat{H}_0$. Typically, the electron–electron interaction is treated in the Hartree approximation, wherein the electrons give rise to an effective electric potential that can be written as a one-body function $W(x)$ that solves the classical Poisson equation. Then, we can find the spectrum of the Hamiltonian $\hat{H}_0 + \hat{H}_{e-e}$ by solving a nonlinear stationary Schrödinger–Poisson system of equations. If we do not distinguish the effective electron masses and dielectric constants of the semiconductors comprising wells and barriers of the SL (a common simplification), the basis states Eq. (5.59) solve the eigenvalue problem of the Kronig–Penney model at a zero external field:

$$\hat{H}_0 \varphi_\nu \equiv \left[\mathcal{E}_c + V_c(x) - \frac{\hbar^2}{2m^*} \nabla_x^2 \right] \varphi_\nu = \mathcal{E}_\nu(\boldsymbol{k}) \varphi_\nu , \tag{5.61}$$

$$V_c(x) = \begin{cases} 0 & \text{if } x \text{ corresponds to a quantum well } W \\ V_c & \text{if } x \text{ corresponds to a barrier B} . \end{cases} \tag{5.62}$$

Here $\mathcal{E}_c$ is the conduction band edge of material W (GaAs, well) and $W(x)$ is the electric potential due to the electron–electron interaction. The eigenvalues have the form

$$\mathcal{E}_\nu(\boldsymbol{k}) = \frac{\hbar^2 \boldsymbol{k}_\perp^2}{2m^*} + \mathcal{E}_\nu(k) , \tag{5.63}$$

and the Bloch functions $\varphi_\nu(x,k)$ satisfy the orthogonality condition

$$\int_{-\infty}^{\infty} \varphi_\mu^*(x,k) \varphi_\nu(x,k') \, dx = \delta_{\mu\nu} \delta(k-k') , \tag{5.64}$$

and the closure condition

$$\int_{-\infty}^{\infty} \varphi_\mu^*(x,k) \varphi_\nu(x',k) \, dk = \delta_{\mu\nu} \delta(x-x') , \tag{5.65}$$

provided the integral of $|\varphi_\nu|^2$ over one SL period is unity.

1) The discussion in this and the following subsection is based on the treatment presented in [5].

5.2.1.1 Wigner Equation

To find a kinetic equation, we start writing equations for the coefficients $a_{\nu,\boldsymbol{k}}(t)$ in the expansion of the wave function

$$\psi(\boldsymbol{x},t)=\sum_{\nu,\boldsymbol{k}} a_{\nu,\boldsymbol{k}}(t)\,\varphi_\nu(\boldsymbol{x},\boldsymbol{k})\equiv\sum_\nu \psi_\nu(\boldsymbol{x},t)\,. \tag{5.66}$$

If we ignore the scattering term $\hat{H}_{\mathrm{sc}}$ in Eq. (5.60), the coefficients $a_{\nu,\boldsymbol{k}}(t)$ solve

$$i\hbar\frac{\partial}{\partial t}a_{\nu,\boldsymbol{k}}=\mathcal{E}_\nu(\boldsymbol{k})a_{\nu,\boldsymbol{k}}-e\sum_{\nu',\boldsymbol{k}'}\langle\nu\boldsymbol{k}|\,W\,|\nu'\boldsymbol{k}'\rangle\,a_{\nu',\boldsymbol{k}'}\,, \tag{5.67}$$

where we have used the self-consistent electric potential W defined above instead of $\hat{H}_{\mathrm{e-e}}$. The equations for the perturbed *band wave functions* ψ_ν of Eq. (5.66) can be obtained from Eq. (5.67) after some algebra with the result

$$\begin{aligned}i\hbar\frac{\partial}{\partial t}\psi_\nu&=-\frac{\hbar^2}{2m^*}\frac{\partial^2}{\partial\boldsymbol{x}_\perp^2}\psi_\nu+\sum_{m=-\infty}^{\infty}\mathcal{E}_{\nu,m}\,\psi_\nu(x+ml,\boldsymbol{x}_\perp,t)\\&\quad-e\sum_{\nu'}\int\Phi_\nu(\boldsymbol{x},\boldsymbol{x}')\,W(\boldsymbol{x}')\psi_{\nu'}(\boldsymbol{x}',t)\,d^3\boldsymbol{x}'\,,\end{aligned} \tag{5.68}$$

$$\Phi_\nu(\boldsymbol{x},\boldsymbol{x}')=\sum_{\boldsymbol{k}}\varphi_\nu(\boldsymbol{x},\boldsymbol{k})\,\varphi_\nu^*(\boldsymbol{x}',\boldsymbol{k})\,, \tag{5.69}$$

$$\mathcal{E}_\nu(k)=\sum_{m=-\infty}^{\infty}\mathcal{E}_{\nu,m}\,e^{imkl}\,. \tag{5.70}$$

Notice that Eq. (5.59) implies

$$\begin{aligned}\Phi_\nu(\boldsymbol{x},\boldsymbol{x}')&=\delta(\boldsymbol{x}_\perp-\boldsymbol{x}'_\perp)\,\phi_\nu(x,x')\,,\\\phi_\nu(x,x')&=\sum_k\varphi_\nu(x,k)\,\varphi_\nu^*(x',k)\,,\end{aligned} \tag{5.71}$$

and the closure condition in Eq. (5.65) yields

$$\sum_\nu\Phi_\nu(\boldsymbol{x},\boldsymbol{x}')=\delta(\boldsymbol{x}-\boldsymbol{x}')\,. \tag{5.72}$$

Thus, $\Phi_\nu(\boldsymbol{x},\boldsymbol{x}')$ can be considered as the projection of the delta function $\delta(\boldsymbol{x}-\boldsymbol{x}')$ onto the band ν.

After a second quantization, the band density matrix is defined by

$$\rho_{\mu,\nu}(\boldsymbol{x},\boldsymbol{y},t)=\langle\hat{\psi}_\mu^\dagger(\boldsymbol{x},t)\,\hat{\psi}_\nu(\boldsymbol{y},t)\rangle\,, \tag{5.73}$$

so that the 2D electron density is (the factor 2 is due to spin degeneracy)

$$n(\boldsymbol{x},t)=2l\sum_{\mu,\nu}\langle\hat{\psi}_\mu^\dagger(\boldsymbol{x},t)\,\hat{\psi}_\nu(\boldsymbol{x},t)\rangle=2l\sum_{\mu,\nu}\rho_{\mu,\nu}(\boldsymbol{x},\boldsymbol{x},t)\,. \tag{5.74}$$

Using Eqs. (5.73) and (5.74), we can derive the following evolution equation for the band density matrix

$$i\hbar\frac{\partial}{\partial t}\rho_{\mu,\nu} + \frac{\hbar^2}{2m^*}\left(\frac{\partial^2}{\partial \mathbf{y}_\perp^2} - \frac{\partial^2}{\partial \mathbf{x}_\perp^2}\right)\rho_{\mu,\nu}$$
$$- \sum_{m=-\infty}^{\infty}[\mathcal{E}_{\nu,m}\rho_{\mu,\nu}(\mathbf{x}, y+ml, \mathbf{y}_\perp, t) - \mathcal{E}_{\mu,m}\rho_{\mu,\nu}(x-ml, \mathbf{x}_\perp, \mathbf{y}, t)]$$
$$+ e\sum_{\nu'}\int W(z)\,[\Phi_\nu(y,z)\rho_{\mu\nu'}(\mathbf{x}, z, t) - \Phi_\mu(z, \mathbf{x})\rho_{\nu'\nu}(z, \mathbf{y}, t)]d^3\mathbf{z} = Q[\rho]\,, \tag{5.75}$$

with $Q[\rho] \equiv 0$ in the absence of scattering. This means that the collision term has been reintroduced after neglecting it at the beginning of Section 5.2.1. It appears that this form of collision term is quite general. Though below, we will once again make simplifying assumptions, as in BGK. The Hartree potential satisfies the Poisson equation

$$\varepsilon\,\frac{\partial^2 W}{\partial x^2} = \frac{e}{l}\,(n - N_D) \tag{5.76}$$

(recall that $e > 0$ and that W is the self-consistent electric potential. The charge density is thus equal to minus the right-hand side of this equation). When considering scattering, the right-hand side of Eq. (5.75) is equal to a non-zero functional of the band density matrix $Q[\rho]$, whose form depends on the closure assumption we have made to close the density matrix hierarchy. In the semiclassical limit, the kernel of the collision term $Q[\rho]$ is usually found by using perturbation theory in the impurity potential, electron–phonon interaction, and so on. For the time being, we shall not try to formulate collision models. Instead, and in order to make contact with the kinetic equations in the semiclassical limit, we shall rewrite Eq. (5.75) in terms of the band Wigner function

$$w_{\mu,\nu}(\mathbf{x}, \mathbf{k}, t) = \int_{\mathbb{R}^3}\rho_{\mu,\nu}\left(\mathbf{x}+\frac{1}{2}\boldsymbol{\xi}, \mathbf{x}-\frac{1}{2}\boldsymbol{\xi}, t\right)\,e^{i\mathbf{k}\cdot\boldsymbol{\xi}}\,d^3\boldsymbol{\xi}\,. \tag{5.77}$$

The evolution equation for the Wigner function is

$$\frac{\partial}{\partial t}w_{\mu,\nu} + \frac{\hbar\mathbf{k}_\perp}{m^*}\cdot\frac{\partial}{\partial \mathbf{x}_\perp}w_{\mu,\nu} + \frac{i}{\hbar}\sum_{m=-\infty}^{\infty}e^{imkl}\left[\mathcal{E}_{\nu,m}w_{\mu,\nu}\left(x+\frac{ml}{2}, \mathbf{x}_\perp, \mathbf{k}, t\right)\right.$$
$$\left.-\mathcal{E}_{\mu,m}w_{\mu,\nu}\left(x-\frac{ml}{2}, \mathbf{x}_\perp, \mathbf{k}, t\right)\right] + \frac{ie}{\hbar}\sum_{\nu'}\int\left[W\left(z+\frac{1}{2i}\frac{\partial}{\partial k}, \mathbf{x}_\perp\right)\right.$$
$$\times\,\phi_\mu(z,x)\,e^{ik(x-z)}w_{\nu',\nu}\left(\frac{x+z}{2}, \mathbf{x}_\perp, \mathbf{k}, t\right) - W\left(z-\frac{1}{2i}\frac{\partial}{\partial k}, \mathbf{x}_\perp\right)$$
$$\left.\times\,\phi_\nu(x,z)\,e^{-ik(x-z)}w_{\mu,\nu'}\left(\frac{x+z}{2}, \mathbf{x}_\perp, \mathbf{k}, t\right)\right]dz = Q_{\mu,\nu}[w]\,, \tag{5.78}$$

in which the collision term is again left unspecified. Notice that the 2D electron density is

$$n(\boldsymbol{x}, t) = \frac{2l}{8\pi^3} \sum_{\mu,\nu} \int_{\mathbb{R}^3} w_{\mu,\nu}(\boldsymbol{x}, \boldsymbol{k}, t)\, d^3\boldsymbol{k}\,, \tag{5.79}$$

because of Eq. (5.74) and the definition in Eq. (5.77). From Eqs. (5.78) and (5.79), we obtain the charge continuity equation

$$\frac{e}{l}\frac{\partial n}{\partial t} + \nabla_x \cdot \boldsymbol{J} = 0\,, \tag{5.80}$$

$$\boldsymbol{J}_\perp = \frac{2e}{8\pi^3} \int \frac{\hbar \boldsymbol{k}_\perp}{m^*} \sum_{\mu,\nu} w_{\mu,\nu}(\boldsymbol{x}, \boldsymbol{k}, t)\, d^3\boldsymbol{k}\,, \tag{5.81}$$

$$\begin{aligned}\frac{\partial J}{\partial x} = \frac{ie}{4\pi^3\hbar} \sum_{\mu,\nu,m} \int e^{imkl} \Big[& \mathcal{E}_{\nu,m} w_{\mu,\nu}\left(x + \frac{ml}{2}, \boldsymbol{x}_\perp, \boldsymbol{k}, t\right) \\ & - \mathcal{E}_{\mu,m} w_{\mu,\nu}\left(x - \frac{ml}{2}, \boldsymbol{x}_\perp, \boldsymbol{k}, t\right)\Big]\, d\boldsymbol{k}\,,\end{aligned} \tag{5.82}$$

provided our collision model satisfies $\int \sum_{\mu,\nu} Q_{\mu,\nu}\, d^3\boldsymbol{k} = 0$.

One difficulty with our formulation is that the Wigner function in Eq. (5.77) is not $2\pi/l$-periodic in k. This can be corrected by using the following definition

$$\begin{aligned} f_{\mu,\nu}(\boldsymbol{x}, \boldsymbol{k}, t) &\equiv \sum_{s=-\infty}^{\infty} w_{\mu,\nu}\left(\boldsymbol{x}, k + \frac{2\pi s}{l}, \boldsymbol{k}_\perp, t\right) = \sum_{j=-\infty}^{\infty} e^{ijkl} l \\ &\times \int \rho_{\mu,\nu}\left(x + \frac{jl}{2}, \boldsymbol{x}_\perp + \frac{1}{2}\boldsymbol{\xi}_\perp, x - \frac{jl}{2}, \boldsymbol{x}_\perp - \frac{1}{2}\boldsymbol{\xi}_\perp, t\right) e^{i\boldsymbol{k}_\perp \cdot \boldsymbol{\xi}_\perp} d^2\boldsymbol{\xi}_\perp\,. \end{aligned} \tag{5.83}$$

To derive this equation, we have used the Poisson summation formula

$$\sum_{j=-\infty}^{\infty} \delta(\xi - jl) = \frac{1}{l} \sum_{s=-\infty}^{\infty} e^{i2\pi\xi s/l}\,, \tag{5.84}$$

together with the definition of Eq. (5.77). From Eqs. (5.79) and (5.83), we obtain the 2D electron density in terms of $f_{\mu,\nu}$

$$n(\boldsymbol{x}, t) = \frac{2l}{8\pi^3} \sum_{\mu,\nu} \int_{-\pi/l}^{\pi/l} \int_{\mathbb{R}^2} f_{\mu,\nu}(\boldsymbol{x}, \boldsymbol{k}, t)\, dk\, d^2\boldsymbol{k}_\perp\,. \tag{5.85}$$

Similarly, the transversal current density can be obtained from Eqs. (5.81) and (5.83):

$$\boldsymbol{J}_\perp = \frac{2e}{8\pi^3} \int_{-\pi/l}^{\pi/l} \int \frac{\hbar \boldsymbol{k}_\perp}{m^*} \sum_{\mu,\nu} f_{\mu,\nu}(\boldsymbol{x}, \boldsymbol{k}, t)\, dk\, d^2\boldsymbol{k}_\perp\,. \tag{5.86}$$

The current density along the growth direction has the form

$$J = \frac{2e\hbar}{8\pi^3 m^*} \sum_{\mu,\nu,s} \int_{-\pi/l}^{\pi/l} \int \left(k + \frac{2\pi s}{l}\right) w_{\mu,\nu}\left(\boldsymbol{x}, k + \frac{2\pi s}{l}, \boldsymbol{k}_\perp, t\right) dk\, d^2\boldsymbol{k}_\perp\,, \tag{5.87}$$

from which we can also derive Eq. (5.82).

To find a Wigner–Poisson description of transport in a single miniband, we sum all the Wigner equations (5.78) over the band indices and use the closure condition in Eq. (5.72), so as to find an equation for $w(\boldsymbol{x}, \boldsymbol{k}, t) = \sum_{\mu,\nu} w_{\mu,\nu}(\boldsymbol{x}, \boldsymbol{k}, t)$

$$\begin{aligned}
&\frac{\partial}{\partial t} w + \frac{\hbar \boldsymbol{k}_\perp}{m^*} \cdot \frac{\partial}{\partial \boldsymbol{x}_\perp} w + \frac{i}{\hbar} \sum_{m=-\infty}^{\infty} e^{imkl} \sum_{\mu,\nu} \left[\mathcal{E}_{\nu,m} w_{\mu,\nu}\left(x + \frac{ml}{2}, \boldsymbol{x}_\perp, \boldsymbol{k}, t\right)\right. \\
&\left. - \mathcal{E}_{\mu,m} w_{\mu,\nu}\left(x - \frac{ml}{2}, \boldsymbol{x}_\perp, \boldsymbol{k}, t\right)\right] + \frac{ie}{\hbar}\left[W\left(x + \frac{1}{2i}\frac{\partial}{\partial k}, \boldsymbol{x}_\perp\right)\right. \\
&\left. - W\left(x - \frac{1}{2i}\frac{\partial}{\partial k}, \boldsymbol{x}_\perp\right)\right] w = \sum_{\mu,\nu} Q_{\mu,\nu}[w]\,.
\end{aligned} \tag{5.88}$$

Let us now assume that only the first miniband is populated and that there are no transitions between minibands, $w(\boldsymbol{x}, \boldsymbol{k}, t) \approx w_{1,1}(\boldsymbol{x}, \boldsymbol{k}, t)$. This approximation is commonly used when describing strongly coupled SLs with wide minibands. Then, Eq. (5.88) becomes

$$\begin{aligned}
&\frac{\partial}{\partial t} w + \frac{\hbar \boldsymbol{k}_\perp}{m^*} \cdot \frac{\partial}{\partial \boldsymbol{x}_\perp} w + \frac{i}{\hbar} \sum_{m=-\infty}^{\infty} e^{imkl} \mathcal{E}_{1,m} \left[w\left(x + \frac{ml}{2}, \boldsymbol{x}_\perp, \boldsymbol{k}, t\right)\right. \\
&\left. - w\left(x - \frac{ml}{2}, \boldsymbol{x}_\perp, \boldsymbol{k}, t\right)\right] + \frac{ie}{\hbar}\left[W\left(x + \frac{1}{2i}\frac{\partial}{\partial k}, \boldsymbol{x}_\perp\right)\right. \\
&\left. - W\left(x - \frac{1}{2i}\frac{\partial}{\partial k}, \boldsymbol{x}_\perp\right)\right] w = Q_{1,1}[w]\,.
\end{aligned} \tag{5.89}$$

This yields the following equation for the periodic Wigner function (5.83)

$$\begin{aligned}
&\frac{\partial}{\partial t} f + \frac{\hbar \boldsymbol{k}_\perp}{m^*} \cdot \frac{\partial}{\partial \boldsymbol{x}_\perp} f + \frac{i}{\hbar} \sum_{m=-\infty}^{\infty} e^{imkl} \mathcal{E}_{m} \left[f\left(x + \frac{ml}{2}, \boldsymbol{x}_\perp, \boldsymbol{k}, t\right)\right. \\
&\left. - f\left(x - \frac{ml}{2}, \boldsymbol{x}_\perp, \boldsymbol{k}, t\right)\right] + \frac{ie}{\hbar}\left[W\left(x + \frac{1}{2i}\frac{\partial}{\partial k}, \boldsymbol{x}_\perp\right)\right. \\
&\left. - W\left(x - \frac{1}{2i}\frac{\partial}{\partial k}, \boldsymbol{x}_\perp\right)\right] f = Q[f]\,.
\end{aligned} \tag{5.90}$$

Here and henceforth, we drop the subscript 1 in the miniband dispersion relation since only the first miniband appears in all our subsequent descriptions. Therefore, $\mathcal{E}_m$ means the m-th Fourier component of the (single) miniband dispersion relation $\mathcal{E}(k)$.

The dispersion relation $\mathcal{E}(k)$ is an even periodic function of k with period $2\pi/l$ that can be written as $\mathcal{E}(k) = \Delta[1 - \cos(kl)]/2$ plus a constant in the tight-binding approximation (Δ denotes the width of the first miniband). Moreover, the field $\boldsymbol{F} = \nabla_x W$ (note that the actual electric field is $-\boldsymbol{F}$) satisfies

$$\varepsilon \left(\frac{\partial F}{\partial x} + \nabla_{x_\perp} \cdot \boldsymbol{F}_\perp \right) = \frac{e}{l} (n - N_D) , \tag{5.91}$$

$$n(x, \boldsymbol{x}_\perp, t) = \frac{l}{4\pi^3} \int_{-\pi/l}^{\pi/l} \int_{\mathbb{R}^2} f(x, \boldsymbol{x}_\perp, k, \boldsymbol{k}_\perp, t) dk \, d^2\boldsymbol{k}_\perp . \tag{5.92}$$

Let us now assume that we can ignore exchange of lateral momentum in scattering processes and that the electron density and the electric potential are functions of x and t only. Then the electric field is directed along the x direction, $\boldsymbol{F} = (F(x,t), 0, 0)$. Defining

$$f(x, k, t) = \frac{1}{2\pi^2} \iint f(x, \boldsymbol{x}_\perp, k, \boldsymbol{k'}_\perp, t) \, d^2\boldsymbol{k'}_\perp d^2\boldsymbol{x}_\perp , \tag{5.93}$$

Eq. (5.92) yields

$$n(x, t) = \frac{l}{2\pi} \int_{-\pi/l}^{\pi/l} f(x, k, t) \, dk , \tag{5.94}$$

and Eq. (5.90) gives the following 1D Wigner equation:

$$\begin{aligned} &\frac{\partial}{\partial t} f + \frac{i}{\hbar} \sum_{m=-\infty}^{\infty} e^{imkl} \mathcal{E}_m \left[f\left(x + \frac{ml}{2}, k, t \right) - f\left(x - \frac{ml}{2}, k, t \right) \right] \\ &+ \frac{ie}{\hbar} \left[W\left(x + \frac{1}{2i} \frac{\partial}{\partial k} \right) - W\left(x - \frac{1}{2i} \frac{\partial}{\partial k} \right) \right] f = Q[f] . \end{aligned} \tag{5.95}$$

In order to complete the specification of our quantum kinetic theory, we need a description of scattering. Scattering processes such as phonon scattering change the energy and momentum of the electrons leading the distribution function toward thermal equilibrium. We can describe these processes by a BGK collision model

$$Q[f] = -\nu(f - f^{\mathrm{FD}}), \tag{5.96}$$

$$f^{\mathrm{FD}}(k;n) = \int_{-\infty}^{\infty} \frac{D_\Gamma(E - \mathcal{E}(k))}{1 + \exp\left(\frac{E-\mu}{k_B T}\right)} \, dE, \tag{5.97}$$

$$D_\Gamma(E) = \frac{2}{(2\pi)^2} \int_{\mathbb{R}^2} \delta_\Gamma\left(\frac{\hbar^2 \mathbf{k}_\perp^2}{2m^*} - E\right) d^2\mathbf{k}_\perp = \frac{m^*}{\pi\hbar^2} \int_0^\infty \delta_\Gamma(E_\perp - E)\, dE_\perp, \tag{5.98}$$

$$\delta_\Gamma(E) = \frac{\sqrt{2}\,\Gamma^3/\pi}{\Gamma^4 + E^4}, \tag{5.99}$$

$$n(x,t) = \frac{l}{2\pi} \int_{-\pi/l}^{\pi/l} f^{\mathrm{FD}}(k;n)\, dk. \tag{5.100}$$

The chemical potential $\mu = \mu(x,t)$ is a function of the exact electron density n of Eq. (5.94) that is calculated by solving Eq. (5.100). With these definitions, the integral of $Q[f]$ over k vanishes, and the equation of charge continuity holds, as in the case of the simpler BGK equation of the previous section.

First, principles formulations of scattering show that collisions include two quantum effects: collision broadening of line widths and the intracollisional field effect. Collision broadening has been included phenomenologically in Eq. (5.97) by means of the coefficient Γ, which measures the finite width of the spectral function in thermal equilibrium due to scattering [6]. As $\Gamma \to 0$, $\delta_\Gamma(E) \to \delta(E)$, $D_\Gamma(E)$ tends to the 2D density of states, $D(E) = m^*\theta(E)/(\pi\hbar^2)$, and f^{FD} tends to the 3D Fermi–Dirac distribution function integrated over the lateral wave vector $\mathbf{k}_\perp$. Our line-width Eq. (5.99) allows us to carry out integrations explicitly by means of the residue theorem. This cannot be done with other standard line-widths such as a Gaussian or a hyperbolic secant. Integration by parts of Eq. (5.97) yields

$$\begin{aligned} f^{\mathrm{FD}}(k;n) &= -k_B T\, D_\Gamma(E - \mathcal{E}(k)) \ln\left[1 + \exp\left(-\frac{E-\mu}{k_B T}\right)\right]\Bigg|_{E=-\infty}^{\infty} \\ &\quad + \frac{m^* k_B T}{\pi\hbar^2} \int_{-\infty}^{\infty} \ln\left[1 + \exp\left(\frac{\mu - E}{k_B T}\right)\right] \delta_\Gamma(E - \mathcal{E}_1(k))\, dE, \end{aligned}$$

where we have used that

$$\frac{dD_\Gamma}{dE} = \frac{m^*}{\pi\hbar^2}\delta_\Gamma(E).$$

Therefore, we have

$$f^{\rm FD}(k;n) = \frac{m^* k_B T}{\pi \hbar^2} \int_{-\infty}^{\infty} \ln\left[1 + \exp\left(\frac{\mu - E}{k_B T}\right)\right] \delta_\Gamma (E - \mathcal{E}(k))\, dE\,. \tag{5.101}$$

Quantum collision operators involve integrals over momentum, space and time, although the integrals over space and time have kernels that become delta functions in the semiclassical limit. Moreover, the collision operators involve the electric field at different times, which is called the intracollisional field effect. As in the case of the BGK model for the Boltzmann equation, our model (5.96) replaces nonlocal effects due to integrals in the collision term by a strongly nonlinear functional of the density n which is local in k, x and t. Even if we include the electric potential in the 3D Fermi function (for example, replacing $E - eW - \mu$ instead of $E - \mu$ in Eq. (5.97)), a redefinition of the chemical potential (replacing the electrochemical potential $\mu + eW$ instead of μ) rids of the electric potential. The price we pay for having a model which is simple enough to calculate reduced equations for the electron density and the electric field is that we must make a severe but intuitively plausible approximation for the intracollisional field effect in our formulation.

5.2.1.2 Equivalent form of the Quantum Kinetic Equation

The Wigner function f is periodic in k. Its Fourier expansion is

$$f(x,k,t) = \sum_{j=-\infty}^{\infty} f_j(x,t)\, e^{ijkl}\,. \tag{5.102}$$

Since $F = \partial W/\partial x$ (*minus* the electric field), defining the local average of a function $g(x)$ as

$$\langle g\rangle_j(x,t) = \frac{1}{jl}\int_{-jl/2}^{jl/2} g(x+s,t)\, ds, \quad \text{which implies}\,, \tag{5.103}$$

$$\frac{\partial}{\partial x}\langle g\rangle_j = \left\langle\frac{\partial g}{\partial x}\right\rangle_j = \frac{g(x+jl/2) - g(x-jl/2)}{jl}\,,$$

it is possible to rewrite Eqs. (5.95) and (5.96) as the following equivalent Wigner equation [7]

$$\frac{\partial f}{\partial t} + \sum_{j=-\infty}^{\infty} \frac{ijl}{\hbar} e^{ijkl}\left(\mathcal{E}_j \frac{\partial}{\partial x}\langle f\rangle_j + e\,\langle F\rangle_j\, f_j\right) = -\nu\,(f - F^{\rm FD})\,. \tag{5.104}$$

Here, the nonzero Fourier coefficients of the dispersion relation are simply $\mathcal{E}_0 = \Delta/2$ and $\mathcal{E}_{\pm 1} = -\Delta/4$ for the tight-binding dispersion relation $\mathcal{E}(k) = \Delta(1 - \cos kl)/2$ (Δ is the miniband width), which yields a miniband group velocity $v(k) =$

$\frac{\Delta l}{2\hbar}\sin kl$. To derive Eq. (5.104), note that

$$\begin{aligned} & W\left(x+\frac{1}{2i}\frac{\partial}{\partial k},t\right)f - W\left(x-\frac{1}{2i}\frac{\partial}{\partial k},t\right)f \\ &= \sum_{j=-\infty}^{\infty}\left[W(x+jl/2,t)-W(x-jl/2,t)\right] f_j e^{ijkl} \\ &= \sum_{j=-\infty}^{\infty} jl\langle F\rangle_j\, f_j\, e^{ijkl}\,, \end{aligned}$$

according to Eq. (5.103).

Integrating Eq. (5.104) over k yields the charge continuity equation

$$\frac{\partial n}{\partial t}+\frac{\partial}{\partial x}\sum_{j=1}^{\infty}\frac{2jl}{\hbar}\langle \mathrm{Im}(\mathcal{E}_{-j} f_j)\rangle_j = 0\,. \tag{5.105}$$

Here, we can eliminate the electron density by using the Poisson equation and then integrate over x, thereby obtaining the nonlocal Ampère's law for the total current density $J(t)$:

$$\varepsilon\frac{\partial F}{\partial t}+\frac{2e}{\hbar}\sum_{j=1}^{\infty} j\langle \mathrm{Im}(\mathcal{E}_{-j} f_j)\rangle_j = J(t)\,. \tag{5.106}$$

For future convenience, let us write the complete system of equations for the Wigner function and the electric potential, Eqs. (5.91), (5.94), (5.100), (5.101) and (5.104):

$$\frac{\partial f}{\partial t}+\sum_{j=-\infty}^{\infty}\frac{ijl}{\hbar}e^{ijkl}\left(\mathcal{E}_j\frac{\partial}{\partial x}\langle f\rangle_j + e\langle F\rangle_j f_j\right) = -\nu\left(f-F^{\mathrm{FD}}\right) \tag{5.107}$$

$$\varepsilon\frac{\partial F}{\partial x} = \frac{e}{l}(n-N_D)\,, \tag{5.108}$$

$$n(x,t) = \frac{l}{2\pi}\int_{-\pi/l}^{\pi/l} f(x,k,t)\,dk = \frac{l}{2\pi}\int_{-\pi/l}^{\pi/l} f^{\mathrm{FD}}(k;n(x,t))\,dk\,, \tag{5.109}$$

$$f^{\mathrm{FD}}(k;n) = \frac{m^* k_B T}{\pi\hbar^2}\int_{-\infty}^{\infty}\ln\left[1+\exp\left(\frac{\mu-E}{k_B T}\right)\right]\frac{\sqrt{2}\,\Gamma^3/\pi}{[E-\mathcal{E}_1(k)]^4+\Gamma^4}\,dE\,. \tag{5.110}$$

5.2.2 Derivation of Reduced Equations for *n* and *F*

It is possible to find a reduced description in terms of electron density and field by using the ideas and perturbation methods presented in the previous section. If we blindly mimic what we did there, we would use the parabolic scaling and find a modified drift-diffusion equation with a current which is linear in the electric field. This equation would not describe interesting phenomena observed in SLs, namely, that under sufficient dc voltage bias in the growth direction, there appear self-sustained oscillations of the electric current. The physical cause for such oscillations is that the electrons move over the whole miniband if the applied field is strong enough, whereas parabolic scaling implies that the electrons move close to the minima of the miniband.

To see why electrons may produce a periodic current if they move through the whole miniband with dispersion relation $\mathcal{E} = \Delta\,(1 - \cos kl)/2$, consider the semiclassical description of an electron moving in an applied constant field F in the absence of scattering:

$$\hbar\,\frac{dk}{dt} = eF\,, \qquad \frac{dx}{dt} = v(k) = \frac{l\Delta}{2\hbar}\,\sin kl\,. \tag{5.111}$$

Then, an electron initially at $(x, k) = (0, 0)$ satisfies $k = eFt/\hbar$ and $x = [\Delta/(2eF)][1 - \cos(eFlt/\hbar)]$, and therefore, it oscillates with the Bloch frequency

$$\omega_B = \frac{eFl}{\hbar}\,. \tag{5.112}$$

Of course, scattering substantially modifies this picture of coherent current oscillations and in fact, only damped Bloch oscillations have been observed in experiments. In superlattices, a few damped Bloch oscillations have been observed in undoped SLs in which carriers have been created by lasers.

5.2.2.1 **Nondimensional Wigner Equation**

We are interested in the *hyperbolic limit* of the Wigner Eq. (5.107), in which the term containing the field is of the same order as the collision term. Equivalently, the Bloch frequency Eq. (5.112) and the collision frequency ν are of the same order,

$$\frac{e[F]l}{\hbar}\,[f] = \nu\,[f] \Longrightarrow [F] = \frac{\hbar\nu}{el}\,. \tag{5.113}$$

From the Poisson equation (5.108), we obtain

$$\frac{\varepsilon\,[F]}{[x]} = \frac{eN_D}{l} \Longrightarrow [x] = \frac{\varepsilon\,[F]l}{eN_D} = \frac{\varepsilon\hbar\nu}{e^2 N_D}\,. \tag{5.114}$$

$[x]$ is the distance over which the field varies an amount $[F]$. If $[v]$ is the scale of electron velocity, the time it takes an electron with speed $[v]$ to move a distance $[x]$ is

$$[t] = \frac{[x]}{[v]} = \frac{\varepsilon\hbar\nu}{e^2 N_D[v]}\,. \tag{5.115}$$

In the semiclassical case, the electron velocity is the group velocity

$$v(k) = \hbar^{-1}\frac{d\mathcal{E}}{dk} \Longrightarrow [v] = \frac{[\mathcal{E}]}{[k]\hbar} = \frac{\Delta\,l}{\hbar}\,, \tag{5.116}$$

which gives the scale $[\nu]$. Then, Eq. (5.115) becomes

$$[t] = \frac{\varepsilon \hbar^2 \nu}{e^2 N_D l \Delta} . \tag{5.117}$$

The first and the second terms in the left hand side of Eq. (5.107) are of order $[f]/[t]$. We shall assume that these terms are small compared to the others, and therefore, that $[f]/[t] \ll \nu [f]$. Equivalently, the dimensionless parameter

$$\epsilon = \frac{e^2 N_D l \Delta}{\varepsilon \hbar^2 \nu^2} \ll 1 \tag{5.118}$$

is very small and the limit $\epsilon \to 0$ is the hyperbolic limit. We have also obtained the following table of magnitudes in hyperbolic scaling:

Table 5.2 also includes representative values of the different scales taken from experiments by Schonburg *et al.* [8]. Using these values $\epsilon = \nu^{-1}/[t] \approx 0.32$, which is relatively small. We now remove the dimensions from Eqs. (5.107)–(5.110) by using the following definitions extracted from Table 5.2:

$$\tilde{n} = \frac{n}{N_D}, \quad \tilde{f} = \frac{f}{N_D}, \quad \tilde{\mathcal{E}} = \frac{\mathcal{E}}{\Delta}, \quad \tilde{\Gamma} = \frac{\Gamma}{\Delta}, \quad \tilde{F} = \frac{eFl}{\hbar \nu},$$
$$\tilde{x} = \frac{e^2 N_D x}{\varepsilon \hbar \nu}, \quad \tilde{k} = kl, \quad \tilde{t} = \frac{e^2 N_D l \Delta t}{\varepsilon \hbar^2 \nu} .$$

The resulting nondimensional equations are:

$$f - F^{FD} + \sum_{j=-\infty}^{\infty} i j e^{ijk} \langle F \rangle_j f_j = -\epsilon \left(\frac{\partial f}{\partial t} + \sum_{j=-\infty}^{\infty} i j e^{ijk} \mathcal{E}_j \frac{\partial}{\partial x} \langle f \rangle_j \right) \tag{5.119}$$

$$\frac{\partial F}{\partial x} = n - 1 , \tag{5.120}$$

$$n(x,t) = \frac{1}{2\pi} \int_{-\pi}^{\pi} f(x,k,t) dk = \frac{1}{2\pi} \int_{-\pi}^{\pi} f^{FD}(k; n(x,t)) dk , \tag{5.121}$$

$$f^{FD}(k;n) = \alpha \int_{-\infty}^{\infty} \ln\left[1 + e^{\beta(\mu - E)}\right] \frac{\sqrt{2}\,\Gamma^3/\pi}{[E - \mathcal{E}_1(k)]^4 + \Gamma^4} dE , \tag{5.122}$$

$$\langle g(x) \rangle = \frac{1}{j\lambda} \int_{-j\lambda/2}^{j\lambda/2} g(x+s)\, ds , \tag{5.123}$$

where we have suppressed tildes and defined the additional dimensionless parameters:

$$\alpha = \frac{m^* k_B T}{\pi \hbar^2 N_D}, \quad \beta = \frac{\Delta}{k_B T}, \quad \lambda = \frac{l}{[x]} = \frac{e^2 l N_D}{\varepsilon \hbar \nu} . \tag{5.124}$$

For the numerical values of Table 5.2, $\alpha = 26.45$, $\beta = 0.62$, $\lambda = 0.24$.

Table 5.2 Hyperbolic scaling

f, n	F	$\mathcal{E}, \mu$	J_n	x	k	t
N_D	$\frac{\hbar \nu}{e l}$	Δ	$\frac{e N_D \Delta}{\hbar}$	$\frac{\varepsilon \hbar \nu}{e^2 N_D}$	$\frac{1}{l}$	$\frac{\varepsilon \hbar^2 \nu}{e^2 N_D l \Delta}$
$4.05 \times 10^{10}\,\mathrm{cm}^{-2}$	23.4 kV/cm	16 meV	157.65 kA/cm^2	21.3 nm	0.2/nm	0.17 ps

5.2.2.2 Derivation of a Reduced System

We shall now derive the reduced equations in the hyperbolic limit by means of the Chapman–Enskog method. With hyperbolic scaling, the method of multiple scales yields inconsistent results. The solution of Eq. (5.119) for $\epsilon = 0$ is calculated in terms of its Fourier coefficients as

$$f^{(0)}(k; F) = \sum_{j=-\infty}^{\infty} \frac{(1 - i j \langle F \rangle_j)\, f_j^{\mathrm{FD}}}{1 + j^2 \langle F \rangle_j^2}\, e^{i j k} . \tag{5.125}$$

Note that $f_0^{(0)} = f_0^{\mathrm{FD}} = n$.

The Chapman–Enskog ansatz for the Wigner function is then

$$f(x, k, t; \epsilon) = f^{(0)}(k; F) + \sum_{m=1}^{\infty} f^{(m)}(k; F)\, \epsilon^m , \tag{5.126}$$

$$\frac{\partial F}{\partial t} + \sum_{m=0}^{\infty} J^{(m)}(F)\, \epsilon^m = J(t) , \quad \frac{\partial n}{\partial t} + \sum_{m=0}^{\infty} \frac{\partial J^{(m)}(F)}{\partial x}\, \epsilon^m = 0 . \tag{5.127}$$

The coefficients $f^{(m)}(k; F)$ depend on the "slow variables" x and t only through their dependence on the electric field and the electron density. The electric field obeys a reduced evolution equation (5.127) in which the functionals $J^{(m)}(F)$ are chosen so that the $f^{(m)}(k; F)$ are bounded, $2\pi/l$-periodic in k and satisfy the compatibility conditions:

$$f_0^{(j)} = \frac{1}{2\pi} \int_{-\pi}^{\pi} f^{(j)}\, d^3 k = n\, \delta_{j0} , \quad j = 0, 1, 2 . \tag{5.128}$$

These conditions are obtained by inserting Eq. (5.119) in Eq. (5.121). After we keep the desired number of terms, Eq. (5.127) is the reduced equation provided by our perturbation procedure.

The exact continuity equation is obtained by integrating Eq. (5.119) over k:

$$\frac{\partial n}{\partial t} + \frac{\partial J_n}{\partial x} = 0 , \quad J_n = 2 \sum_{j=1}^{\infty} j \langle \mathrm{Im}(\mathcal{E}_{-j} f_j) \rangle_j , \tag{5.129}$$

where we have used that

$$\left\langle \frac{1}{2\pi} \int_{-\pi}^{\pi} e^{i j k} f\, dk \right\rangle_j = \langle f \rangle_{-j} ,$$

and that $\mathcal{E}$ and f are real valued functions. By inserting $n = 1 + \partial F/\partial x$ in Eq. (5.129), we obtain the Ampère's law

$$\frac{\partial F}{\partial t} + J_n = J\,, \tag{5.130}$$

where $J(t)$ is the total current density.

Inserting Eqs. (5.126) and (5.127) in Eq. (5.119), we find the hierarchy:

$$\mathcal{L} f^{(1)} = -\left.\frac{\partial f^{(0)}}{\partial t}\right|_0 + \sum_{j=-\infty}^{\infty} i j \mathcal{E}_j e^{ijk} \frac{\partial}{\partial x} \langle f^{(0)} \rangle_j\,, \tag{5.131}$$

$$\mathcal{L} f^{(2)} = -\left.\frac{\partial f^{(1)}}{\partial t}\right|_0 + \sum_{j=-\infty}^{\infty} i j \mathcal{E}_j e^{ijkl} \frac{\partial}{\partial x} \langle f^{(1)} \rangle_j - \left.\frac{\partial}{\partial t} f^{(0)}\right|_1\,, \tag{5.132}$$

and so on. Here,

$$\mathcal{L} u(k) \equiv u(k) + i \sum_{j=-\infty}^{\infty} j \langle F \rangle_j u_j e^{ijk}\,, \tag{5.133}$$

and the subscripts 0 and 1 in the right-hand side of these equations mean that $\partial F/\partial t$ is replaced by $J - J^{(0)}(F)$ and by $-J^{(1)}(F)$, respectively.

The compatibility condition (5.128) implies that

$$\int_{-\pi}^{\pi} \mathcal{L} f^{(m)} dk = 0 \tag{5.134}$$

for $m \geq 1$. Using Eq. (5.128) as solvability conditions for the linear hierarchy of equations, we obtain

$$J^{(m)} = 2 \sum_{j=1}^{\infty} j \big\langle \mathrm{Im}\,(\mathcal{E}_{-j} f_j^{(m)}) \big\rangle_j\,, \tag{5.135}$$

which can also be obtained by insertion of Eq. (5.119) in Eq. (5.129). For the tight binding dispersion relation, $\mathcal{E}_j = \delta_{k0}/2 - (\delta_{j1} + \delta_{j,-1})/4$, and therefore, Eqs. (5.128) and (5.135) yield

$$J^{(0)}(F) = \left\langle \frac{2\langle F \rangle_1}{1 + \langle F \rangle_1^2} \frac{f_1^{\mathrm{FD}}}{4} \right\rangle_1\,. \tag{5.136}$$

Thus, the leading order of the Ampère's law Eq. (5.127) is

$$\frac{\partial F}{\partial t} + \left\langle \frac{2\langle F \rangle_1}{1 + \langle F \rangle_1^2} \frac{f_1^{\mathrm{FD}}}{4} \right\rangle_1 = J(t)\,, \tag{5.137}$$

in which f_1^{FD} is a function of n that has to be, in general, numerically determined. Using that $\mathcal{E}(k)$ is even, we have

$$f_j^{\mathrm{FD}} = \alpha\,\mathcal{I}_j(\mu)\,, \tag{5.138}$$

$$\mathcal{I}_j = \frac{1}{\pi}\int_0^{\pi}\int_{-\infty}^{\infty} \ln\left(1 + e^{\beta[\mu-(1-\cos k)/2]}\right) \frac{\sqrt{2}\,\Gamma^3\cos(j\,k)/\pi}{\left[E - \frac{1-\cos k}{2}\right]^4 + \Gamma^4}\,dk\,dE\,. \tag{5.139}$$

Using $f_0^{\mathrm{FD}} = n$, we find $n = \alpha\,\mathcal{I}_0(\mu)$. By numerically solving this equation, we can obtain the chemical potential as a function of electron density: $\mu(n)$. Then, we can write Eq. (5.138) as a function of n:

$$f_j^{\mathrm{FD}} = n\,\mathcal{M}_j(n)\,, \quad \mathcal{M}_j(n) = \frac{\mathcal{I}_j(\mu(n))}{\mathcal{I}_0(\mu(n))}\,. \tag{5.140}$$

$\mathcal{M}_j(n)$ can be explicitly calculated in the limit as $\Gamma \to 0$ and $\beta\mu \to -\infty$. In this limit, we can approximate $\ln(1+x) \sim x$ and the previous integrals are proportional to modified Bessel functions:

$$\mathcal{M}_j(n) \sim \frac{e^{\beta(\mu-1/2)}\int_0^{\pi} dk\,\cos(j\,k)\,e^{\beta\cos k/2}}{e^{\beta(\mu-1/2)}\int_0^{\pi} dk\,\cos k\,e^{\beta\cos k/2}} = \frac{I_j(\beta/2)}{I_0(\beta/2)}\,,$$

which yields

$$f_1^{\mathrm{FD}} \sim n\,\frac{I_1(\beta/2)}{I_0(\beta/2)}\,. \tag{5.141}$$

The same result is obtained if we use Boltzmann statistics instead of Fermi–Dirac in the expression (5.123) for local equilibrium and set $\Gamma \to 0$. By inserting Eq. (5.140) into Eq. (5.137) and using the Poisson equation (5.120) to eliminate n, we find

$$\frac{\partial F}{\partial t} + \frac{1}{4}\left\langle v_{ES}(\langle F\rangle_1)\left(1 + \frac{\partial F}{\partial x}\right)\mathcal{M}_1\left(1 + \frac{\partial F}{\partial x}\right)\right\rangle_1 = J(t)\,, \tag{5.142}$$

$$v_{ES}(F) = \frac{2F}{1+F^2}\,. \tag{5.143}$$

Except for the local averages which disappear if $\lambda \to 0$ (semiclassical limit), this is a first order hyperbolic partial differential equation. The drift velocity has the Esaki–Tsu form (5.143) with a peak velocity $I_1(\beta/2)/[4I_0(\beta/2)]$ in the limit as $\Gamma \to 0$, $\beta\mu \to -\infty$. In the semiclassical limit and returning to dimensional units, the drift velocity is

$$v_d(F) = \frac{l\Delta\,I_1(\beta/2)}{4\hbar\,I_0(\beta/2)}\,\frac{2eFl\hbar\nu}{(\hbar\nu)^2 + (eFl)^2}\,.$$

Ignoring local averages in Eq. (5.142), we have a hyperbolic equation very similar to that describing the Gunn effect for vanishing diffusion. We know that such equations may develop shock waves for constant J. Thus, we need to calculate the following term in the expansion (5.127) in order to regularize possible shocks. To find the first-order correction in Eq. (5.127), we first solve Eq. (5.131) and then find $J^{(1)}$. The result is

$$f^{(1)}(k;F) = \sum_{j=-\infty}^{\infty} \frac{e^{ijk}}{1+ij\langle F\rangle_j} \left[\frac{\frac{df_j^{\mathrm{FD}}}{dn}\frac{\partial J^{(0)}}{\partial x}}{1+ij\langle F\rangle_j} + \frac{ij f_j^{\mathrm{FD}}}{(1+ij\langle F\rangle_j)^2} \right.$$
$$\left. \times (J - \langle J^{(0)}\rangle_j) + \frac{\partial}{\partial x} \sum_{m=-\infty}^{\infty} im\mathcal{E}_m \left\langle \frac{f_{j-m}^{\mathrm{FD}}}{1+i(j-m)\langle F\rangle_{j-m}} \right\rangle_m \right], \tag{5.144}$$

which simplifies to

$$f^{(1)}(k;F) = \sum_{j=-\infty}^{\infty} \frac{e^{ijk}}{1+ij\langle F\rangle_j} \left[\frac{\frac{df_j^{\mathrm{FD}}}{dn}\frac{\partial J^{(0)}}{\partial x}}{1+ij\langle F\rangle_j} + \frac{ij f_j^{\mathrm{FD}}}{(1+ij\langle F\rangle_j)^2} \right.$$
$$\times (J - \langle J^{(0)}\rangle_j) - \frac{i}{4}\frac{\partial}{\partial x} \left\langle \frac{f_{j-1}^{\mathrm{FD}}}{1+i(j-1)\langle F\rangle_{j-1}} \right\rangle_1$$
$$\left. - \frac{i}{4}\frac{\partial}{\partial x} \left\langle \frac{f_{j+1}^{\mathrm{FD}}}{1+i(j+1)\langle F\rangle_{j+1}} \right\rangle_1 \right], \tag{5.145}$$

in the particular case of the tight binding dispersion relation $\mathcal{E}_{\pm 1} = -1/4$. In this particular case, we now obtain $J^{(1)}$ from Eq. (5.135) (here, $'$ means differentiation with respect to n):

$$J^{(1)} = \mathcal{N}^{(1)}\left(F, \frac{\partial F}{\partial x}\right) - \left\langle D^{(1)}\left(F, \frac{\partial F}{\partial x}, \frac{\partial^2 F}{\partial x^2}\right)\right\rangle_1 - \langle A^{(1)}\rangle_1 J(t), \tag{5.146}$$

$$A^{(1)} = \frac{1-3\langle F\rangle_1^2}{(1+\langle F\rangle_1^2)^3}\frac{n\mathcal{M}_1}{2}, \tag{5.147}$$

$$\mathcal{N}^{(1)} = \left\langle A^{(1)} \left\langle \left\langle n\mathcal{M}_1 \frac{\langle F\rangle_1}{2(1+\langle F\rangle_1^2)} \right\rangle_1 \right\rangle_1 \right\rangle_1 - \left\langle \frac{B^{(1)}}{1+\langle F\rangle_1^2} \right\rangle_1, \tag{5.148}$$

$$D^{(1)} = \frac{1}{8(1+\langle F\rangle_1^2)} \left(\frac{\partial^2 \langle F\rangle_1}{\partial x^2} - C^{(1)} \right), \tag{5.149}$$

$$B^{(1)} = \left\langle \frac{\langle F\rangle_2 n\mathcal{M}_2}{(1+4\langle F\rangle_2^2)^2}\frac{\partial\langle F\rangle_2}{\partial x} \right\rangle_1 + \frac{\langle F\rangle_1}{4}\left\langle \frac{n\mathcal{M}_2(1-4\langle F\rangle_2^2)}{(1+4\langle F\rangle_2^2)^2}\frac{\partial\langle F\rangle_2}{\partial x} \right\rangle_1$$
$$- \frac{\langle F\rangle_1 (n\mathcal{M}_1)'}{2(1+\langle F\rangle_1^2)} \left\langle n\mathcal{M}_1 \frac{1-\langle F\rangle_1^2}{(1+\langle F\rangle_1^2)^2}\frac{\partial\langle F\rangle_1}{\partial x} \right\rangle_1, \tag{5.150}$$

$$C^{(1)} = \left\langle \frac{(n\mathcal{M}_2)'}{1+4\langle F\rangle_2^2}\frac{\partial^2 F}{\partial x^2} \right\rangle_1 - 2\langle F\rangle_1 \left\langle \frac{(n\mathcal{M}_2)'\langle F\rangle_2}{1+4\langle F\rangle_2^2}\frac{\partial^2 F}{\partial x^2} \right\rangle_1$$
$$+ \frac{4(n\mathcal{M}_1)'\langle F\rangle_1}{1+\langle F\rangle_1^2} \left\langle \frac{(n\mathcal{M}_1)'\langle F\rangle_1}{1+\langle F\rangle_1^2}\frac{\partial^2 F}{\partial x^2} \right\rangle_1. \tag{5.151}$$

By inserting this current and $J^{(0)}$ given by Eqs. (5.136) in (5.127), we obtain Ampère's law which we may write with dimensional units as [7]

$$\varepsilon \frac{\partial F}{\partial t} + \frac{e\Delta}{4\hbar} \left\langle n \mathcal{M}_1 \left(\frac{n}{N_D} \right) \frac{2\hbar \nu e l \langle F \rangle_1}{(\hbar \nu)^2 + \langle e l F \rangle_1^2} \right\rangle_1 + \epsilon \frac{e N_D \Delta}{\hbar} \mathcal{N}^{(1)} =$$
$$\epsilon \frac{e N_D \Delta}{\hbar} \langle D^{(1)} \rangle_1 + (1 + \epsilon \langle A^{(1)} \rangle_1) J(t) . \tag{5.152}$$

Here, the terms Eqs. (5.147)–(5.151) contain the appropriate dimensionless combinations of dimensional variables. If the electric field and the electron density do not change appreciably over two SL periods, λ in Eq. (5.124) is very small and the spatial averages can be ignored. Then, the *nonlocal* QDDE Eq. (5.152) becomes the *local* generalized DDE (GDDE) obtained from the semiclassical theory [9].

With a Boltzmann distribution as the local equilibrium and in the semiclassical limit $\lambda \to 0$, the nondimensional reduced system of equations is

$$\begin{aligned}
&\frac{\partial F}{\partial t} + \frac{I_1}{4 I_0} n \, v_{ES}(F) \left[1 + \epsilon \frac{I_1}{2 I_0} \frac{1 - 3F^2}{(1 + F^2)^3} \right. \\
&\quad \left. - \epsilon \left(\frac{I_2}{I_1} \frac{5 - 4F^2}{(1 + 4F^2)^2} - \frac{I_1}{2 I_0} \frac{5 - 7F^2}{(1 + F^2)^3} \right) \frac{\partial F}{\partial x} \right] \\
&= \frac{\epsilon}{8(1 + F^2)} \left[1 - \frac{I_2}{I_0} \frac{1 - 2F^2}{1 + 4F^2} + \frac{4 I_1^2}{I_0^2} \frac{F^2}{(1 + F^2)^2} \right] \frac{\partial^2 F}{\partial x^2} \\
&\quad + \left(1 + \frac{\epsilon I_1}{2 I_0} \frac{1 - 3F^2}{(1 + F^2)^3} n \right) J(t) , \\
&n = 1 + \frac{\partial F}{\partial x} .
\end{aligned} \tag{5.153}$$

In the limit of large temperatures, $k_B T \gg \max(\Delta, \pi \hbar^2 N_D / m^*)$, $I_1 / I_0 \sim \beta/4 = \Delta/(4 k_B T)$, the terms in square brackets in Eq. (5.153) are approximately one and we obtain the following simpler drift-diffusion equation:

$$\frac{\partial F}{\partial t} + \frac{\beta}{16} v_{ES}(F) \left(1 + \frac{\partial F}{\partial x} \right) = \frac{\epsilon}{8(1 + F^2)} \frac{\partial^2 F}{\partial x^2} + J(t) . \tag{5.154}$$

Here, we have also ignored the correction of order ϵ to the total current density. Written in dimensional units, Eq. (5.154) becomes

$$\varepsilon \frac{\partial F}{\partial t} + F \mu_{ES}(F) \left(\varepsilon \frac{\partial F}{\partial x} + \frac{e N_D}{l} \right) - D_{ES}(F) \, \varepsilon \frac{\partial^2 F}{\partial x^2} = J(t) , \tag{5.155}$$

$$\mu_{ES}(F) = \frac{e \Delta^2 l^2}{8 k_B T \hbar^2 \nu_e} \frac{1}{1 + \left(\frac{e F l}{\hbar \nu_e} \right)^2} , \tag{5.156}$$

$$D_{ES}(F) = \frac{k_B T}{e} \mu_{ES}(F) . \tag{5.157}$$

Equation (5.157) is a nonlinear Einstein relation linking the diffusion coefficient $D_{ES}(F)$ and the electron mobility $\mu_{ES}(F)$.

The boundary conditions for the QDDE Eq. (5.152) (which contains triple spatial averages) need to be specified on the intervals $[-2l, 0]$ and $[Nl, Nl + 2l]$, not just at the points $x = 0$ and $x = Nl$, as in the case of the parabolic GDDE. Similarly, the initial condition has to be defined on the extended interval $[-2l, Nl + 2l]$. For realistic values of the parameters representing a strongly coupled SL under dc voltage bias, the numerical solution of the QDDE yields a stable self-sustained oscillation of the current [7] in quantitative agreement with experiments [8]. Details of the numerical procedure can be found in [10].

Figure 5.2 shows the evolution of the current during the self-sustained oscillations that appear when the QDDE Eq. (5.152) and the Poisson equation (5.120) are solved for boundary conditions $\varepsilon \partial F/\partial t + \sigma F = J$ at each point of the intervals $[-2l, 0]$ and $[Nl, Nl + 2l]$ and appropriate dc voltage bias. The contact conductivity σ is selected so that σF intersects $e N_D v_M V(F/F_M)/l$ on its decreasing branch, as in the theory of the Gunn effect [5]. Parameter values correspond to a 157-period 3.64 nm GaAs/0.93 nm AlAs SL at 5 K, with $N_D = 4.57 \times 10^{10}\,\text{cm}^{-2}$, $\nu_i = 2\nu_e = 18 \times 10^{12}$ Hz under a dc voltage bias of 1.62 V [8]. Cathode and anode contact conductivities are 2.5 and 0.62 $\Omega^{-1}\text{cm}^{-1}$, respectively.

Our numerical solution shows that the current and the field profile become stationary for $\phi < 0.75$ (1.2 V). For larger values of the dimensionless voltage ϕ, the initial field profile evolves toward a stable time-periodic solution for which J oscillates with time and the field profile shows recycling and motion of a pulse

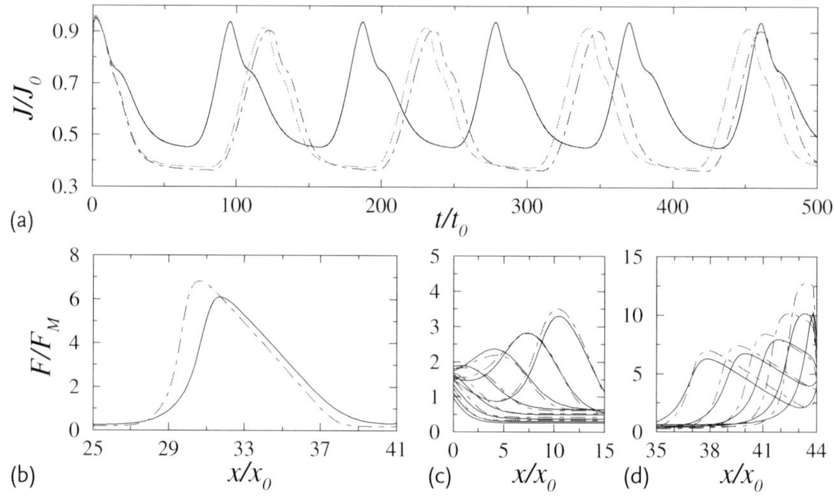

Fig. 5.2 (a) Current ($J_0 = e v_M N_D/l$) vs. time ($t_0 = \varepsilon F_M/J_0$) during self-oscillations for a voltage biased GaAs/AlAs SL, as described by the QDDE (solid line), the QDDE with $\Gamma = 0$ (long-dashed line) and by the GDDE (dot-dashed line). (b) Comparison between the fully developed dipole wave for the QDDE (solid line) and the dipole wave for the GDDE (dashed line). (c) Dipole wave at different times during the stage in which it is shed from the injecting contact. (d) Same as (c) for the stage in which the dipole disappears at the anode, located at $Nl/x_0 \approx 44$. Parameter values are $x_0 = v_M t_0 = 16$ nm, $t_0 = 0.43$ ps, $J_0 = 6.07 \times 10^5$ A/cm^2, $\phi = 1$. From [10].

from $x = 0$ to the SL end. Figure 5.2 shows the self-oscillations of the current for 1.62 V ($\phi = 1$) and the corresponding field pulse at different times. In this figure, we compare the solution of the GDDE corresponding to the semiclassical BGK–Poisson kinetic equation and the solution of the QDDE for $\Gamma = 0$ (no collision broadening) and for $\Gamma = 18$ meV, which is of the same order as the collision frequencies.

Self-oscillations in the QDDE have a frequency of $\nu_Q = 25.5$ GHz, faster than in the GDDE, $\nu_G = 20.6$ GHz (relative frequency $(\nu_Q - \nu_G)/\nu_G = 23.8\%$). When $\Gamma = 0$, the frequency is $\nu_{Q^*} = 21$ GHz, and the relative frequency $(\nu_{Q^*} - \nu_G)/\nu_G = 1.94\%$.

Collision broadening shortens the period of the current oscillations and therefore it reinforces the effects of the nonlocal terms in the QDDE due to quantum effects. We observe that the field profile of the dipole wave during self-oscillations is sharper in the case of the GDDE than in the case of the QDDE. The local spatial averages appearing in the QDDE have a smoothing effect on the sharp gradients of the electric field. This smoothing effect produces rounder and smaller dipole waves in the QDDE, as compared to the same solution for the GDDE. The equal-area rule, as in the theory of the Gunn effect, hints that smaller waves are faster [11], resulting in a slightly larger frequency for the self-oscillations in the QDDE (37.6 GHz) than in the case of the GDDE (36.8 GHz).

5.3 Concluding Remarks

In this chapter, we have reviewed the kinetic theory basis of the drift-diffusion models that are widely used in this book. General ideas have been presented in the context of simple BGK models and parabolic scaling in which collision terms dominate and force the distribution function to be close to local equilibrium. The Einstein relation connects drift and diffusion in the resulting equation for the electron density and the electric field. Both multiple scales and the Chapman–Enskog method yield the same limiting drift-diffusion equation. In the second part of this chapter, we have applied the same general ideas to the case of a strongly coupled semiconductor superlattice with a single populated miniband. We have derived the corresponding Wigner–Poisson–BGK system of equations with a collision broadened local Fermi–Dirac distribution. In the hyperbolic limit in which the collision and Bloch frequencies are of the same order and dominate all other frequencies, the Chapman–Enskog perturbation method yields a quantum drift-diffusion equation for the field. Numerical solutions of this equation exhibit self-sustained oscillations of the current due to recycling and motion of charge dipole domains [7].

There are interesting extensions and results related to the topics in this chapter. In the semiclassical limit, Cebrián, Bonilla and Carpio have directly solved the Boltzmann–Poisson–BGK system of equations for a strongly coupled miniband superlattice by using a weighted particle numerical method [12]. This is a deterministic method that directly yields the distribution function. The results show that the limiting drift-diffusion equation found by using the Chapman–Enskog method is an excellent description of many superlattices for which there are measurements of self-sustained current oscillations. Of course for other parameter ranges far from the hyperbolic limit, the drift-diffusion equations may not be such suitable approximations and the direct numerical solution of the kinetic equation may be preferable.

For strongly coupled SLs having two populated minibands, Bonilla, Barletti and Alvaro [13] have introduced a description based on a periodic version of the Kane Hamiltonian and derived the corresponding Wigner–Poisson–BGK system of equations. The collision model comprises two terms, a BGK term trying to bring the Wigner matrix closer to a broadened Fermi–Dirac local equilibrium at each miniband, and a scattering term that brings down electrons from the upper to the lower miniband. By using the Chapman–Enskog method, they have derived quantum drift-diffusion equations for the miniband populations which contain generation-recombination terms. As it should be, the recombination terms vanish if there is no inter-miniband scattering and the off-diagonal terms in the Hamiltonian are zero. These terms may represent a Rashba spin–orbit interaction for a lateral superlattice. For a lateral superlattice under dc voltage bias in the growth direction, numerical solutions of the corresponding quantum drift-diffusion equations show self-sustained current oscillations due to periodic recycling and motion of electric field pulses. The periodic changes of the spin polarization and spin polarized current indicate that this system acts as a spin oscillator.

An interesting development is related to Bloch oscillations in strongly coupled superlattices. Restricting ourselves to semiclassical descriptions, the usual collision models contain BGK terms Eq. (5.96) with a local equilibrium such as Eq. (5.101) (usually without collision broadening). The local equilibrium only depends on the electron density n. An initial distribution function rapidly approaches exponentially the leading order term in the Chapman–Enskog expansion, and then n and the field slowly evolve to their stable values on a much longer time scale, as described in this chapter. In the longer scale corresponding to a hydrodynamic (diffusive) regime, the only oscillations of the current that may be described by the usual models are the Gunn-type self-sustained oscillations in this chapter. However, what happens if we introduce a richer local equilibrium that also depends on the electron current density and the electron mean energy? This local equilibrium may exhibit Bloch oscillations at the Bloch frequency and a slower hydrodynamic regime in which electron density, field and the envelope of the Bloch oscillations evolve. Bonilla and Carretero [14] have proposed a BGK collision model based on these ideas. Consider the following local equilibrium instead of Eq. (5.101):

$$f^{1D\alpha} = \frac{m^* \Delta}{2\pi \tilde{\beta} \hbar^2} \ln\left(1 + e^{\tilde{\mu} + \tilde{u}\, kl - \tilde{\beta} + \tilde{\beta} \cos kl}\right) , \tag{5.158}$$

where the nondimensional functions $\tilde{\mu}$, $\tilde{u}$ and $\tilde{\beta}$ are obtained in terms of the electron density, current density and mean energy,

$$n = \frac{l}{2\pi} \int_{-\pi/l}^{\pi/l} f \, dk \,, \tag{5.159}$$

$$J_n = \frac{e}{2\pi} \int_{-\pi/l}^{\pi/l} v(k) \, f \, dk \,, \tag{5.160}$$

$$E = \frac{l}{2\pi n} \int_{-\pi/l}^{\pi/l} \mathcal{E}(k) \, f \, dk \,, \tag{5.161}$$

by solving:

$$\frac{l}{2\pi} \int_{-\pi/l}^{\pi/l} f^{1D\alpha} \, dk = n \,, \tag{5.162}$$

$$\frac{e}{2\pi} \int_{-\pi/l}^{\pi/l} v(k) \, f^{1D\alpha} \, dk = (1 - \alpha_j) J_n \,, \tag{5.163}$$

$$\frac{l}{2\pi n} \int_{-\pi/l}^{\pi/l} \left[\frac{\Delta}{2} - \mathcal{E}(k) \right] f^{1D\alpha} \, dk = \alpha_e E_0 + (1 - \alpha_e) E \,. \tag{5.164}$$

Then, the source terms in the moment equations for n, J_n and E are:

$$\int_{-\pi/l}^{\pi/l} Q_e(f) \, dk = 0 \,, \tag{5.165}$$

$$\frac{e}{2\pi} \int_{-\pi/l}^{\pi/l} v(k) \, Q_e(f) \, dk = -\nu_e \alpha_j J_n \,, \tag{5.166}$$

$$\frac{l}{2\pi n} \int_{-\pi/l}^{\pi/l} \mathcal{E}(k) \, Q_e(f) \, dk = -\nu_e \alpha_e (E - E_0) \,. \tag{5.167}$$

The restitution coefficients α_j and α_e, $0 \leq \alpha_{j,e} \leq 1$, measure the fraction of momentum and of energy lost in inelastic collisions. Obviously, for $\alpha_{e,j} = 0$, the collisions are elastic. If we integrate the semiclassical version of Eq. (5.107),

$$\frac{\partial f}{\partial t} + v(k) \frac{\partial f}{\partial x} + \frac{eF}{\hbar} \frac{\partial f}{\partial k} = Q_e(f) \equiv -\nu_e (f - f^{1D\alpha}) \,, \tag{5.168}$$

times 1, $v(k)$ or $\mathcal{E}(k)$ over k and use Eqs. (5.165)–(5.167), we obtain the moment equations:

$$\frac{e}{l}\frac{\partial n}{\partial t} + \frac{\partial J_n}{\partial x} = 0\,, \tag{5.169}$$

$$\frac{\partial J_n}{\partial t} + \frac{e}{l}\frac{\partial}{\partial x}(n\,\langle v(k)^2\rangle) - \frac{e^2 n F}{\hbar l}\,\langle v'(k)\rangle = -\nu_e \alpha_j J_n\,, \tag{5.170}$$

$$\frac{\partial(nE)}{\partial t} + \frac{\partial}{\partial x}(n\,\langle v(k)\mathcal{E}(k)\rangle) - F\,J_n = -\nu_e\alpha_e\,(E - E_0)\,. \tag{5.171}$$

Here,

$$\langle\phi(k)\rangle = \frac{\int_{-\pi/l}^{\pi/l} \phi(k)\, f\, dk}{\int_{-\pi/l}^{\pi/l} f\, dk}\,. \tag{5.172}$$

The source terms in the right-hand sides of these equations try to make J_n and E equal to zero (no current) and E_0 (the lattice energy), respectively. Equation (5.169) is the charge continuity equation. If $\mathcal{E}(k) = \Delta(1 - \cos kl)/2$ (tight-binding dispersion relation) and we seek x-independent solutions, we find that n is constant and

$$\frac{\partial S}{\partial t} - \frac{eFl}{\hbar}\,C = -\nu_e\alpha_j\,S\,, \tag{5.173}$$

$$\frac{\partial C}{\partial t} + \frac{eFl}{\hbar}\,S = -\nu_e\alpha_e\left(C - 1 + \frac{2E_0}{\Delta}\right)\,, \tag{5.174}$$

provided that we define $C = \langle\cos kl\rangle$ and $S = \langle\sin kl\rangle$. F is also constant and therefore, C and S (or, equivalently, J_n and E) oscillate at the Bloch frequency $\omega_B = eFl/\hbar$ in the limit of elastic collisions $\alpha_{e,j} = 0$. Bonilla and Carretero have derived hydrodynamic equations for n, F and the envelope of the Bloch oscillations in the hyperbolic limit [14]. These equations can be used to discuss the stability of the Bloch oscillations which is a topic that has attracted the attention of researchers for many years.

References

1 Toda, M., Kubo, R., and Saitô, N. (1983) *Statistical Physics I*, Springer, Berlin.

2 Bhatnagar, P.L., Gross, E.P., and Krook, M. (1954) A model for collision processes in gases. I. Small amplitude processes in charged and neutral one-component systems. *Phys. Rev.*, **94**, 511–525.

3 Cercignani, C., Illner, R., and Pulvirenti, M. (1994) *The Mathematical Theory of Dilute Gases*, Springer, New York.

4 Sze, S.M. (1981) *Physics of Semiconductor Devices*, John Wiley & Sons, Inc., New York.

5 Bonilla, L.L. and Grahn, H.T. (2005) Nonlinear dynamics of semiconductor superlattices. *Rep. Prog. Phys.*, **68**, 577–683. Copyright (2005), Institute of Physics.

6 Wacker, A. (2002) Semiconductor superlattices: A model system for nonlinear transport. *Phys. Rep.*, **357**, 1–111.

7 Bonilla, L.L. and Escobedo, R. (2005) Wigner–Poisson and nonlocal drift-diffusion model equations for semiconductor superlattices. *Math. Models Methods Appl. Sci.*, **15**(8), 1253–1272.

8 Schomburg, E., Blomeier, T., Hofbeck, K., Grenzer, J., Brandl, S., Lingott, I., Ignatov, A.A., Renk, K.F., Pavelev, D.G., Koschurinov, Y., Melzer, B.Y., Ustinov, V.M., Ivanov, S.V., Zhukov, A., and Kopev, P.S. (1998) Current oscillations in superlattices with different miniband widths. *Phys. Rev. B*, **58**, 4035–4038.

9 Bonilla, L.L., Escobedo, R., and Perales, A. (2003) Generalized drift-diffusion model for miniband superlattices. *Phys. Rev. B*, **68**, 241304(R).

10 Escobedo, R. and Bonilla, L.L. (2006) Numerical methods for a quantum drift-diffusion equation in semiconductor physics. *J. Math. Chem.*, **40**, 3–13.

11 Bonch-Bruevich, V.L., Zvyagin, I.P., and Mironov, A.G. (1975) *Domain Electrical Instabilities in Semiconductors*. Consultants Bureau, New York.

12 Cebrián, E., Bonilla, L.L., and Carpio, A. (2009) Self-sustained current oscillations in the kinetic theory of semiconductor superlattices. *J. Comput. Phys.*, **228**, 7689–7705.

13 Bonilla, L.L., Barletti, L., and Álvaro, M. (2008) Nonlinear electron and spin transport in semiconductor superlattices. *SIAM J. Appl. Math.*, **69**, 494–513.

14 Bonilla, L.L. and Carretero, M. (2008) Nonlinear electron transport in nanostructures, In: *Numerical Analysis and Applied Mathematics*, (eds T.E. Simos, G. Psihoyios and Ch. Tsitouras), American Institute of Physics Proceedings, Vol. 1048, Melville, New York, pp. 9–12.

6
Electric Field Domains in Bulk Semiconductors I: the Gunn Effect

6.1
Introduction

In this chapter, we will study pattern formation and oscillatory phenomena involving recycling and motion of charge dipole waves in bulk semiconductors. We focus on cases where the space charge instability is due to negative differential mobility (see Chapter 1). These phenomena were first observed in bulk n-type GaAs, for which the effective electron drift velocity is an $\mathcal{N}$-shaped function of the electric field: it increases linearly for low fields, it has a local maximum at about 3.2 kV/cm, and then it decays to a nearly constant value as $E \to \infty$. When planar contacts are attached to an n-GaAs sample and an appropriate dc voltage bias is applied across them, self-sustained oscillations of the current with typical frequencies in the microwave range appear. In an elegant experiment, Gunn demonstrated that these oscillations are accompanied by periodic recycling and motion of charge density dipole waves, which are pulses of the electric field [1]. Each wave is nucleated near the cathode, it moves towards the anode and as it disappears, a new wave is nucleated at the cathode. The Gunn effect is exploited in Gunn diodes, which are electronic devices used in the communication industry as microwave sources.

After Gunn's discovery, many materials were shown to have current self-oscillations accompanied by recycling of dipole waves. Basically, all of these materials possess intrinsic current–voltage characteristics that are $\mathcal{N}$-shaped when the samples have one-dimensional geometry and are subjected to dc voltage bias conditions. The physical mechanisms causing the oscillations are often very different and a great deal of research was required in order to clarify these mechanisms. Additionally, some effort was made in the 1960s and 1970s to understand the mathematics behind the Gunn effect. Essentially, several basic facts on infinite semiconductors under current bias were found and many numerical simulations of one-dimensional drift-diffusion models were performed. Further understanding of the mathematics of self-oscillations has been achieved more recently. In this chapter, we present asymptotic and numerical analyses of the Kroemer model for the Gunn effect in n-GaAs [2]. We will also clarify which features of the analysis are model independent.

Nonlinear Wave Methods for Charge Transport. Luis L. Bonilla and Stephen W. Teitsworth

ISBN: 978-3-527-40695-1

This chapter is organized as follows. There are two basic mechanisms responsible for $\mathcal{N}$-shaped current–voltage characteristics: intervalley transfer, which results in negative differential conductivity, and impurity trapping of electrons, which results in negative differential carrier concentration. The first mechanism yields Kroemer's model which we analyze in detail in the rest of this chapter. (Instabilities mediated by carrier trapping are taken up in Chapter 7.) We first give a heuristic explanation about why $\mathcal{N}$-shaped current–voltage characteristics may give rise to charge dipole waves. The stationary solutions of Kroemer's model under voltage bias are constructed and their linear stability is examined in Section 6.3 under conditions of both current and voltage bias. Section 6.4 discusses the onset of the current oscillations under voltage bias. Sections 6.3 and 6.4 are of a rather technical character and can be omitted in a first reading of this chapter. The main results of this chapter are presented in Sections 6.5 and 6.6: asymptotic analyses of the Gunn effect using simple phase plane considerations and matched asymptotic expansions. Section 6.5 considers a $\mathcal{N}$-shaped electron velocity curve as a function of the electric field, whereas the theory for a velocity curve which saturates at high fields is presented in Section 6.6. In both cases, useful explicit formulas are obtained in the small diffusivity limit where shock wave dynamics provides a good approximation. This chapter also reviews work in other materials presenting Gunn-type self-oscillations. However, the extension of these methods to several space dimensions and other diverse open problems are discussed in Chapter 9.

6.2
$\mathcal{N}$-shaped Current-Field Characteristics and Kroemer's Model

The Gunn effect and similar oscillatory phenomena occur if an $\mathcal{N}$-shaped current-field characteristic originates from the bulk properties of a semiconductor. This is suggested by the following simple heuristic argument. Let us consider the one-dimensional charge continuity and Poisson equations for the electric field, F, and charge density, ρ:

$$\frac{\partial \rho}{\partial t} + \frac{\partial J_q}{\partial x} = 0 ,$$

$$\frac{\partial F}{\partial x} = \frac{\rho}{\kappa \epsilon_0} .$$

Here, $\kappa\epsilon_0$ is the dielectric constant and J_q is the charge current density, provided free electrons are the majority carriers. We assume that J_q is an $\mathcal{N}$-shaped function of F, and let $F = F_0 =$ constant, $\rho = 0$, be a stationary state of this system of equations, such that $dJ_q(F_0)/dF < 0$. Thus, F_0 lies in a region of negative differential resistance. To keep the argument as simple as possible, we temporarily ignore bias and boundary conditions. Let us examine the dynamical stability of an initial condition corresponding to a sinusoidal charge dipole disturbance of the previous stationary state:

$$\frac{\partial F}{\partial x} = \rho = \epsilon \sin x \, \theta(\pi - |x|),$$

with $\epsilon \ll 1$. $\theta(x)$ is the Heaviside step function. If $F = F_0 + \epsilon f$, $\rho = \epsilon r$, the previous system of equations can be written as

$$\frac{\partial r}{\partial t} = -\frac{dJ_q(F_0)}{dF}\frac{\partial f}{\partial x},$$
$$\frac{\partial f}{\partial x} = \frac{r}{\kappa\epsilon_0}, \quad \text{which yields}$$
$$\frac{\partial r}{\partial t} = -\frac{dJ_q(F_0)}{dF}\frac{r}{\kappa\epsilon_0},$$

up to terms of order ϵ. As $-J_q'(F_0) > 0$, the absolute value of the initial charge disturbance grows in time. Thus, an $\mathcal{N}$-shaped current-field characteristic may result in the generation of charge dipole waves and it is known that such waves play a major role in the Gunn effect. This heuristic argument is suggestive but, if the aim is to develop precise predictions and estimates, *it is not a substitute for rigorous analysis.*

Let us describe two physical mechanisms responsible for a $\mathcal{N}$-shaped current-field characteristic curve. Suppose $J_q = en(F)\mathcal{V}(F)$, where e, $n(F)$ and $\mathcal{V}(F)$ denote the positive magnitude of electron charge, the field-dependent electron density and velocity, respectively. Clearly,

$$\frac{dJ_q}{dF} = en\frac{d\mathcal{V}}{dF} + e\mathcal{V}\frac{dn}{dF}$$

may be negative provided that either $d\mathcal{V}/dF < 0$ (negative differential velocity or mobility) or $dn/dF < 0$ (negative differential carrier concentration). Both possibilities are found in real semiconductor systems: the first possibility occurs, for example, in materials where intervalley transfer in k-space occurs for high electric fields, that is, under conditions of carrier heating. The second mechanism occurs in semiconductors possessing field-dependent carrier trapping dynamics, for example, due to shallow donor or acceptor impurities. This is discussed in the next chapter. Here, we focus on the first of these physical mechanisms.

6.2.1 Intervalley Transfer Mechanism

Intervalley transfer occurs in semiconductors with non-equivalent valleys k-space corresponding to the conduction band. At sufficiently high electric fields, electrons sitting in the lowest valley have their energy raised high enough that they may scatter by phonon emission to another valley's state of lower energy, without changing the total electron concentration. This is called *intervalley transfer* of hot electrons and it yields an electron drift velocity which is a $\mathcal{N}$-shaped function of the local electric field. In turn, the local current density becomes a $\mathcal{N}$-shaped function of the local electric field. Figure 6.1 illustrates the intervalley transfer mechanism proposed by Ridley, Watkins and Hilsum [3] in 1961, two years before the Gunn effect was discovered experimentally by Gunn in 1963 [4]. The electron energy, E, is a function of the electron wavevector, $\boldsymbol{k}$, and reveals two energy minima (two valleys). The lower valley is located at the center of Brillouin zone ($\boldsymbol{k} = 0$) and is characterized by a relatively small effective mass, m_c (for III–V compounds, this minimum is termed a Γ-minimum). The side minima have a larger mass, m_s (these minima are now known to be L valleys for GaAs). An important parameter is the energy splitting between these valleys, Δ. Under equilibrium conditions and if the crystal temperature T is much smaller than Δ/k_B, all the electrons populate the lowest valley. There, they have a small mass and a large mobility, μ_l. When an electric field, $\boldsymbol{F}$, is applied, the electron temperature, T_e, may exceed T. Then, the electrons are able to transfer to the higher minima. This typically occurs via emission of an optical phonon (i. e., lattice vibrational mode) with large k-vector. In the higher valley, the effective mass is larger and their mobility, μ_s, is lower. Thus, the average mobility of the electrons may decrease. This can lead to an overall electron velocity which decreases as the local electric field increases for a certain range of electric field values.

More precisely, we may quantify this mechanism as follows. We assume that the timescale for momentum and energy changes due to interaction with the electric field are faster than electron scattering processes with phonons or impurities. We also assume that the characteristic timescale for intervalley transitions is much longer than electron momentum and energy relaxation times within each valley. Then, the effective electron temperature is the same for all valleys while the populations are different. As the intervalley transition time is $\tau_i \approx 10^{-11}$ s or less for III–V compounds, the description we give below holds for materials displaying Gunn oscillations of frequencies up to 100 GHz (whose corresponding periods are of the order of or larger than τ_i).

Let n_c and n_s be the electron concentration at the central and side valleys respectively. If the total concentration $n = n_c + n_s$ is constant, the steady state current

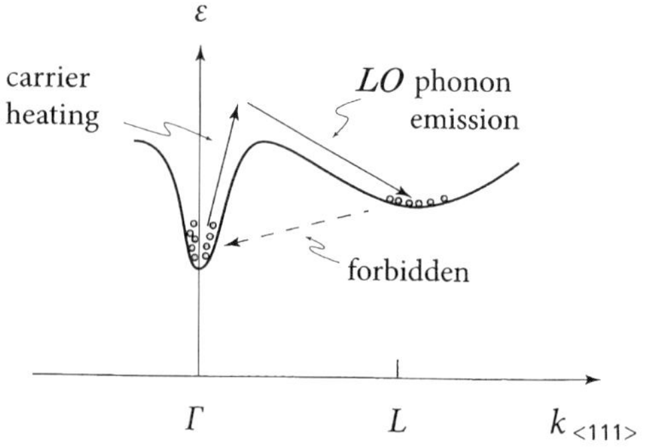

Fig. 6.1 Energy band structure schematic of GaAs indicating the key dynamical processes leading to the Gunn effect.

density can be written as

$$J = e\left(\mu_c n_c + \mu_s n_s\right) F \equiv en\mathcal{V}(F) , \tag{6.1}$$

where we have introduced the absolute value of the average drift velocity

$$\mathcal{V}(F) \equiv \frac{\left(\mu_c n_c + \mu_s n_s\right)}{n_c + n_s} F \approx \frac{\mu_c}{1 + n_s/n_c} F . \tag{6.2}$$

Here, we suppose the inequality $\mu_c \gg \mu_s$ that usually holds for III–V compounds. Notice that the $\mathcal{V} - F$ characteristic is equivalent to the current–voltage characteristic at constant n. The ratio between the lower and upper valley populations is determined by the energy splitting and the electron temperature:

$$\frac{n_s}{n_c} = R e^{-\frac{\Delta}{k_B T_e}} . \tag{6.3}$$

Here, $R = \mathcal{N}_s (m_s/m_c)^{\frac{3}{2}}$ is the ratio between the density of states of the valleys c and s ($\mathcal{N}_s$ is the number of upper valleys), and T_e is the effective electron temperature which is greater than T due to electric field heating. For *GaAs*, there are four nonequivalent upper valleys (*L*-valleys), $n_s = 4$, $m_c = 0.067 m_0$, $m_s = 0.55 m_0$, so that $R \approx 94$. To find $\mathcal{V}(F)$, Eqs. (6.1)–(6.3) should be supplemented by a calculation of the electron temperature as a function of the field. In general, such a calculation is performed by approximately solving the Boltzmann equation; this has been considered by Butcher [5]. The result depends on particular scattering mechanisms. For an estimate, one can use the simplest energy balance equation

$$e F \mathcal{V} = \frac{3}{2} k_B \frac{T_e - T}{\tau_E} , \tag{6.4}$$

assuming that the *intravalley* energy relaxation time, τ_E, is independent of the field. Inserting Eq. (6.2) in Eq. (6.4), we obtain the following equation for the electron temperature as a function of the field:

$$T_e = T + \frac{2e\tau_E\mu_c}{3k_B} \frac{F^2}{1 + R e^{-\frac{\Delta}{k_B T_e}}} . \tag{6.5}$$

$\mathcal{V}(F)$ is now found substituting the result in Eq. (6.2). Depending on the parameters $\mathcal{V} - F$ characteristic can show a negative slope. For example, for *GaAs* at room temperature, one can set $\tau_E = 1.2 \times 10^{-12}$ s, $\mu_c = 6000\,\mathrm{cm^2/V\,s}$ and $\Delta = 0.31$ eV. Then, Eqs. (6.5) and (6.2) yield the electron velocity curve depicted in Figure 6.2a. Parameters of importance in this curve are the threshold field, F_M, defining the onset of the region with negative differential mobility, $d\mathcal{V}/dF < 0$, the maximum electron velocity, $\mathcal{V}_M$, and the peak-to-valley velocity ratio, $\mathcal{V}_M/\mathcal{V}_m$ ($\mathcal{V}_m$ is the minimum velocity at $F > F_M$). For several III–V compounds, the parameters Δ, F_M and $\mathcal{V}_M$ are presented in Table 6.1 at different crystal temperatures.

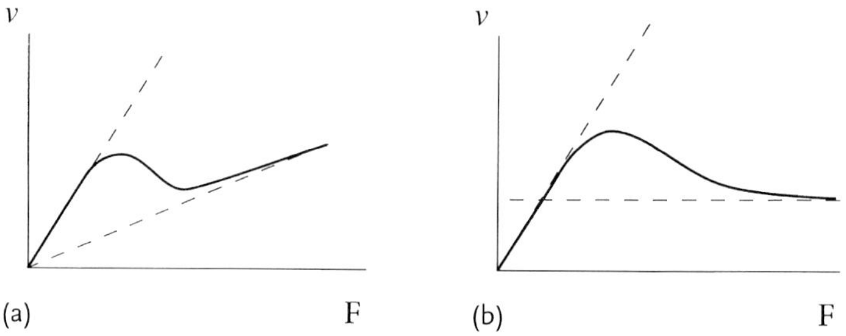

Fig. 6.2 Electron drift velocity curves. (a) N-shaped velocity curves with three branches. (b) Saturating velocity curve with only two branches.

Table 6.1 Parameters related to intervalley electron transfer effect.

Semiconductor	T (K)	Δ (eV)	F_M (kV/cm)	$\mathcal{V}_M$ (10^7 cm/s)
GaAs	300	0.31	3.2	2.2
InP	300	0.53	10.5	2.5
InAs [a]	300	1.25	1.6	3.6
InSb [a]	77	0.41	0.6	5.0

a under pressure

6.2.2
Kroemer's Drift-Diffusion Model

A simple and powerful model that is able to explain the Gunn effect in n-GaAs was proposed by Kroemer in 1964 [2]. It consists of the charge continuity equation, Poisson equation, a constitutive relation for the electron flux, bias, boundary and initial conditions. The unknowns are the electron concentration, $n(\boldsymbol{r}, t)$, and the electric field, $\boldsymbol{F}(\boldsymbol{r}, t)$ (or equivalently, the electric potential). Peculiarities of the semiconductor material (intervalley transfer) are considered in the constitutive relation which links the electron flux to the electron density and the electric field.

The equations governing the model are

$$\frac{\partial n}{\partial t} + \nabla \cdot \boldsymbol{I} = 0\,, \tag{6.6}$$

$$\nabla \cdot \boldsymbol{F} = -\frac{e}{\kappa \epsilon_0}(n - n_0)\,. \tag{6.7}$$

Here, $\kappa\epsilon_0$ is the dielectric constant, n_0 is the doping (which equals the electron concentration in equilibrium), and $\boldsymbol{I}$ is the electron flux. The constitutive relation for the latter is

$$\boldsymbol{I} = \boldsymbol{\mathcal{V}}(F)n - \nabla[D(F)\,n]\,. \tag{6.8}$$

The average velocity, $\mathcal{V}$, and the diffusion coefficient, D, are known functions of the field. Notice that, insofar as the diffusion coefficient is field dependent, there are two space dependent effects in Eq. (6.6): Fick's law, $-D\,\nabla n$, and the so-called *thermo-diffusion of hot electrons* described by the term $-n\nabla D = -n(dD/dT_e)\,\nabla T_e$. The electron temperature may be found as a function of the electric field as we have explained before. In this chapter, we consider electrons with isotropic properties. This means that the electron velocity, $\boldsymbol{\mathcal{V}}$, will always be directed opposite to the field, $\boldsymbol{F}$. The drift velocity $|\boldsymbol{\mathcal{V}}|$ is an N-shaped nonlinear function of the electric field $|\boldsymbol{F}|$ due to the intervalley transfer mechanism explained above. For n-type GaAs, $\mathcal{V}$ has a maximum $\mathcal{V}_M \approx 2.2 \times 10^7$ cm/s at $F_M \approx 3.2\,\mathrm{kV/cm}$. The minimum of $\mathcal{V}$ occurs (if at all) at such high electric fields that a curve $\mathcal{V}(F)$ saturating at high fields is a very good approximation to reality.

We will consider that the electric field is irrotational,

$$\nabla \times \boldsymbol{F} = 0\,. \tag{6.9}$$

Then, the field equals minus the gradient of the electric potential. It is convenient to eliminate the electron concentration in Eq. (6.6) by means of Eq. (6.7). The result is Ampere's law

$$\boldsymbol{j} = -\frac{\kappa\epsilon_0}{e}\frac{\partial \boldsymbol{F}}{\partial t} + \boldsymbol{I}\,, \quad \text{with} \tag{6.10}$$

$$\nabla \cdot \boldsymbol{j} = 0\,,$$

$$\boldsymbol{I} = \left(n_0 - \frac{\kappa\epsilon_0}{e}\nabla\cdot\boldsymbol{F}\right)\boldsymbol{\mathcal{V}}(\boldsymbol{F}) - n_0\nabla D(\boldsymbol{F}) + \frac{\kappa\epsilon_0}{e}\,\nabla\left[D(\boldsymbol{F})\nabla\cdot\boldsymbol{F}\right]. \tag{6.11}$$

Here, $e\boldsymbol{j}$ is the total (electronic plus displacement) current density.

6.2.3 Boundary Conditions

We should now provide the boundary conditions needed to obtain a mathematically well-posed problem for Eqs. (6.10) and (6.11). These conditions are the bias condition (for voltage bias, the electric potential is assumed zero at the cathode and $V > 0$ at the anode) and proper conditions in the contact regions. Models of varying degrees of complexity and accuracy have been studied in great detail – see, for example, [6]. Here, we focus on relatively simple models that are known to give good agreement with available data. Plausible boundary conditions are:

- Fixed electron density at the contact at $x \in \Sigma$: $n|_\Sigma = n_0$ (Ohmic contact). This is a Neumann-type condition for the electric field.
- Phenomenological relation for a $n^+ - n$ contact: The normal components of the diffusive and electronic current densities are proportional, so that $\hat{n}\cdot\nabla[D\,n(\boldsymbol{x},t)] = \zeta_c\,\hat{n}\cdot\boldsymbol{I}$ where $\hat{n}$ denotes the unit normal vector to the contact–sample interface surface and ζ_c is a phenomenological contact parameter which is generally field-dependent. This is a mixed boundary con-

dition for the electric field, a fact that is easily seen if the diffusion current density is written as $-\boldsymbol{I} + n\,\boldsymbol{\mathcal{V}}$ and Poisson's equation is used.

- Contact-field characteristics. It is assumed that there is a definite functional relation between the normal components of the electric field and the electronic current density. In the simplest case, this relation is linear, $\hat{n} \cdot \boldsymbol{F} = er_c\hat{n} \cdot \boldsymbol{I}$, and r_c is interpreted as the resistivity of the contact. This is a Dirichlet-type condition for the electric field and may be understood from the previous boundary condition as follows: we apply the previous condition to the n^+ side of the contact where n_0 is so high that $n \approx n_0$ and $\boldsymbol{\mathcal{V}} \approx -\boldsymbol{F}$.
- Ideal metal–semiconductor contact (Schottky barrier): $-\hat{n}\cdot\boldsymbol{I} = \mathcal{V}_c\, n|_\Sigma - I_c^m$, where $\mathcal{V}_c$ and I_c^m are the contact parameters introduced in [6]. This is a mixed boundary condition for the electric field. Depending on the parameters in the boundary condition and the voltage bias, different stationary or time-dependent attractors are stable solutions of the model [6, 7].

We shall mostly consider the linear contact-field characteristics in our analyses, as it is mathematically simple yet representative of real materials. Often, nonlinear contact-field characteristics can be derived for the other types of boundary conditions, and their study follows along the same lines as explained here.

6.2.4 Nondimensionalization

The first step in the study of nonlinear spatiotemporal structures should be to rewrite the governing equations and conditions in dimensionless form. This should be done so that a relevant small parameter can be easily spotted in the dimensionless equations.

Since our main interest is the Gunn effect, we should adopt F_M and $\mathcal{V}_M$ as the units of electric field and velocity, respectively. Clearly, the doping n_0 is a reasonable unit for the electron density, which varies on the order of n_0 during the Gunn oscillations. Now, we need units for length, time and current density. There are different candidates for the unit of length. One of these would be the sample length defined as l. However, we would like to have an *intrinsic* length scale so that we will not consider using the size of the semiconductor sample as a unit for the time being. The right-hand side of the Poisson equation (6.7) is of the order of $en_0/\kappa\epsilon_0$. In the left side, the variation of the electric field is of order F_M, so that the length scale over which this variation balances $en_0/\kappa\epsilon_0$ is

$$l_1 = \frac{\kappa\epsilon_0 F_M}{en_0}\,. \tag{6.12}$$

The time scale is obtained by imposing that the electron drift balances Maxwell's displacement current in Ampère's law. This is the same as balancing $\partial n/\partial t$ and the drift current in the charge continuity equation (6.6). The result is that the time scale is set by the time it takes an electron to move a distance l_1 at its characteristic velocity $\mathcal{V}_M$:

$$t_1 = \frac{l_1}{\mathcal{V}_M} = \frac{\kappa\epsilon_0 F_M}{e n_0 \mathcal{V}_M} . \tag{6.13}$$

Thus, we are led to the following definitions for the dimensionless variables:

$$\boldsymbol{E} \equiv -\frac{\boldsymbol{F}}{F_M}, \quad \boldsymbol{v}(\boldsymbol{E}) \equiv \frac{\boldsymbol{\mathcal{V}}(-F_M \boldsymbol{E})}{\mathcal{V}_M}, \quad \nu \equiv \frac{n}{n_0}, \quad \boldsymbol{X} \equiv \frac{\boldsymbol{x}}{l_1}, \quad \tau \equiv \frac{t}{t_1} . \tag{6.14}$$

We have defined the dimensionless electric field to be opposite to the standard dimensional electric field. This definition causes the dimensionless electron velocity to have the same direction as the dimensionless electric field. Then, the electrons and charge density waves will move in the direction of the field, that is, from left to right on a one-dimensional geometry. This sign convention is often used in the literature on the Gunn effect. Inserting Eq. (6.14) in Eqs. (6.10) and (6.11), we obtain

$$\boldsymbol{J} = \frac{\partial \boldsymbol{E}}{\partial \tau} + (1 + \nabla \cdot \boldsymbol{E})\ \boldsymbol{v}(\boldsymbol{E}) - \delta\ \nabla[\mathcal{D}(\boldsymbol{E})(1 + \nabla \cdot \boldsymbol{E})] , \tag{6.15}$$

$$\nabla \cdot \boldsymbol{J} = 0 , \tag{6.16}$$

where we have used the following definitions

$$\boldsymbol{J} \equiv -\frac{\boldsymbol{j}}{e n_0 \mathcal{V}_M}, \quad \mathcal{D}(\boldsymbol{E}) \equiv \frac{D(-F_M \boldsymbol{E})}{D_0}, \quad \delta \equiv \frac{D_0}{\mathcal{V}_M l_1} . \tag{6.17}$$

Here, $D_0 = D(F_M \boldsymbol{I}/|\boldsymbol{I}|)$ gives the order of magnitude of the diffusion coefficient at typical values of the electric field. It is interesting to note that $1/\delta$ is the product of the Reynolds number (velocity times length divided by viscosity) and the Schmidt number (viscosity divided by diffusivity), which is sometimes called the Bodenstein number in fluid mechanics [8]. For n-GaAs at room temperature, $\delta = 0.01$ is a small parameter, which justifies our choice of units.

Assuming linear current-field contact characteristics, the boundary conditions at the cathode and anode, respectively, are:

$$x \in \Sigma_c : \quad \boldsymbol{E} \cdot \hat{n} = \rho\,(\nu \boldsymbol{v} - \delta \nabla[\mathcal{D}\nu]) \cdot \hat{n} \quad \text{and} \quad \varphi = 0 , \tag{6.18}$$

$$x \in \Sigma_a : \quad \boldsymbol{E} \cdot \hat{n} = \rho\,(\nu \boldsymbol{v} - \delta \nabla[\mathcal{D}\nu]) \cdot \hat{n} \quad \text{and} \quad \varphi = \phi\, L , \tag{6.19}$$

where $\boldsymbol{E} = \nabla\varphi$, $L = l/l_1$ is the dimensionless distance between the contacts, and $\rho = r_c e n_0 \mathcal{V}_M / F_M$ is the dimensionless contact resistivity, and $\phi = V/(F_M l)$ is the dimensionless applied field. We have assumed that the resistivity of both contacts is the same in order to simplify the formulas. Instead of using a dimensionless voltage, it is convenient to adopt an average electric field, ϕ, as our bias parameter. The reason is that L is typically large (from 5 to 200 for n-GaAs), and we want to keep our control parameter ϕ of order one when describing the Gunn instability.

Notice that our nondimensionalization has left us with a system of equations and conditions depending only on three dimensionless parameters: ϕ (our control parameter for voltage bias conditions), L (the ratio of the distance between contacts to the dielectric length l_1), and δ (a reciprocal Bodenstein number; it may be also interpreted as the ratio between the length, $D_0/\mathcal{V}_M$, and l_1). Additional dimensionless parameters such as the aspect ratio may appear depending on the spatial geometry of the semiconductor. In a typical Gunn diode, ϕ is of order one, while $\delta \ll 1$ and $L \gg 1$. The crucial simplification that we will use in what follows is that $L \gg 1$. In dimensional units, $n_0 l \gg \kappa\epsilon_0 F_M/e$. As we shall see, in this limit, it is always possible to obtain the Gunn effect for ϕ larger than a critical value. Then, moving charge dipole waves are much smaller than the distance between contacts, which allows for a simple asymptotic description thereof.

For the rest of this section, we will consider a semiconductor sample with attached planar contacts, as shown in Figure 6.3. Then, $\nu = \nu(x,\tau)$, $\boldsymbol{E} = (E(x,\tau),0,0)$, $\varphi = \int_0^x E(s,\tau)\,ds$, $\boldsymbol{j} = (J(\tau),0,0)$, and the patterns depend on the coordinate x directed along the electron flux $\boldsymbol{J}$. Contacts are at $x = 0$ (a cathode) and $x = L$ (an anode), L is the sample length. The one-dimensional Kroemer model is now given by the following nondimensional equations:

$$\frac{\partial \nu}{\partial \tau} + \frac{\partial}{\partial x}\left(\nu v - \delta \frac{\partial}{\partial x}[\mathcal{D}\nu]\right) = 0\,, \tag{6.20}$$

$$\frac{\partial E}{\partial x} = \nu - 1\,, \tag{6.21}$$

or, equivalently, by Ampére's equation and the bias condition

$$\frac{\partial E}{\partial \tau} + v(E)\left(\frac{\partial E}{\partial x} + 1\right) - \delta\,\frac{\partial}{\partial x}\left[\mathcal{D}(E)\left(\frac{\partial E}{\partial x} + 1\right)\right] = J\,, \tag{6.22}$$

$$\frac{1}{L}\int_0^L E(x,\tau)\,dx = \phi\,, \tag{6.23}$$

plus the boundary conditions:

$$E = \rho\left(J - \frac{\partial E}{\partial \tau}\right) \quad \text{at} \quad x = 0, L\,. \tag{6.24}$$

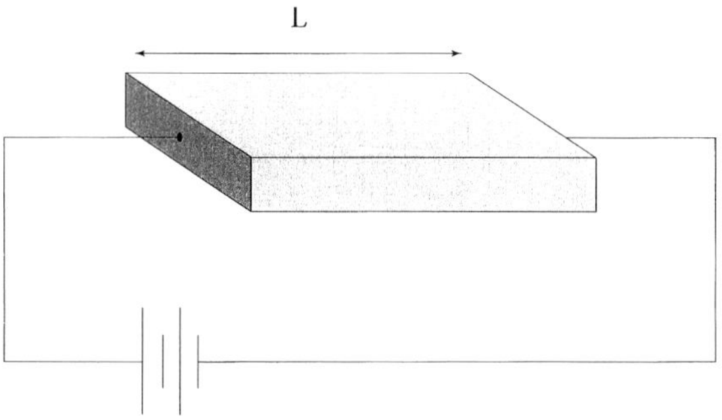

Fig. 6.3 Sample schematic showing a rectangular sample sandwiched between two planar contacts.

We shall assume that the straight line E/ρ intersects the drift velocity curve $v(E)$ on its second (decreasing) branch, at a field $E_c > 1$. This condition is necessary for our model to describe the Gunn effect. Thus, we need $\rho > 1$ or a sufficiently large resistivity of the contact, $r_c > F_M/(en_0 V_M)$ in dimensional units. To perform specific calculations, we need to specify the functions $v(E)$ and $\mathcal{D}(E)$. Since the results do not vary qualitatively so long as $v(E)$ is $\mathcal{N}$-shaped and $\mathcal{D}(E) > 0$, we shall assume that

$$\mathcal{D}(E) = 1\,, \quad \text{and therefore:}$$

$$\frac{\partial E}{\partial \tau} + v(E)\left(\frac{\partial E}{\partial x} + 1\right) - \delta\,\frac{\partial^2 E}{\partial x^2} = J\,. \tag{6.25}$$

In the general case $\mathcal{D}(E) > 0$, the convective term is $[v(E) + \delta\,\mathcal{D}'(E)]\frac{\partial E}{\partial x}$. This slightly complicates the formulas to be obtained and we omit the needed straightforward modifications. To perform actual calculations, we shall adopt the following $\mathcal{N}$-shaped dimensionless electron velocity with parameters B and K due to Kroemer [9]:

$$v(E) = \frac{v_0\,E}{E_0}\left[\frac{1 + B\left(\frac{E}{E_0}\right)^K}{1 + \left(\frac{E}{E_0}\right)^K}\right],$$

$$E_0 = \left[\frac{2B}{K - 1 - (K+1)B - R}\right]^{\frac{1}{K}}, \quad \frac{v_0}{E_0} = \frac{(K-1)(1-B) - R}{[(K+1)(1-B) - R]\,B},$$

$$R = \sqrt{[K - 1 - (K+1)B]^2 - 4B}\,, \quad 0 < B < \left(\frac{K-1}{K+1}\right)^2. \tag{6.26}$$

Another important case that may be considered is a velocity curve which saturates to a constant value B as the field increases past its maximum value. An example is

$$v(E) = \frac{v_0\,E}{E_0}\left[\frac{1 + B\left(\frac{E}{E_0}\right)^{K-1}}{1 + \left(\frac{E}{E_0}\right)^K}\right],$$

$$E_0 = \left[K(1-B) - 1\right]^{\frac{1}{K}},$$

$$v_0 = \frac{K(1-B)}{[K(1-B) - 1]^{1-\frac{1}{K}} + B}, \quad 0 \le B < 1 - \frac{1}{K}. \tag{6.27}$$

In both cases, $K > 1$, E_0, and v_0 have been chosen so that $v(E)$ has a local maximum $v(1) = 1$. Figure 6.2b shows a schematic graph of the saturating velocity-field characteristic.

6.3
Stationary Solutions and Their Linear Stability in the Limit $L \gg 1$

Stationary solutions are easy to construct in the asymptotic limit of samples which are much larger than the dielectric length, $n_0 l \gg \kappa\epsilon_0 F_M/e$, or $L \gg 1$ in dimensionless units. We shall also assume that the dimensionless diffusivity δ is zero. How one includes boundary layers at the anode and a description of the inner structure of shock waves will be explained later. In this limit, it is also possible to study the linear stability properties of stationary solutions with great generality. Then, the onset of the Gunn effect can be analyzed by means of amplitude equations describing the evolution of small disturbances about a linearly unstable stationary solution. This will be done in the next section. The discussion in this section and the next section follows that given in [11] and [12].

Let us assume that the 1D electric field, E, and the current density, J, are time-independent. Then, Eqs. (6.23)–(6.25) (with $\delta = 0$) become

$$\frac{dE}{dx} = \frac{J - v(E)}{v(E)}, \tag{6.28}$$

$$E(0) = \rho J, \tag{6.29}$$

$$\Phi(J) \equiv \frac{1}{L}\int_0^L E(x;J)\,dx = \phi\,. \tag{6.30}$$

Here, we have denoted by $E(x;J)$ the solution of Eqs. (6.28) and (6.29) corresponding to a given positive J. In the case of current bias, this current is specified a priori. In the case of voltage bias, the current J is determined as a function of the applied bias ϕ by Eq. (6.30). The qualitative behavior of $E(x;J)$ and of $\Phi(J)$ are seen by the analysis of the one-dimensional phase diagram of Eq. (6.28).

6.3.1
Stationary States and Their Linear Stability under *Current Bias*

We briefly digress to discuss the case of current bias dynamics. This case is interesting for academic reasons, but less so for practical reasons since in experimental systems it is almost always the voltage bias that is controlled rather than the current level. Nonetheless, the case of current-bias in the Gunn system has been studied by several groups. For technical details on the analysis, one may refer to [10]. Here, we summarize the key qualitative results. For sufficiently small or large currents, one finds a single stable state for the complete system. On the other hand, for current values that fall within the NDR range, there are three distinct spatially uniform solutions. The solutions on the lower and upper branches are stable for infinite samples, whereas the middle solution is unstable. It is possible to construct fronts between the upper and lower states and these are typically found to move in the direction of the current bias with a speed approximately equal to the drift velocity. Under current bias, these fronts are typically stable. The eigenvalue equation determining their linear stability can be transformed into a stationary Schrödinger equation by a change of variable and, by translation invariance of the front, it has a zero eigenvalue whose eigenfunction is simply the space derivative of the front. Since the front is monotonically increasing or decreasing, its space derivative does not vanish (except at infinity), and therefore it corresponds to the ground state of the Schrödinger equation. All other eigenvalues are then negative which implies linear stability of the front. An example of this procedure is worked out in Chapter 7 for a closely related example.

6.3.2
Construction of the Stationary Solution and of $\Phi(J)$ under Voltage Bias

In this subsection, we construct the spatially nonuniform stationary solutions, equivalent to the fixed point of the dynamical system Eqs. (6.22)–(6.24). Let us assume that the curve $v(E)$ is N-shaped and that $v_m < J < 1$ in our dimensionless units; see Figure 6.2a. Then $J - v(E)$ has three zeroes $E_1(J) < E_2(J) < E_3(J)$, which are the critical points of Eq. (6.28). Let us define $y = x/L$ and $\epsilon = 1/L$. Then, Eq. (6.28) may be rewritten as

$$\frac{dE}{dy} = \frac{J - v(E)}{\epsilon v(E)} . \tag{6.31}$$

If y is kept fixed and $\epsilon \to 0$, $J \approx v(E)$. Thus, $E(y; J)$ should be approximately constant or close to a piecewise constant function of y, approximately equal to one of the stable zeroes of $J - v(E)$: $E_1(J)$ or $E_3(J)$, except on transition layers whose width is of order 1, and therefore, it is small compared to L. Which zero is selected depends on the boundary condition Eq. (6.29). Near the cathode at $y = 0$, E varies according to Eq. (6.28) on the spatial scale x. Provided the contact resistivity is large, if the field at the cathode, $E(0; J) = \rho J$, is to the left (resp. right) of $E_2(J)$, then $E(x; J)$ tends to $E_1(J)$ [resp. $E_3(J)$] as x increases. Of particular interest is the critical value $J = J_c$ for which $E(0; J)$ coincides with $E_2(J)$:

$$E_2(J_c) = \rho J_c . \tag{6.32}$$

This relation holds only for resistivities $\rho \in (1, E_m/v_m)$, which we assume here, as explained in Section 6.1. For $0 < J < J_c$, $E \to E_1(J)$ as $x \gg 1$ whereas, if $J > J_c$, then $E \to E_3(J)$ as $x \gg 1$:

1. For $0 < J < J_c$, the stationary solution is $E(y; J) = E_1(J)$. At $y = 0$ there is a boundary layer obeying Eq. (6.28), with the boundary condition Eq. (6.29) at $x = 0$ and $E \to E_1(J)$ as $x \to \infty$. See Figure 6.4b.
2. For $J > J_c$, the stationary solution is $E(y; J) = E_3(J)$. At $y = 0$, there is a boundary layer obeying Eq. (6.28), with the boundary condition Eq. (6.29) at $x = 0$ and $E \to E_3(J)$ as $x \to \infty$.

In these two cases, we calculate the bias by means of Eq. (6.30), $\Phi(J) = E_i(J) + O(\epsilon) \sim \phi$, $i = 1, 3$. Then, we have

$$J = v(\phi) + O(\epsilon) .$$

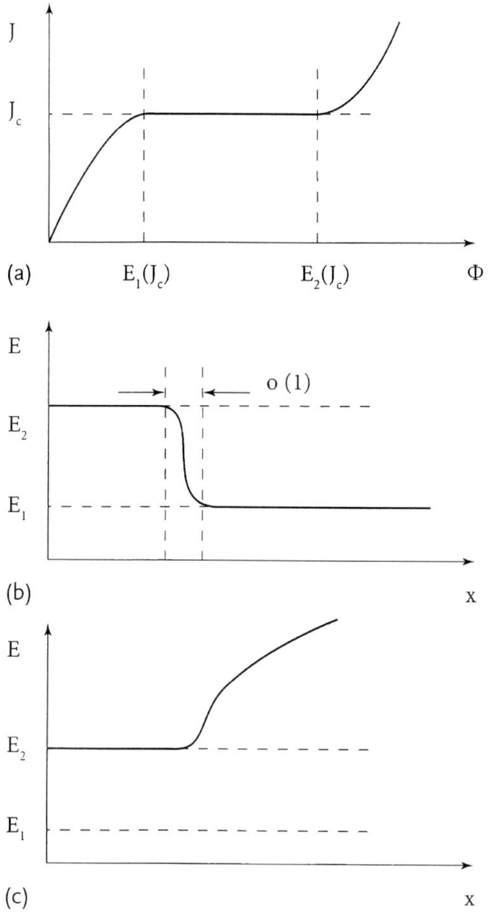

Fig. 6.4 (a) Spatial profile of total current density vs. average electric field (voltage divided by length) when $0 < J < J_c$ and the v–E curve has a three-branch form. For $E_1(J_c) < \Phi < E_3(J_c)$, J is almost flat, $J \approx J_c$. (b) Spatial profile of electric field vs. position when $0 < J < J_c$ and the v–E curve has a three-branch form. (c) Spatial profile of electric field vs. position when $0 < J < J_c$ and the v–E curve has a two-branch form.

This means that the characteristic current–voltage curve follows the first branch of the velocity curve $v(E)$ if $0 < \phi < E_1(J_c)$ and $0 < J < J_c$, it is flat with $J = J_c$ and $E_1(J_c) < \phi < E_3(J_c)$, and it follows the third branch of $v(E)$ if $\phi > E_3(J_c)$ and $J > J_c$. If we use the bias as our control parameter, then we see that $J \sim v(\phi)$ outside the interval $E_1(J_c) < \phi < E_3(J_c)$. This overall behavior is shown in Figure 6.4a.

If $E_1(J_c) < \phi < E_3(J_c)$, $E(y; J)$ should be E_1 or E_3 only on part of the sample $0 < y < 1$. This can be achieved if J is very close to J_c so that $E \approx E_2(J_c)$ close to the cathode $y = 0$. In particular, we have

- If $E_1(J_c) < \phi < E_2(J_c)$,

$$E(y;J) \sim E_2(J_c)\theta\left(\frac{\phi - E_1(J_c)}{E_2(J_c) - E_1(J_c)} - y\right) + E_1(J_c)\theta\left(y - \frac{\phi - E_1(J_c)}{E_2(J_c) - E_1(J_c)}\right).$$

- If $E_2(J_c) < \phi < E_3(J_c)$,

$$E(y;J) \sim E_2(J_c)\theta\left(\frac{E_3(J_c) - \phi}{E_3(J_c) - E_2(J_c)} - y\right) + E_3(J_c)\theta\left(y - \frac{E_3(J_c) - \phi}{E_2(J_c) - E_2(J_c)}\right).$$

$\theta(x)$ is the Heaviside step function, 1 if $x > 0$ and 0 if $x < 0$. The discontinuities at $y = 0$ and inside the sample are resolved by inserting appropriate boundary layers. These field profiles have been depicted in Figure 6.4b and c and a more precise characterization (including boundary layers) is given in Lemma 6.1.

Lemma 6.1

Let $E_1(J_c) < \phi < E_2(J_c)$ and fix the value of E at the transition point X such that $E(X;J) = 1$. Then, provided that $\{X, (L - X)\} \gg 1$, for each value of $X \in (0, L)$, ϕ and J are uniquely determined by the asymptotic formulas:

$$\phi = E_1 + (E_2 - E_1)\frac{X}{L} + \left(\frac{L - X}{v_1' L} + \frac{X}{v_2' L}\right)(J - J_c)$$
$$+ \frac{1}{L}\int_{E_2}^{E_0} \frac{(s - E_2)\,v(s)}{J_c - v(s)}\,ds + \frac{1}{L}\int_{E_0}^{E_1} \frac{(s - E_1)\,v(s)}{J_c - v(s)}\,ds + o\left(\frac{1}{L}\right), \qquad (6.33)$$

with $E_0 = 1$ and

$$J - J_c \sim -\frac{c_L \mid v_2' \mid}{1 - \rho v_2'} \exp\left\{-\frac{\mid v_2' \mid X}{J_c}\right\}. \qquad (6.34)$$

Also, the position-dependent electric field can be written as

$$E(x;J) \sim E_2 - c_L \exp\left\{-\frac{v_2'(x - X)}{J}\right\} \quad (\text{as} \quad (x - X) \to -\infty), \qquad (6.35)$$

$$c_L = (E_2 - E_0)\exp\left\{\int_{E_0}^{E_2}\left[\frac{1}{s - E_2} - \frac{v_2'}{v(s) - J} - \frac{v_2'}{J}\right] ds\right\}, \qquad (6.36)$$

and

$$E(x;J) \sim E_1 - c_R \exp\left\{-\frac{v_1'(x - X)}{J}\right\} \quad (\text{as} \quad (x - X) \to +\infty), \qquad (6.37)$$

$$c_R = (E_0 - E_1)\exp\left\{-\int_{E_1}^{E_0}\left[\frac{1}{s - E_1} - \frac{v_1'}{v(s) - J} - \frac{v_1'}{J}\right] ds\right\}. \qquad (6.38)$$

In the right-hand sides of Eqs. (6.35) and (6.37), $v_k' \equiv v'(E_k(J))$, $k = 1,2$ and $E_0 = 1$. $E(x; J)$ in these equations refers to the field in the transition layer $(x - X) = O(1)$: $(x - X) \to -\infty$ (resp. $\to +\infty$) means leaving the transition layer towards the left (resp. right). Notice that c_L and c_R are both positive.

Another part of the lemma pertains to the case where $E_2(J_c) < \phi/L < E_3(J_c)$. Then, let us fix $E(X; J) = E_0$, where E_0 is a given number in the interval $E_2(J_c) < E_0 < E_3(J_c)$. Then, for each value of $X \in (0, L)$, such that $\{X, (L - X)\} \gg 1$, ϕ and J are uniquely determined by the asymptotic formulas (6.33)–(6.38) where $E_3(J)$ replaces $E_1(J)$ everywhere. Notice that c_L and c_R are now both negative. •

The proof of the lemma is elementary and it follows directly from Eq. (6.31); see [11].

6.3.3 Linear Stability of the Stationary Solution under Voltage Bias

To analyze the linear stability of the steady state, we study the evolution of a small disturbance about the steady state of the form

$$\begin{aligned} J(\tau) &= J + \varepsilon\, \tilde{j}(\tau)\,, \\ E(x,\tau) &= E(x) + \varepsilon\, \tilde{e}(x,\tau)\,, \quad 0 < \varepsilon \ll 1\,, \end{aligned} \tag{6.39}$$

as $\tau \to +\infty$. Inserting Eq. (6.39) into Eqs. (6.23)–(6.25) yields

$$\mathcal{L}\tilde{e} - \tilde{j} = 0\,. \tag{6.40}$$

$$\tilde{e}(0,\tau) + \rho\left(\frac{\partial \tilde{e}(0,\tau)}{\partial \tau} - \tilde{j}(\tau)\right) = 0\,, \quad \tau > 0\,, \tag{6.41}$$

and

$$\int_0^L \tilde{e}(x,\tau)\, dx = 0\,. \tag{6.42}$$

In Eq. (6.40), $\mathcal{L}$ is the linear operator

$$\mathcal{L}\tilde{e} \equiv \frac{\partial \tilde{e}}{\partial \tau} + \frac{\partial [v(E)\tilde{e}]}{\partial x} + v'(E)\tilde{e}\,. \tag{6.43}$$

Equations (6.40)–(6.42) can be solved by separation of variables:

$$\tilde{j}(\tau) = \hat{j}\, e^{\lambda\tau}\,, \quad \tilde{e}(x,\tau) = \hat{e}(x;\lambda)\, e^{\lambda\tau}\,. \tag{6.44}$$

Insertion of Eq. (6.44) into Eqs. (6.40)–(6.42) yields

$$\frac{\partial[v(E)\,\hat{e}]}{\partial x} + [v'(E) + \lambda]\,\hat{e} = \hat{j}\,, \tag{6.45}$$

$$Z(\lambda) \equiv \frac{1}{L}\int_0^L \frac{\hat{e}(x;\lambda)}{\hat{j}}\,dx = 0\,, \tag{6.46}$$

$$\hat{e}(0;\lambda) = \frac{\rho\hat{j}}{1+\rho\lambda}\,. \tag{6.47}$$

It is important to note that we have written Eq. (6.46) in terms of the differential impedance of the system. As is well known from circuit analysis, the zero of the impedance $Z(\lambda)$ with largest the real part determines the linear stability of the steady state. In particular, if the zero with the largest real part has a positive real part, the state is linearly unstable. We have numerically evaluated the neutral stability curve (corresponding to the zero with the largest real part being purely imaginary) for the steady state in the parameter space ϕ *vs* L for different values of the resistivity ρ. The results are shown in Figure 6.5. The discontinuities in the slope of the in Figure 6.5 are due to the crossing of different zeroes as L increases. Notice that above a certain $L = L_m$, there are two values of the voltage for each L, ϕ_α and ϕ_ω, such that the steady state is linearly stable for ϕ outside $(\phi_\alpha, \phi_\omega)$. At the voltages ϕ_α and ϕ_ω, we expect that time-periodic solutions bifurcate from the steady state. We shall calculate these bifurcating branches of periodic solutions in Section 6.4 for semiconductors of finite length. In the limit $L \gg 1$, many modes become unstable almost simultaneously, as we shall see below, and a different calculation of the bifurcating branches is needed; see Section 6.4.

Remark 6.1

Notice in Figure 6.5 that ϕ_α rapidly tends to a constant value $E_1(J_c)$ as L increases. This suggests that the results we will obtain in the asymptotic limit $L \to \infty$ may be of practical applicability even for moderate L near $\phi = \phi_\alpha$.

Remark 6.2

Restoring dimensional units, we notice that a necessary condition for the stationary solution to become unstable is

$$\frac{e n_0 l}{\kappa\epsilon_0 F_M} > L_m\,, \quad \text{that is,} \quad n_0 l > 4.6\times 10^{10}\,\text{cm}^{-2}\,,$$

for n-GaAs. This inequality is the so-called NL stability criterion. Lower bounds for $n_0 l$ (equivalently for L_m) can be found rigorously or by the heuristic arguments first introduced by Kroemer [2]. We do not delve further into these matters due to their predominantly technical character.

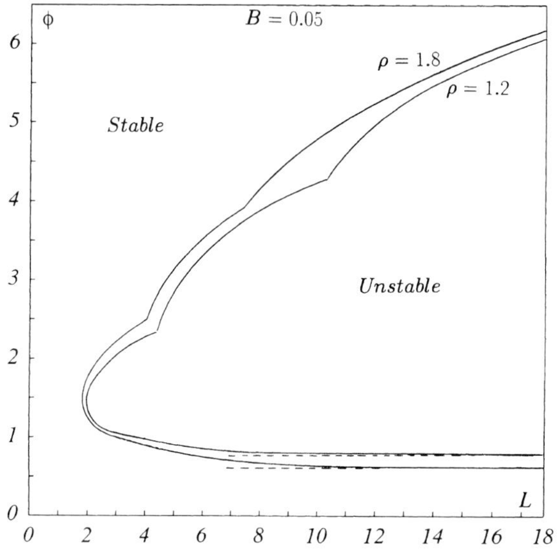

Fig. 6.5 Phase diagram of bifurcations for Gunn effect: Nondimensional voltage divided by nondimensional sample length L (i.e., nondimensional average field $\Phi(J) = \phi$) vs. L (from [11]).

For long semiconductors, the linear stability of the steady state can be ascertained without resorting to numerical calculations. The main result is the following. Let us consider a stationary state with a step-like electric field profile as in Lemma 6.1, with bias $\phi \sim E_1(J_c)$. If the bias is such that the electric field takes values on the second (decreasing) branch of $v(E)$ for only a small fraction $O(\ln L/L)$ of its length, then the stationary state may become unstable. More precisely, we shall prove:

Lemma 6.2

Asymptotically as $L \to \infty$, the steady state is linearly unstable for voltages $\phi \in (\phi_\alpha, \phi_\omega)$, where:
(a) ϕ_α is given by Eq. (6.33) with $X = X_c$,

$$X_c \sim \frac{J}{|\,v_2'\,|} \ln \left\{ \frac{c_L\, v_2'^2\, L}{J v_1'(1-\rho\, v_2')(E_2 - E_1)} \right\} = \frac{J}{|\,v_2'\,|} \ln L + O(1)\,. \tag{6.48}$$

Consider the case of voltages slightly over ϕ_α, $\phi = \phi_\alpha + \delta\phi$, such that $\delta\phi$ satisfies the inequality:

$$\frac{1}{(\ln L)^2} \ll \delta\phi L \ll 1 \quad (\text{as} \quad \ln L \to \infty)\,, \tag{6.49}$$

and $X = X_c + \delta X$ (with $\delta X/L \sim \delta\phi/[2(E_2 - E_1)] > 0$). Then, many eigenvalues of the form $\lambda = i\Omega_n + \alpha_n + i\omega_n$, $(\alpha_n, \omega_n) \ll \Omega_n \ll 1$, become unstable nearly simultaneously, where

$$\Omega_n \sim \frac{n\,J\,\pi}{X_c}, \quad n = \pm 1, \pm 3, \pm 5, \ldots, O(X_c\sqrt{\delta X}), \tag{6.50}$$

$$\alpha_n \sim \frac{|\,v_2'\,|\;\delta X}{X_c} - \frac{J\left(f_2 - \frac{f_1^2}{2}\right)\Omega_n^2}{X_c}, \tag{6.51}$$

$$\omega_n \sim \frac{J\,f_1\,\Omega_n}{X_c}. \tag{6.52}$$

Here, the f_k's are defined as:

$$f_1 = -\rho - \frac{v_1' - v_2'}{v_1'\,v_2'} - \phi_1, \tag{6.53}$$

$$f_2 = (\rho + \phi_1)\left(\rho + \frac{v_1' - v_2'}{v_1'\,v_2'}\right) + \frac{v_1' - v_2'}{v_1'\,v_2'} - v_1' - \phi_2, \tag{6.54}$$

where

$$\phi_k = \frac{(-1)^{k+1}}{k!(E_2 - E_1)} \int_{-\infty}^{+\infty} \frac{\partial E(x;J)}{\partial x}\left(\frac{x - X + E(x;J) - E_2}{J}\right)^k dx, \; k = 1, 2, \cdots \tag{6.55}$$

(b) ϕ_ω is given by the formulas (6.33) and (6.48) with E_3 replacing E_1. With this replacement, formulas (6.50)–(6.55) hold for voltages $\phi = \phi_\omega - \delta\phi$, $X = X_c + \delta X$ $(\delta X/L \sim \delta\phi/[2(E_3 - E_2)] > 0)$. •

Note that the α_n and ω_n are unrelated to the subscripts in ϕ_α and ϕ_ω.

Proof [11]. As said in Lemma 6.1, the steady state differs appreciably from the piecewise constant profile $E = E_2\,\theta(X - x) + E_1\,\theta(x - X)$ only in a transition layer of $O(1)$ width centered at $x = X$. A narrower boundary layer of width $O(\delta)$ is located at $x = L$ and will be ignored. Near the critical biases ϕ_α and ϕ_ω, we have $1 \ll X \ll L$. Noting that $\partial E/\partial x$ solves Eq. (6.45) with $\hat{j} = \lambda = 0$, it is convenient to write the solution of Eqs. (6.45)–(6.47) in the following form:

$$\hat{e}(x;\lambda) = \hat{e}_p(x;\lambda) + h(x;\lambda)\,\frac{\partial E(x;J)}{\partial x}, \tag{6.56}$$

where $\hat{e}_p(x;\lambda)$ obeys Eq. (6.45) with the natural boundary condition

$$\hat{e}_p(0;\lambda) = \frac{\hat{j}}{v_2' + \lambda} \sim \hat{e}_p(X - \Delta x;\lambda) \quad (\text{as} \quad \Delta x \to \infty). \tag{6.57}$$

Any solution of Eq. (6.45), $\hat{e}_p(x;\lambda)$ in particular, satisfies

$$\hat{e}_p(X + \Delta x;\lambda) \sim \frac{\hat{j}}{v_1' + \lambda} \quad (\text{as} \quad \Delta x \to \infty). \tag{6.58}$$

The second term in Eq. (6.56) solves Eq. (6.45) with $\hat{j} = 0$ and an appropriately modified boundary condition. It is given by

$$h(x;\lambda) = h(X;\lambda)\,\exp\left[-\lambda\,\frac{x - X + E(x;J) - E_0}{J}\right],$$
$$h(X;\lambda) \sim -\frac{(1-\rho\,v_2')\,J\,\hat{j}}{(1+\rho\lambda)\,(v_2'+\lambda)\,c_L\,v_2'}\,\exp\left[-\frac{(v_2'+\lambda)\,X + (E_0 - E_2)\lambda}{J}\right]. \tag{6.59}$$

Here, E_0 is as defined in Lemma 6.1. As $(x - X) \to \infty$, the second term in Eq. (6.56) becomes exponentially small, so that $\hat{e} \sim \hat{e}_p$ in $(X + \Delta x, L]$, and

$$\int_{X+\Delta x}^{L} \hat{e}\,dx \sim \frac{(L - X - \Delta x)\,\hat{j}}{v_1' + \lambda} \sim \frac{L\,\hat{j}}{v_1' + \lambda}\,. \tag{6.60}$$

For λ to be a zero of $Z(\lambda)$ in Eq. (6.46), the integral of $\hat{e}$ on $[0, X + \Delta x)$ has to be of the same order as Eq. (6.60). The contribution of $\hat{e}_p$ to this integral is, according to Eqs. (6.57) and (6.58), $O(X)$, much smaller than needed. Thus, it is the contribution coming from the second term of Eq. (6.56) that should balance Eq. (6.60). This yields:

$$Z(\lambda) \sim \frac{L}{v_1'+\lambda} - \frac{(1-\rho\,v_2')\,J\,K(\lambda)}{(1+\rho\lambda)\,(v_2'+\lambda)\,c_L\,v_2'}\,\exp\left[-\frac{(v_2'+\lambda)\,X}{J}\right], \tag{6.61}$$

$$K(\lambda) = \int_0^{X+\Delta x} \frac{\partial E}{\partial x}\,\exp\left[-\lambda\,\frac{x - X + E(x;J) - E_2}{J}\right] dx$$
$$= (E_2 - E_1)\,(1 + \phi_1\,\lambda + \phi_2\,\lambda^2 + \ldots)\,, \tag{6.62}$$

where in the last equality in Eq. (6.62), we have used the definitions from Eq. (6.55).

By means of Eqs. (6.61) and (6.62), we can rewrite the impedance condition $Z(\lambda) = 0$ as follows:

$$\frac{(E_2 - E_1)\,(1-\rho v_2')\,J\,v_1'}{c_L\,v_2'^2}\,F(\lambda)\,\exp\left\{-\frac{(v_2'+\lambda)\,X}{J} - \ln L\right\} \sim -1\,,$$
$$F(\lambda) = 1 + f_1\,\lambda + f_2\,\lambda^2 + \ldots\,, \tag{6.63}$$

where f_1 and f_2 are given by Eqs. (6.53) and (6.54), respectively. Equation (6.61) holds for $\mathrm{Re}\,\lambda + v_2' < 0$ (the impedance $Z(\lambda)$ does not have zeros with $\mathrm{Re}\,\lambda + v_2' > 0$, for otherwise, the contribution of the high- field region $0 < x < X + \Delta x$ to the impedance is, at most, $X/(v_2' + \lambda)$, which cannot cancel the term of order L/λ due to the low-field region in the rest of the sample for $X < X_c$) and then the integrand in Eq. (6.62) tends exponentially to zero as $|x - X| \to \infty$. This means that we can replace the endpoints of the integral Eq. (6.62) by $\pm\infty$. At the instability threshold, $\lambda = i\Omega$, and $|K(i\Omega)| < |K(0)| = E_2 - E_1$ for nonzero real Ω. Thus, $K(i\Omega)$ is of order 1 and Eq. (6.61) implies that $\mathrm{Re}\,\lambda = 0$ at $X \sim X_c(\Omega)$, where $X_c(\Omega)$ is given by

$$X_c(\Omega) = \frac{J}{|\,v_2'\,|} \ln \left\{ \frac{c_L L \,|\, v_2'(v_2' + i\Omega)(1 + i\Omega)\,|}{J\,(1 - \rho\, v_2')\,|\,(v_1' + i\Omega) K(i\Omega)\,|} \right\}, \quad \text{and} \tag{6.64}$$

$$\frac{\Omega\, X_c(\Omega)}{J} \sim 2n\pi + \arg\left[\frac{(v_1' + i\Omega) K(i\Omega)}{(|v_2'| - i\Omega)(1 + i\Omega)} \right], \quad n = 0, \pm 1, \pm 2, \ldots \tag{6.65}$$

Here, $X_c(\Omega)$ in Eq. (6.64) is an increasing function of Ω. For ϕ close to the threshold value ϕ_α, the voltage is an increasing function of X and therefore ϕ_α corresponds to $X_c(\Omega)$ with the lowest possible value, $\Omega = 0$. Since $K(0) = E_1 - E_2$, we obtain $X_c = X_c(0)$ given by Eq. (6.48). Similarly, Eq. (6.65) yields Eq. (6.50) when we set $\Omega = 0$ in its right-hand side. Analogous formulas hold for ϕ_ω with E_3 instead of E_1.

To obtain Eqs. (6.51) and (6.52), we first set $x = 0$ in Eq. (6.35), so that the boundary condition yields

$$\rho J \sim E_2(J) - c_L e^{v_2' X/J}\,.$$

Therefore, a small variation of X, $X = X_c + \delta X$, $\delta X \ll 1$, provokes a much smaller variation of J:

$$\delta J = -\frac{c_L v_2'}{[\rho - E_2'(J)]J} \exp\left(\frac{v_2' X_c}{J}\right) \delta X \sim \frac{v_1'(E_2 - E_1)}{L}\,\delta X\,, \tag{6.66}$$

where we have used Eq. (6.48) and $E_2'(J) = 1/v_2'$. We have $\delta J \ll X_c \delta J \ll \delta X$, so that the variations of J due to δX are negligible compared to δX itself. When $\phi = \phi_\alpha + \delta\phi$, $\delta\phi \ll 1$, we have $\lambda = i\Omega_n + \alpha_n + i\omega_n$, with $(\alpha_n, \omega_n) \ll \Omega_n \ll 1$, where Ω_n is given by Eq. (6.50) and α_n and ω_n vanish if $\delta\phi = 0$. We now substitute $\lambda = i\Omega_n + \alpha_n + i\omega_n$ and Eq. (6.50) in Eq. (6.63) with $X = X_c + \delta X$, use Eq. (6.48), and finally equate real and imaginary parts of the resulting expression. We obtain Eqs. (6.51) and (6.52).

□

Discussion. What is the meaning of the formulas compiled in Lemma 2? If we have $\phi = \phi_\alpha + \delta\phi$, $\delta\phi \ll 1$, from the bias condition,

$$\phi = E_1 + (E_2 - E_1)\frac{X}{L} + \frac{1}{L}\int_{X-\Delta x}^{X} [E(x) - E_2]dx + \frac{1}{L}\int_{X}^{X+\Delta x} [E(x) - E_1]dx + o(1),$$

we obtain the relation between δX and $\delta\phi$:

$$\delta\phi \sim E_1' \delta J + (E_2 - E_1)\frac{\delta X}{L} \sim 2(E_2 - E_1)\frac{\delta X}{L}\,. \tag{6.67}$$

Similarly, we find $\phi = \phi_\omega - \delta\phi$, with $\delta\phi \sim 2(E_3 - E_2)\delta X/L$. The separation between two consecutive eigenvalues given by Eq. (6.51) is

$$\alpha_{2m+1} - \alpha_{2m-1} \sim \frac{8m\pi^2 J^3 \left(f_2 - \frac{f_1^2}{2}\right)}{X_c^3} = m\, O\left(\frac{1}{(\ln L)^3}\right) \ll \frac{1}{(\ln L)^2}, \tag{6.68}$$

and it vanishes faster than $(\ln L)^{-2}$. Thus, for a bias variation $\delta\phi$ satisfying Eq. (6.49), Eq. (6.51) indicates that many oscillatory modes N of vanishing frequency Eq. (6.50) become unstable, with

$$N = O(\sqrt{L\,\delta\phi}\,\ln L)\,, \tag{6.69}$$

obtained by setting $\alpha_N = 0$. In other words, for $\phi = \phi_\alpha + \delta\phi$ (resp. $\phi = \phi_\omega - \delta\phi$) with $\delta\phi$ in the range Eq. (6.49), a quasicontinuum of eigenvalues with vanishing frequencies Eq. (6.50) crosses the imaginary axis. The resulting bifurcations are composites of quasicontinuum sets of Hopf bifurcations which will be analyzed in Section 6.4.

Remark 6.3

From Eqs. (6.39), (6.44), (6.48)–(6.50), (6.56)–(6.59), we find that the eigenfunction corresponding to the eigenvalue with zero real part obeys:

$$\begin{aligned}\frac{\tilde{e}(x,\tau)}{\hat{j}} &= \frac{\tilde{e}(x; i\Omega_n)}{\hat{j}} e^{i\Omega_n \tau} \\ &\sim \frac{L}{v_1'(E_2 - E_1)} \frac{\partial E}{\partial x} \exp\left[i\Omega_n\left(\tau - \frac{x - X + E(x;J) - E_2}{J}\right)\right]\end{aligned} \tag{6.70}$$

on $0 < x < X_c + \Delta x$, and

$$\frac{\tilde{e}(x,\tau)}{\hat{j}} = \frac{\hat{e}(x; i\Omega_n)}{\hat{j}} e^{i\Omega_n \tau} \sim \frac{e^{i\Omega_n \tau}}{v_1' + i\Omega_n}, \quad \text{on} \quad X_c + \Delta x < x < L\,. \tag{6.71}$$

Here, $1 \ll \Delta x \ll X_c = O(\ln L)$, and $X = X_c + \delta X$ such that $\operatorname{Re}\lambda = 0$, $\lambda \sim i\Omega_n$. Equation (6.70) indicates that the disturbance of the electric field represents a wave that travels with speed J to the right while its amplitude grows exponentially with x [recall Eq. (6.35)].

Remark 6.4

Equation (6.51) is a relation between the growth rate due to the increment of the bifurcation parameter measured from its critical value, δX, and the square of the frequency (which is the imaginary part of the eigenvalue). This equation relates the Fourier coefficients of the terms in a diffusion equation where the "spatial" variable corresponds to a slow time scale $\chi = \tau\sqrt{\delta X}$ and the "time" variable corresponds to a slower time scale $\sigma = (\delta X/X_c)\,\tau$. (There is an authentic spatial scale associated with χ because the field disturbance is a wave traveling with speed J on the interval $0 < x < X_c + \Delta x$. See the interpretation of the amplitude equation in Section 6.4). Then, the coefficient of $-\Omega_n^2$ plays the role of an effective diffusivity and it appears as such in the nonlinear amplitude equation that we will derive in Section 6.4. For values of the resistivity in the interval $\rho \in (E_M/v_M, E_m/v_m)$, and typical values of B, the effective diffusivity is always positive. Consistency requires $\alpha_n \ll \Omega_n$, which implies $\Omega_n \ll X_c = O(\ln L)$, as stated in Lemma 6.2.

6.4 Onset of the Gunn Effect

This section follows [12].

6.4.1 The Linear Inhomogeneous Problem and Secular Terms

Here, we carry out a perturbation calculation of the oscillatory branch that bifurcates from the steady state both for finite L and for $L \to \infty$. Then, it is important to know under which conditions the solution of the linear nonhomogeneous problem associated to Eqs. (6.40)–(6.42) (at the critical voltage) is bounded and periodic in time. In other words, we will find the precise conditions under which no secular terms (i. e., terms that become unbounded in time) are present in our calculation. In this section, we consider the simpler case of finite L where only two complex conjugate zeros of the impedance Eq. (6.46), $\lambda = \pm i\,\omega$, cross the imaginary axis at the critical voltage. The linear nonhomogeneous problem is

$$\mathcal{L}\tilde{e} - \tilde{j} = f(x)\,e^{i\,\omega\,\tau}\,,$$

$$\tilde{e}(0,\tau) + \rho\left[\frac{\partial \tilde{e}(0,\tau)}{\partial \tau} - \tilde{j}(\tau)\right] = g\,e^{i\,\omega\,\tau}\,,$$

$$\int_0^L \tilde{e}(x,\tau)\,dx = 0\,, \tag{6.72}$$

where $f(x)\,e^{i\,\omega\,\tau}$ and $g\,e^{i\,\omega\,\tau}$ are source terms.

In Eq. (6.72), the operator $\mathcal{L}$ is given by Eq. (6.43) and the coefficients are calculated for $E = E(x; J)$, the steady state at the critical voltage. We now find the conditions that ensure absence of secular terms. The approach here is a classic application of the method of multiple scales discussed in a previous chapter. Insertion of the ansatz

$$\tilde{j}(\tau) = \hat{j}\,e^{i\,\omega\,\tau}\,, \quad \tilde{e}(x,\tau) = \hat{e}(x)\,e^{i\,\omega\,\tau}\,, \tag{6.73}$$

in Eq. (6.72) yields,

$$\frac{\partial[v(E)\hat{e}]}{\partial x} + [i\omega + v'(E)]\hat{e} = \hat{j} + f(x)\,,$$
$$\hat{e}(0) = \frac{\rho\,\hat{j} + g}{1 + i\rho\,\omega}\,,$$
$$\int_0^L \hat{e}(x)\,dx = 0\,. \tag{6.74}$$

We now solve the first two equations and insert the solution in the third one. The result is that a sum of three terms is zero. The first term, proportional to $\hat{j}$, is zero because it is the solution of the homogeneous problem Eq. (6.72) with $f \equiv 0$ and $g \equiv 0$. Then, the sum of the remaining terms must also be zero for a bounded periodic solution of the type Eq. (6.73) to be possible:

$$\int_0^L \frac{dx}{v(E(x))} \left\{ \frac{g\,v(\rho J)}{1 + i\rho\omega} \exp\left[-\int_0^x \frac{i\omega + v'(E(z))}{v(E(z))}\,dz \right] + \int_0^x f(y) \exp\left[-\int_y^x \frac{i\omega + v'(E(z))}{v(E(z))}\,dz \right] dy \right\} = 0\,. \tag{6.75}$$

If the left side of Eq. (6.75) is not zero, the solution of Eq. (6.72) has to be proportional to $\tau\, e^{i\,\omega\,\tau}$, thereby yielding secular terms (unbounded as $\tau \to \infty$) in the perturbation theory of which the linear problem Eq. (6.72) is a part. We will call Eq. (6.75) the nonresonance condition, and we will use it extensively in the bifurcation calculations that follow.

6.4.2
Hopf Bifurcation

Let $E = E(x;\phi)$ be the steady state corresponding to a bias ϕ and $J = J(\phi)$ the corresponding current. Let $E_0(x)$ and J_c be the corresponding field and current at the critical bias ϕ_c where the steady state ceases to be linearly stable (ϕ_c is either ϕ_α or ϕ_ω of Section 6.3). We want to construct the time periodic solutions that bifurcate from $E(x)$ at $\phi = \phi_c$. Let us define the small parameter ϵ such that $\epsilon^2\varphi$ denotes the deviation from the critical bias ϕ_c:

$$\phi = \phi_c + \epsilon^2\,\varphi,\quad \epsilon \ll 1\,, \tag{6.76}$$

(the sign of $\varphi = \pm 1$ will be determined later). Then, the corresponding stationary field and current are

$$E(x;\epsilon) = E_0(x) + \epsilon^2\,\varphi\,E_2(x) + O(\epsilon^4)\,,$$
$$J = J_0 + \epsilon^2\,\varphi\,J_2 + O(\epsilon^4)\,. \tag{6.77}$$

Clearly, $E_2(x) = \partial E(x;\phi_c)/\partial\phi$, $J_2 = dJ(\phi_c)/d\phi$. To calculate the bifurcating oscillatory solution, we shall assume the usual Hopf multiscale ansatz:

$$\begin{aligned} E(x,\tau;\epsilon) &= E_0(x) + \epsilon\, E^{(1)}(x,\tau,T) + \epsilon^2[E^{(2)}(x,\tau,T) + \varphi E_2(x)] \\ &\quad + \epsilon^3 E^{(3)}(x,\tau,T) + O(\epsilon^4)\,, \\ J(\tau;\epsilon) &= J_0 + \epsilon\, J^{(1)}(\tau,T) + \epsilon^2[J^{(2)}(\tau,T) + \varphi J_2] + \epsilon^3 J^{(3)}(\tau,T) + O(\epsilon^4)\,; \\ \tau &= \tau,\ T = \epsilon^2\tau\,. \end{aligned} \tag{6.78}$$

The slow time scale $T = \epsilon^2\tau$ is chosen so as to eliminate the secular terms that first appear in the equation for $E^{(3)}$ (see below). By inserting Eq. (6.78) into Eqs. (6.25) (with $\delta = 0$), (6.23) and (6.24) and equating like powers of ϵ, we obtain the following hierarchy of equations for the $E^{(k)}$ and $J^{(k)}$'s:

$$\mathcal{L}_c E^{(1)} - J^{(1)} = 0\,, \tag{6.79}$$

$$E^{(1)}(0,\tau,T) + \rho\left[\frac{\partial E^{(1)}(0,\tau,T)}{\partial\tau} - J^{(1)}(\tau,T)\right] = 0\,, \tag{6.80}$$

$$\int_0^L E^{(1)}(x,\tau,T)\,dx = 0\,, \tag{6.81}$$

$$\mathcal{L}_c E^{(2)} - J^{(2)} = -\frac{1}{2}\left[\frac{\partial}{\partial x} v' + v''\right](E^{(1)})^2\,, \tag{6.82}$$

$$E^{(2)}(0,\tau,T) + \rho\left[\frac{\partial E^{(2)}(0,\tau,T)}{\partial\tau} - J^{(2)}(\tau,T)\right] = 0\,, \tag{6.83}$$

$$\int_0^L E^{(2)}(x,\tau,T)\,dx = 0, \tag{6.84}$$

$$\begin{aligned} \mathcal{L}_c E^{(3)} - J^{(3)} &= -\left[\frac{\partial}{\partial T} + \varphi\left(\frac{\partial}{\partial x} v' E_2 + v'' E_2\right)\right] E^{(1)} \\ &\quad - \left[\frac{\partial}{\partial x} v' + v''\right] E^{(1)} E^{(2)} - \frac{1}{6}\left[\frac{\partial}{\partial x} v'' + v'''\right](E^{(1)})^3\,, \end{aligned} \tag{6.85}$$

$$E^{(3)}(0,\tau,T) + \rho\left[\frac{\partial E^{(3)}(0,\tau,T)}{\partial\tau} - J^{(3)}(\tau,T)\right] = -\rho\,\frac{\partial E^{(1)}}{\partial T}\,, \tag{6.86}$$

$$\int_0^L E^{(3)}(x,\tau,T)\,dx = 0\,. \tag{6.87}$$

Here, $\mathcal{L}_c$ is the operator Eq. (6.43) evaluated at $\epsilon = 0$. The argument of the function v and of its derivatives here and in what follows is thus $E = E_0(\cdot)$. Terms such as $[\frac{\partial}{\partial x} v' + v''](E^{(1)})^2$ mean $\frac{\partial[v'(E^{(1)})^2]}{\partial x} + v''(E^{(1)})^2$, so that the operator $[\frac{\partial}{\partial x} v' + v'']$ acts on whatever function of x follows it.

The solution of Eqs. (6.79)–(6.81) is

$$J^{(1)}(\tau, T) = A(T)e^{i\omega\tau} + \text{cc}\,,$$
$$E^{(1)}(x, \tau, T) = A(T)\, e^{i\omega\tau}\, \psi(x) + \text{cc}\,, \tag{6.88}$$

where $\psi(x)$ is the solution of Eqs. (6.45)–(6.47) corresponding to the eigenvalue $\lambda = i\omega$, and cc means the complex conjugate of the preceding term. We now insert Eq. (6.88) in Eq. (6.82) and solve the resulting linear nonhomogeneous equation with the boundary condition Eq. (6.83). We find

$$J^{(2)}(\tau, T) = \nu_0 |A(T)|^2 + \nu_2\, A(T)^2\, e^{i2\omega\tau} + \text{cc}\,,$$
$$E^{(2)}(x, \tau, T) = \xi_0(x)|A(T)|^2 + \xi_2(x)\, A(T)^2 e^{i2\omega\tau} + \text{cc}\,, \tag{6.89}$$

where

$$\xi_n(x)\, v(E_0(x)) = \nu_n \left[\frac{\rho\, \Theta_n(0,x)}{1 + i n \rho \omega} + \int_0^x \Theta_n(y,x)\, dy \right] - \frac{1+\delta_{n0}}{2}$$
$$\int_0^x dy\, \Theta_n(y,x) \left[\frac{\partial}{\partial y} v' + v'' \right] \left[\psi(y)^2 \left(\frac{\bar{\psi}(y)}{\psi(y)} \right)^{\delta_{n0}} \right], \tag{6.90}$$

$$\nu_n = \frac{(1+\delta_{n0}) \int_0^L \frac{dx}{v} \int_0^x dy\, \Theta_n(y,x) \left[\frac{\partial}{\partial y} v' + v'' \right] \left[\psi(y)^2 \left(\frac{\bar{\psi}(y)}{\psi(y)} \right)^{\delta_{n0}} \right]}{2 \int_0^L \frac{dx}{v} \left[\frac{\rho\, \Theta_n(0,x)}{1 + i n \rho \omega} + \int_0^x \Theta_n(y,x)\, dy \right]}\,, \tag{6.91}$$

$$\text{and}\quad \Theta_n(y,x) \equiv \exp\left[- \int_y^x \frac{v'(E_0(z)) + i n \omega}{v(E_0(z))}\, dz \right], \tag{6.92}$$

where n is an integer [$n = 0, 2$ in Eqs. (6.90) and (6.91)]. In Eq. (6.89), we have omitted terms that decay exponentially to zero in the fast time scale τ. δ_{n0} is the Kronecker delta, equal to 1 for $n = 0$, and equal to zero otherwise.

To find the equations for the amplitude $A(T)$, we insert Eqs. (6.88) and (6.89) in the right side of Eq. (6.85). Then, for the solution of this problem to be bounded and time periodic in the fast time scale τ, we need to impose the nonresonance condition (6.75). The result is the following amplitude equation for $A(T)$:

$$\frac{dA}{dT} = \varphi\, \lambda_1\, A - \gamma\, A\, |A|^2\,, \tag{6.93}$$

with

$$\lambda_1 = -\frac{1}{\mathcal{D}} \int_0^L \frac{dx}{v} \int_0^x dy\, \Theta_1(y,x) \left[\frac{\partial}{\partial y} v' + v''\right] (\psi\, E_2)\,, \tag{6.94}$$

$$\gamma = \frac{1}{\mathcal{D}} \int_0^L \frac{dx}{v} \int_0^x dy\, \Theta_1(y,x) \left\{ \left[\frac{\partial}{\partial y} v' + v''\right] (\xi_0 \psi + \xi_2 \bar{\psi}) + \frac{1}{2}\left[\frac{\partial}{\partial y} v'' + v'''\right] \psi\, |\psi|^2 \right\}, \tag{6.95}$$

where

$$\mathcal{D} = \int_0^L \frac{dx}{v} \left\{ \left[\frac{\rho}{1 + i\rho\omega}\right]^2 v(\rho J)\, \Theta_1(0,x) + \int_0^x dy\, \Theta_1(y,x)\, \psi(y) \right\}. \tag{6.96}$$

Notice that $\lambda_1 = \partial\lambda/\partial\phi$ at $\phi = \phi_c$. Thus, Re $\lambda_1 > 0$ at the lower critical bias ϕ_α and Re $\lambda_1 < 0$ at the upper critical bias ϕ_ω.

Equation (6.93) has the following periodic solution:

$$A(T) = R\, e^{i\varphi\mu(T - T_0)}, \quad T_0 = \text{constant}\,,$$
$$R = \sqrt{\frac{\varphi \operatorname{Re} \lambda_1}{\operatorname{Re} \gamma}}\,,$$
$$\mu = \operatorname{Im} \lambda_1 - \frac{\operatorname{Im} \gamma \operatorname{Re} \lambda_1}{\operatorname{Re} \gamma}\,. \tag{6.97}$$

We now show that the stability of the solution (6.97) depends on the sign of Re γ only. Let Re $\gamma > 0$. Then, the bifurcating solution is asymptotically stable (except for the constant phase shift T_0). In fact, if Re $\lambda_1 > 0$, the solution (6.97) exists for $\varphi = 1$ (therefore $\phi > \phi_c = \phi_\alpha$) and it is reached for large positive times starting from any initial condition different from $A = 0$. If Re $\lambda_1 < 0$, the solution (6.97) exists for $\varphi = -1$ (therefore, $\phi < \phi_c = \phi_\omega$) and it is again reached for large positive times starting from any initial condition different from $A = 0$. Similarly, we can show that the bifurcating solution is unstable when Re $\gamma < 0$. This is the usual "principle of exchange of stabilities" between the trivial stationary solution and the bifurcating branch of oscillatory solutions: *a supercritical solution (bifurcating toward the side where the steady state is unstable) is stable, whereas a subcritical solution (bifurcating toward the side where the steady state is stable) is unstable.*

We have numerically evaluated the coefficient Re γ for $\phi = \phi_\alpha$ and L near the minimum length at $\rho = 1.8$ [13]. The result is that: (i) Re $\gamma < 0$ (subcritical bifurcation) near $L = L_m$; (ii) for intermediate lengths, $L \in (L_1, L_2)$, Re $\gamma > 0$ (supercritical bifurcation); (iii) whereas for $L > L_2$, again Re $\gamma < 0$ (subcritical bifurcation, which agrees with the asymptotic result for $\ln L \gg 1$ obtained below). At L_1 and

L_2, there are codimension-2 degenerate Hopf bifurcations with $\mathrm{Re}\,\gamma = 0$, which correspond to transitions from subcritical to supercritical and from supercritical back to subcritical Hopf bifurcations. The corresponding amplitude equations can be derived as explained in Chapter 2.

6.4.3
Amplitude Equation for $\ln L \gg 1$

For very long semiconductors, Lemma 6.2 indicates that many modes Eq. (6.69) become unstable shortly after the voltage crosses its critical value, Eq. (6.49). A convenient representation of the electric field and the current on the bifurcating oscillatory branch for $\ln L \gg 1$ is given by

$$E(x,\tau,T;\epsilon) - E_0(x) = \epsilon \sum_{n\,\mathrm{odd}} A_n(T)\, e^{i\Omega_n\tau}\, \psi_n(x) + O(\epsilon^2)\,,$$
$$J(\tau,T;\epsilon) - J_0 = \epsilon \sum_{n\,\mathrm{odd}} A_n(T)\, e^{i\Omega_n\tau}\,. \qquad (6.98)$$

Here, $\psi_n(x) = \hat{e}(x; i\,\Omega_n)/\hat{j}$ as in Eqs. (6.70) and (6.71), with Ω_n given by Eq. (6.50). The sums are over all odd integers and they include both unstable modes with n of the order of N given by Eq. (6.69), as well as linearly stable modes with larger n. Since the solutions have to be real, $A_{-n} = \bar{A}_n$ in Eq. (6.98) and in what follows, where an overbar denotes complex conjugation. Equations (6.98) are obtained by solving Eqs. (6.23)–(6.25) for $E^{(1)}$ and $J^{(1)}$, with exclusion of linearly stable modes corresponding to eigenvalues whose real part is $O(1)$.

The linear parts of the amplitude equations for A_n would determine their evolution if all the A_n were small. They straightforwardly follow from the dispersion relation (6.51):

$$\frac{\partial A_n}{\partial T} = \frac{|v_2'|\varphi}{2\,X_c\,(E_2 - E_1)}\, A_n - \frac{J\left(f_2 - \frac{f_1^2}{2}\right)}{X_c} \left(\frac{n J \pi}{\epsilon X_c}\right)^2 A_n\,, \qquad (6.99)$$

where $n = \pm 1, \pm 3, \ldots, O(\epsilon X_c)$. The nonlinear terms of the amplitude equation, missing in Eq. (6.99), couple different A_n's and are essential when these are of order 1. A direct derivation using the method of multiple scales is straightforward, though their form can be advanced from the following considerations:

1. The nonlinear terms are cubic in the A_n's. This is similar to Eq. (6.93) and for the same reason: resonant terms first appear in the equations for $E^{(3)}$ that contain products of three factors of the form Eq. (6.98).
2. The coefficient of each trinomial $A_p A_q A_r$ appearing in the amplitude equation for A_n is independent of (p, q, r, n). This is a consequence of the limit $\ln L \to \infty$: $\psi_n(x) \sim \hat{e}(x; 0)/\hat{j}$, as $n = O(\epsilon X_c)$, which implies $\Omega_n \ll 1$. It may be shown that, as far as a determination of the leading order of the nonlinear term in the amplitude equations goes, the integer n does not appear in the right-hand sides of the equations for $E^{(2)}$ and $E^{(3)}$. Then, the

nonlinear term in the amplitude equation for A_n has the form

$$-\gamma_\infty \sum_{p,q \text{ odd}} A_p A_q A_{n-p-q} \,. \tag{6.100}$$

3. The coefficient of the sum Eq. (6.100) is $\gamma_\infty = (3X_c)^{-1} \lim_{L\to\infty}(X_c \gamma)$, where γ is given by Eq. (6.95). This result follows from the complete amplitude equation with initial condition $A_n(0) = \bar{A}_{-n}(0) = \delta_{n1}$. Clearly, $A_1 = \bar{A}_{-1}$ is the only excited mode for $T > 0$, and its evolution should obey Eq. (6.93). The factor 3 in γ_∞ represents the different ways of obtaining $A_1|A_1|^2$ from the sum in Eq. (6.100). An alternative direct derivation and an explicit evaluation of γ_∞ may be obtained by using the method of multiple scales; see Appendix C in [12].

By combining Eqs. (6.99) and (6.100), we obtain the leading order approximation to the amplitude equation in the limit $\ln L \to \infty$:

$$X_c \frac{\partial A_n}{\partial T} = (\alpha\varphi - \beta\Omega_n^2)\, A_n - \Gamma\, L^2 \sum_{p,m \text{ odd}} A_p\, A_m\, A_{n-p-m} \,, \tag{6.101}$$

where the coefficients

$$\alpha = \frac{|v_2'|}{2\,(E_2 - E_1)} > 0\,,$$

$$\beta = J\left(f_2 - \frac{f_1^2}{2}\right) > 0\,,$$

$$\Gamma = -\frac{v_2'^2}{12\, J\, v_1'^2\, (E_2 - E_1)^2} \sim \frac{\gamma_\infty}{L^2} < 0\,, \tag{6.102}$$

are independent of n.

We can rewrite Eq. (6.101) as the following equation for

$$u(\chi, \sigma) = \frac{J^{(1)}(\tau, T)}{L}\,, \quad \chi = \epsilon\tau\,, \quad \sigma = \frac{\epsilon^2 \tau}{X_c} : \tag{6.103}$$

$$\frac{\partial u}{\partial \sigma} = \beta \frac{\partial^2 u}{\partial \chi^2} + (\alpha\varphi - \Gamma\, u^2)\, u\,; \tag{6.104}$$

$$u\left(\chi + \frac{X_c\,\epsilon}{J}, \sigma\right) = -u(\chi, \sigma)\,. \tag{6.105}$$

The antiperiodicity condition Eq. (6.105) results from Eqs. (6.98) and (6.48). Notice that Eq. (6.104) is a reaction-diffusion equation where both “time” σ and “space” χ are slow time scales. The utility of this equation over other approaches is that it provides an effective normal form for this bifurcation. Our result $\Gamma < 0$ indicates that $\varphi = -1$ and the bifurcation is subcritical. Then, any solution of Eqs. (6.104) and (6.105) blows up in finite “time” σ. This means that our perturbation scheme breaks down and presumably strong derivatives or discontinuities in the current (and consequently in the field) are created. These may then evolve into solitary waves with a back shock or into traveling monopole fronts like those discussed in later sections.

Had a result $\Gamma > 0$ been found, a different situation would have occurred. In this case, $\varphi = 1$ and the stable supercritical solution of Eqs. (6.104) and (6.105) would have been a τ-independent steady state of large period $2\epsilon X_c/J \gg 1$. In the phase plane $(u, \frac{\partial u}{\partial \chi})$, such a solution corresponds to a closed orbit that: (i) spends a long time χ near the saddle points $(\pm\sqrt{\alpha/\Gamma}, 0)$, and (ii) jumps from one saddle to the other along the heteroclinic orbits connecting them. One period of $u(\chi)$ is approximately a front connecting $(-\sqrt{\alpha/\Gamma}, 0)$ and $(\sqrt{\alpha/\Gamma}, 0)$ followed after $\chi \sim \epsilon X_c/J$ by the "opposite" front connecting back $(\sqrt{\alpha/\Gamma}, 0)$ and $(-\sqrt{\alpha/\Gamma}, 0)$. In the original variables, we have

$$E(x,\tau;\epsilon) - E_0(x) \sim \frac{\epsilon c_L v_2' e^{|v_2'|\frac{x-X_c}{J}}}{J v_1'(E_2 - E_1)} \sum_{n \text{ odd}} \hat{u}_n \exp\left[-\frac{i n\pi(x - X_c - J\tau)}{X_c}\right], \tag{6.106}$$

for $0 < x < X_c + \Delta x$. Here, $\hat{u}_n$ is the Fourier coefficient of the antiperiodic solution of Eqs. (6.104) and (6.105). $E(x,\tau;\epsilon) - E_0(x)$ decreases exponentially to zero as $(x - X_c - \Delta x)$ grows. One period of $E(x,\tau;\epsilon)$ may thus be described as follows. The front connecting the two saddle points mentioned above moves away from $x = 0$ with speed J and an amplitude that grows exponentially with x. After reaching $x \sim X_c$, two things happen: (i) the front enters $(X_c + \Delta x, L]$ and is exponentially attenuated as it moves further; (ii) at $x = 0$ another front with slope of different sign to the first one is created and starts moving. When this second front reaches $x \sim X_c$, a new front like the first one is created at $x = 0$ and the period is completed. This situation is akin to a *confined* Gunn effect: the oscillations of the current are due to dipole waves which decay after penetrating a distance $x \sim X_c$ in the semiconductor.

The confined Gunn effect has been observed in numerical simulations despite the result that $\Gamma < 0$ in Eq. (6.102). A possible explanation is that $|v_2'|$ can be quite small for $\phi \approx \phi_\alpha$. Then, the subdominant terms that we have neglected in our derivation might be of the same order as those kept here. A supercritical situation would then be a possible outcome.

6.5
Asymptotics of the Gunn Effect for Long Samples and N-shaped Electron Velocity

We now explain the theory of the Gunn effect in the limit of long samples, that is, samples much larger than the dielectric length, $n_0 l \gg \kappa\epsilon_0 F_M/e$, or $L \gg 1$ in dimensionless units. In this limit, the dipole waves have a much smaller size than the sample length. This theory answers questions such as

- What is the actual velocity of dipole waves during each instant of a Gunn oscillation?

- Why are dipole waves stable?
- Can we predict the evolution of the current density and the electric field profiles from a reduced description?

The answers to these questions require a theory that surpasses the numerical simulations of the Kroemer model. The asymptotic theory will be first explained in the case of a $\mathcal{N}$-shaped drift velocity, where scalings are simplest. Then, in Section 6.6 we shall explain how to modify the theory when the drift velocity saturates at high electric fields, as in Eq. (6.27).

6.5.1 General Formulation of Asymptotics for $\delta = O(1)$

An asymptotic theory of the Gunn oscillations was first presented in the case of $\delta \ll 1$ and large L [14]. The general theory for arbitrary δ was presented in [7] for the contact boundary conditions given in [6]. Here, we shall adapt these calculations to the case of a cathode obeying Ohm's law as in [14]. We will assume that $J_c = E_c/\rho = v(E_c)$ has $E_c > 1$ belonging to the second branch of $v(E)$. This boundary condition results in a self-sustained current oscillation due to periodic generation of dipole waves at the cathode that move towards $x = L$ and disappear there, as shown in Figure 6.6. If the cathode resistivity ρ is so small that either (roughly speaking; see [14] for precise details): (i) $E_c < 1$ belonging to the first branch of $v(E)$ or (ii) $J = E/\rho$ and $J = v(E)$ do not intersect and $E > \rho v(E)$, then the self-sustained current oscillations are due to the periodic generation and motion of charge accumulation layers which are charge *monopoles*, not dipoles; see

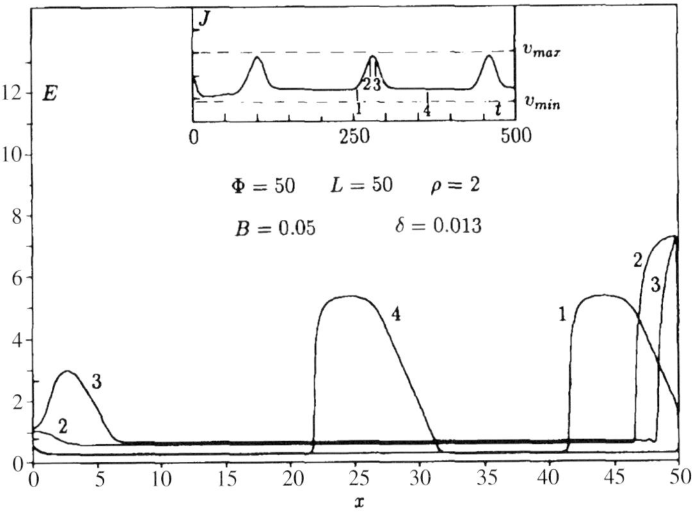

Fig. 6.6 Current vs. time and electric field snapshots showing the traveling wave state for the three branch v–E curve (from [14]).

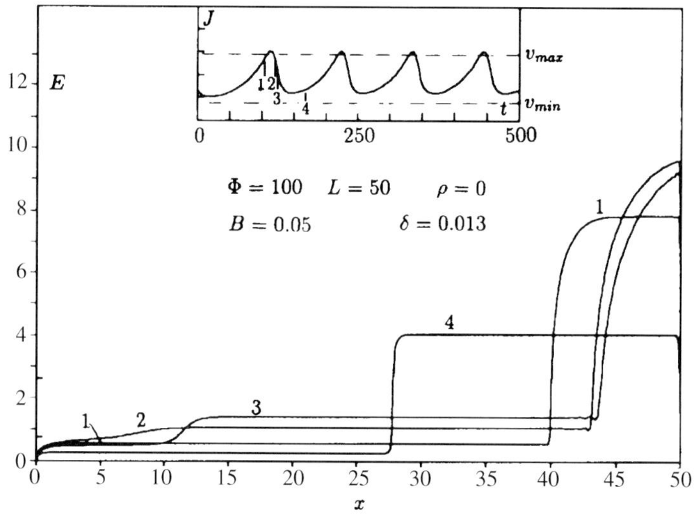

Fig. 6.7 Current-time and snapshots of the field corresponding to traveling waves for a zero resistivity cathode in the three-branch model (from [14]).

Figure 6.7 and their asymptotic description in [14]. In this chapter, we will not discuss monopole waves. They are also important for self-sustained current oscillations in semiconductor superlattices and will be discussed in Chapter 8.

Let us assume that the drift velocity is linear at small fields, it has a local maximum $v(1) = 1$ followed by a local minimum $0 < v(E_m) = v_m < 1$. Then, for larger fields, $v(E) \sim Bv'(0)\,E$ as $E \to \infty$, with $0 < B < 1$. To perform actual calculations, we shall use Kroemer's expression (6.26) for a $\mathcal{N}$-shaped $v(E)$. Suppose that $\phi > \phi_a$ and the dimensionless diffusivity is of order 1. We want to describe one period of the fully developed Gunn effect corresponding to repeated generation and motion of charge dipoles. Figure 6.6 depicts electric field profiles at different times of the current oscillations as marked in the inset. Let us assume that initially, there is a single high-field domain moving towards the anode at $x = L$. The proper time scale is the time the dipole takes to reach the anode, which is of order L. We now rescale time and length as

$$s = \frac{\tau}{L}, \quad y = \frac{x}{L}, \quad \epsilon = \frac{1}{L} \ll 1 . \tag{6.107}$$

Note that the scaling of time is different from the previous discussion of this chapter. Equations (6.23) and (6.25) for E and J become

$$\frac{\partial E}{\partial s} + v(E)\frac{\partial E}{\partial y} - \epsilon\delta\frac{\partial^2 E}{\partial y^2} = \frac{J - v(E)}{\epsilon}, \tag{6.108}$$

$$\int_0^1 E(y, s)\, dy = \phi . \tag{6.109}$$

6.5.1.1 A Single Dipole Wave

The outer limit of Eq. (6.108) as $\epsilon \to 0$ yields

$$v(E) = J\,. \tag{6.110}$$

Thus, the electric field profile is formed out of the zeros of $v(E) - J$. Let these be $E_i(J)$, $i = 1, 2, 3$, with $E_1 < E_2 < E_3$, if $v_m < J < 1$. Outside this interval, we still define E_1 or E_3 in the obvious way. Simple linear stability shows that E_1 and E_3 are stable constant solutions of Eq. (6.22) on the real line (J constant), whereas E_2 is unstable. Letting $Y_+(s)$ and $Y_-(s)$ denote the boundaries of the dipole wave at time s, a dipole wave – equivalently, a high-field domain – may be constructed by means of the stable constant solutions of Eq. (6.110):

$$\begin{aligned} E(y,s) &= E_3(J) \quad \text{if} \quad Y_+(s) < y < Y_-(s)\,, \\ E(y,s) &= E_1(J) \quad \text{otherwise}\,. \end{aligned} \tag{6.111}$$

By inserting this profile in Eq. (6.109), we find the wave width:

$$E_1(J) + E_3(J)\,[Y_- - Y_+] = \phi + O(\epsilon)\,, \tag{6.112}$$

provided that we know the current density J.

To find $Y_\pm(s)$ and $J(s)$, we should consider the inner structure of the dipole wave. Let us assume that the front and the back of the dipole wave depend on an inner coordinate χ, on a frame of reference which moves at speed $c_\pm$:

$$E = E(\chi; c_\pm), \quad \chi = \frac{y - Y_\pm}{\epsilon}\,, \quad \frac{dY_\pm}{ds} = c_\pm\,. \tag{6.113}$$

Then, Eq. (6.108) becomes

$$[v(E) - c_\pm]\,\frac{\partial E}{\partial \chi} - \delta\,\frac{\partial^2 E}{\partial \chi^2} = J - v(E)\,, \tag{6.114}$$

or, equivalently,

$$\frac{dE}{d\chi} = F\,, \tag{6.115}$$

$$\frac{dF}{d\chi} = \frac{[v(E) - c_\pm]\,F + v(E) - J}{\delta}\,. \tag{6.116}$$

Solutions corresponding to the forefront and the backfront of a dipole wave should satisfy the matching conditions:

$$E(-\infty; c_+) = E_1(J), \quad \text{and} \quad E(+\infty; c_+) = E_3(J)\,, \tag{6.117}$$

$$E(-\infty; c_-) = E_3(J), \quad \text{and} \quad E(+\infty; c_-) = E_1(J)\,. \tag{6.118}$$

For a fixed value of $J(s) = J$, Eq. (6.114) together with the conditions (6.117) select a unique velocity $c_+(J)$. The corresponding trajectory in the phase plane of

Eqs. (6.115) and (6.116) is a heteroclinic orbit connecting the saddle points $(E_1, 0)$ and $(E_3, 0)$ with $dE/d\chi > 0$. Similarly, $c_-(J)$ is uniquely selected in such a way that there is a separatrix connecting the saddle $(E_3, 0)$ to $(E_1, 0)$ with $dE/d\chi < 0$. These heteroclinic trajectories and the functions $c_\pm$ may be found numerically for any $J \in (v_m, 1)$. In the asymptotic limit $\delta \to 0$, these constructions can be done explicitly, as shown in Section 6.5.2.

We have shown how a dipole wave can be constructed asymptotically. The field inside the wave is $E_3(J)$, it is $E_1(J)$ outside, its back moves at speed $c_+(J)$, and its front at speed $c_-(J)$. Notice that the backfront and the forefront waves are stable solutions for Eq. (6.22) considered on the complete real line, $-\infty < x < +\infty$, for a fixed value of J. To determine the evolution of the dipole wave, all we need is to find $J(s)$. For this, we shall use the bias condition (6.112). This condition may be written as

$$Y_- - Y_+ \sim \frac{\phi - E_1(J)}{E_3(J) - E_1(J)} .$$

The derivative with respect to the slow time s of the left side is $c_- - c_+$, while the derivative of the right side is proportional to dJ/ds. Thus, we find the following equation for J:

$$\frac{dJ}{ds} = A(J)\,[c_+(J) - c_-(J)] , \qquad (6.119)$$

$$A(J) = \frac{(E_3 - E_1)^2}{\frac{\phi - E_1}{v_1'} + \frac{E_3 - \phi}{v_3'}} > 0 , \qquad (6.120)$$

where we have used that $v(E_i(J)) = J \Rightarrow dE_i/dJ = 1/v'(E_i)$, $v(E_i(J)) \equiv v_i'(J)$, $i = 1, 2, 3$.

Equation (6.119) implies that the current density evolves exponentially fast towards the unique stable fixed point J^* such that $c_+ = c_-$. Then, the current remains flat while the dipole wave advances without changing its size towards the anode. Simple phase plane considerations [see Eq. (6.146) later in Section 6.6] show that the value $c_+ = c_- = J^*$, and that J^* may be found from the so-called *equal area rule*:

$$\int_{E_1}^{E_3} [v(E) - J]\, dE = 0 . \qquad (6.121)$$

The stage described thus far lasts until the dipole wave reaches the anode at $x = L$. Then, $Y_- = 1$ and the current density obeys $dJ/ds = A(J)\, c_+(J)$. Thus, it starts increasing until a new wave is shed at the cathode. To explain how this occurs, we need to analyze the boundary layers at the contacts which we have thus far avoided. However, before doing that, let us anticipate possible situations in which several dipoles may coexist.

Remark 6.5

Notice that Eq. (6.119) explains why a dipole wave far from the contacts should move at constant speed and without changing its size. This situation was observed in numerical simulations over the years, but was not mathematically understood until more recently [7].

6.5.1.2 Several Dipole Waves

The previous equation for the evolution of the current may be generalized to the case of more than one dipole wave. Let us assume that we have N dipole waves moving inside the semiconductor. Let us denote them by a number starting with the first wave which is closest to the receiving contact. The forefront and the back-front of the jth wave will be denoted by $Y_-^{(j)}$ and $Y_+^{(j)}$, respectively. If the first wave has arrived at the anode, we identify $Y_-^{(1)} = 1$; we take $Y_+^{(N)} = 0$ if the last wave is still attached to the cathode. The construction of these waves is exactly as explained above, but now Eq. (6.112) should be replaced by

$$E_1(J) + E_3(J) \sum_{j=1}^{N} \left[Y_-^{(j)} - Y_+^{(j)} \right] = \phi + O(\epsilon) , \qquad (6.122)$$

Time differentiation of this equation yields the following evolution equation for the current:

$$\frac{dJ}{ds} = A(J)\,[n_+ c_+(J) - n_- c_-(J)] . \qquad (6.123)$$

$n_\pm$ is the number of wavefronts which move at speed $c_\pm$. We have set $dY/ds = 0$ if the corresponding wavefront coincides with one of the contacts. Clearly, $n_+ - n_-$ may only be 0, 1 or -1. For example, N dipole waves moving far from the contacts have $n_+ = n_- = N$, so that $J(s) \to J^*$ as the time elapses. When the first wave arrives at the anode, $n_+ - n_- = 1$, and so on.

6.5.1.3 Shedding Waves at the Cathode

After the single dipole wave arrives at the anode, the current starts increasing ,as described by Eq. (6.123), with $n_+ = 1$, $n_- = 0$. When J reaches a critical value, the boundary layer near $y = 0$ becomes unstable and it sheds a new dipole wave. To understand this event, we need to consider the boundary layers at the contacts in more detail. They are described by Eq. (6.22). While the dipole wave is advancing and the current is increasing according to Eq. (6.123), $J = J(s)$, the boundary layers are described by Eq. (6.114) with $c_\pm = 0$, $\chi = y/\epsilon$, or $\chi = (y-1)/\epsilon$. The boundary conditions (6.24) may be approximated by $E = \rho\, J(s)$ at $\chi = 0$. Thus, the boundary layer at $y = 0$ (resp. at $y = 1$) corresponds to the separatrix entering the saddle $(E_1(J), 0)$ from the vertical line $E = \rho\, J(s)$ (resp. leaving the saddle and intersecting the vertical line) on the phase plane $(E, dE/d\chi)$. There are two possible phase portraits (and therefore, two different boundary layer configurations) depending on the value of the current:

- $0 < J < J_c$. [Recall that $J_c = E_c/\rho = v(E_c)$ has $E_c > 1$ belonging to the second branch of $v(E)$]. Then, $E_1(J) < E(0) < E_2(J)$, the field decreases at the injecting boundary layer (at $y = 0$) and it increases at the receiving boundary layer (at $y = 1$).
- $J > J_c$. Then, $E(0) > E_2(J)$ and the field in the injecting boundary layer should, in principle, increase towards $E_3(J)$. At the receiving boundary layer, the field should decrease from $E_3(J)$ (corresponding to the exiting dipole wave) to $E = \rho J$.

As the current follows Eq. (6.123) with $n_+ = 1$, $n_- = 0$, it reaches J_c at $s = s_1$ and increases past it. Then, the field configuration at the injecting boundary layer should abruptly change to accommodate the previous description. This change occurs in a fast scale given by $\sigma = (s - s_1)/\epsilon$. The result is that J reaches a maximum slightly higher than J_c, and a new dipole wave is shed at $y = 0$. *Notice that this description implicitly assumes that $J^* < J_c$. This implies an upper bound for the contact resistivity, which should fulfill the restrictions $1 < \rho < E_2(J^*)/J^*$.*

Other boundary conditions influence the process of shedding new waves differently. For example, mixed boundary conditions corresponding to a $n^+ - n$ contact are equivalent to a multivalued current-field relation at the contact having the same extrema as $v(E)$. In these contacts, when the current increases past J_{cM} (the maximum of the contact curve), a wavefront moving with speed c_- is shed. The current then may slowly decrease on the s-scale until a backfront is shed (moving with speed c_+) when J decreases past J_{cM} (the minimum of the contact curve on its third branch). In this case, the process of shedding dipole waves is much slower: different n_+ and n_- may have to be used in Eq. (6.123) to fully describe it. The situation is similar in Ohmic contacts: the dipole shedding process is slow. Now, the critical current at which forefronts and backfronts are shed is $J_c \sim 1$.

6.5.1.4 Overall Gunn Oscillation

Now we can piece together our previous results and describe one period of a Gunn oscillation. Let us start with a single dipole wave moving far from the contacts. As previously explained, Eq. (6.123) with $n_+ = n_- = 1$ implies that J tends to J^* and the wave moves at constant speed $c_+ = c_-$ until it reaches the anode. Then, $n_+ = 1$, $n_- = 0$ and the current increases past J_c. A new dipole wave is shed quickly and the current satisfies Eq. (6.123) with $n_+ = 2$, $n_- = 1$. Now, J should tend to the new fixed point $J_{2,1}$ such that $2c_+ = c_-$. There are two possible situations:

- If $J_{2,1} < J_c$ [or $\rho < E_2(J_{2,1})/J_{2,1}$], the current decreases during this stage to $J_{2,1}$. After some time, the old dipole has completely left the semiconductor sample. Then, $n_+ = n_- = 1$, and we are back at the first stage of the oscillation with only one dipole wave. This is the usual Gunn effect. Figure 6.6a depicts the $J(s)$ curve in this case and Figure 6.6b shows the corresponding electric field profiles at the times marked in Figure 6.6a.
- If $J_{2,1} > J_c$ [$E_2(J_{2,1})/J_{2,1} < \rho < E_2(J^*)/J^*$], the current increases after nucleation of the new dipole. Numerical simulations show that additional

dipoles are shed at $y = 0$. It is possible to find chaotic oscillations of the current in narrow bias intervals at appropriate values of the contact resistivity.

6.5.2
Explicit Formulation of Asymptotics for $\delta \to 0+$

In the limit $\delta \to 0+$, we can obtain explicit expressions for $c_\pm(J)$ and for the corresponding wavefronts. The forefront (moving depletion layer) joining $(E_3(J), 0)$ and $(E_1(J), 0)$ may be approximated by the exact solution of Eqs. (6.115) and (6.116)

$$E(\chi) = -\chi\,, \quad F(\chi) = -1\,, \quad c_-(J) = J\,, \tag{6.124}$$

which holds for any value of δ plus two corner layers of width $O(\sqrt{\delta})$ (on the χ scale) at $\chi = \pm(E_3 - E_1)/2$. The width of this wavefront on the χ length scale is $(E_3 - E_1) + O(\sqrt{\delta})$, which yields $y - Y_i(s) = O(\epsilon)$ on the large length scale. The velocity of this wavefront is $c_-(J) = J$.

The other wavefront of the dipole wave (moving accumulation layer) can be constructed by matched asymptotic expansions in the limit $\delta \ll 1$. Its velocity $c_+(J)$ and shape depend on whether J is larger or smaller than J^*. Let us assume that $J > J^*$. The forefront is composed of a shock wave joining two field values E_- and E_+ (at least one of them should be equal to E_i, $i = 1, 3$) plus a tail region which moves rigidly with the same velocity as the shock. The inner structure of the shock wave (for very small but not zero δ) can be a quite complicated triple-deck set of boundary layers [15]. Let $V(E_+, E_-)$ be the velocity of a shock wave at $\chi = 0$ such that $E(0\pm; V(E_+, E_-)) = E_\pm$. Let us define the inner variable $\xi = \chi/\delta$, and assume

$$E = E^{(0)}(\xi) + \delta E^{(1)}(\xi) + o(\delta)\,.$$

By inserting this ansatz in Eq. (6.114), we obtain

$$[v(E^{(0)}) - V]\,\frac{\partial E^{(0)}}{\partial \xi} - \frac{\partial^2 E^{(0)}}{\partial \chi^2} = 0\,. \tag{6.125}$$

This equation should be integrated with the matching conditions

$$E^{(0)}(\xi) \sim E_\pm \quad \text{and} \quad \frac{\partial E^{(0)}}{\partial \xi} \sim 0 \quad \text{as} \quad \xi \to \pm\infty\,. \tag{6.126}$$

With these conditions, we can integrate Eq. (6.125) once to obtain

$$\int_{E_-}^{E^{(0)}} [v(E) - V]\,dE - \frac{\partial E^{(0)}}{\partial \xi} = 0\,, \tag{6.127}$$

and therefore, the following equal area rule yields the shock velocity V:

$$\int_{E_-}^{E_+} [v(E) - V(E_+, E_-)]\, dE = 0\,. \tag{6.128}$$

The inner structure of the shock (to this order) is given by integrating Eq. (6.127).

The backfront is a trajectory satisfying Eq. (6.117). Equation (6.128) cannot hold for all $J \in (v_m, 1)$ if $E_- = E_1$ and $E_+ = E_3$. Then, we either use $E_- = E_1(J)$, or $E_+ = E_3(J)$, and join the field value at the other side of the shock, E_+ or E_-, to E_3 or E_1, respectively, by means of a solution of the outer equation

$$[v(E) - V(E_+, E_-)]\frac{\partial E}{\partial \chi} = J - v(E)\,. \tag{6.129}$$

We now have:

1. If $J \in (v_m, J^*)$, $E_+ = E_3(J)$ whereas E_- is calculated as a function of J by imposing the condition that the tail region to the left of the shock wave moves rigidly with it

   $$V(E_3, E_-) = v(E_-)\,. \tag{6.130}$$

   Equations (6.128) and (6.130) (with $E_+ = E_3$), we find both E_- and $c_+ = v(E_-) > J$ as functions of J. To the left of the shock wave (in the tail region), the field satisfies the (approximate) boundary value problem Eq. (6.129) with the boundary conditions

   $$E(-\infty) = E_1(J)\,, \quad E(0) = E_-(J)\,. \tag{6.131}$$

2. If $J \in (J^*, v_M)$, $E_- = E_1(J)$ whereas E_+ is calculated as a function of J by imposing the condition that the tail region to the right of the shock wave moves rigidly with it

   $$V(E_+, E_1) = v(E_+)\,. \tag{6.132}$$

   By simultaneously solving Eqs. (6.128) and (6.132) (with $E_- = E_1$), we find both E_+ and $c_+ = v(E_+) < J$ as functions of J. To the right of the shock wave (in the tail region), the field satisfies the boundary value problem (6.129) with the boundary conditions

   $$E(0) = E_+(J)\,, \quad E(\infty) = E_3(J)\,. \tag{6.133}$$

At $J = J^*$, we have $E_- = E_1$, $E_+ = E_3$, and $c_+ = J^*$. It is not hard to prove that $c_+(J)$ is a decreasing function. The functions $c_\pm(J)$ are similar to those depicted in Chapter 3.

If $c_+ \neq J$, the shock width is $\chi = O(\delta)$. However, at the fixed point $c_+(J^*) = J^*$, $E = E_i(J)$, $i = 1, 3$, at one side of the shock. Then, the coefficient of the convective term in Eq. (6.125) vanishes as we leave the shock at the convenient side. This widens the inner structure of the shock wave to $O(\sqrt{\delta})$. The total width of the backfront (including tail regions) is $O(1)$ on the χ scale and $O(\epsilon)$ on the y scale. In the limit $\delta \ll 1$, the structure of the backfront is thus one-sided: the wavefront is a discontinuity preceded or followed by a tail region.

6.6
Asymptotics of the Gunn Effect for Long Samples and Saturating Electron Velocity

6.6.1
A Single Dipole Wave

We shall now consider that the drift velocity saturates at high electric fields, as in Eq. (6.27). Then, the current density and its corresponding electric field profile during one period of the Gunn oscillation are similar to those shown in Figure 6.6 for the case of a three-branched $v(E)$. The only difference is that the flat top of the dipole wave is absent: far from the contacts, the dipole wave is a triangle. To asymptotically describe this oscillation, we adapt the construction of pulses in the convective Cahn–Allen system [16] to the Kroemer model as follows. Let us assume that, initially, there is a single high-field domain moving towards the anode at $x = L$. If the electron velocity curve saturates at high electric fields, a dipole wave cannot have a flat top at field $E = E_3(J)$, for E_3 is now infinity. Thus, the electric field of a dipole wave will be a spike moving on a background with field $E = E_1(J)$. To describe its motion, we need different time and space scales from Eq. (6.107), valid for the case of a $v(E)$ with three branches. To find them, we need a more precise characterization of the instantaneous positions of the dipole backfront and forefront, $X_+(t)$ and $X_-(t)$, respectively. A precise definition is

$$E(X_+(\tau),\tau) = 1\,, \quad \text{with} \quad \frac{\partial E}{\partial x}(X_+(\tau),\tau) > 0\,, \tag{6.134}$$

$$E(X_-(\tau),\tau) = 1\,, \quad \text{with} \quad \frac{\partial E}{\partial x}(X_-(\tau),\tau) < 0\,. \tag{6.135}$$

Instead of using the field at the maximum of $v(E)$ in Eqs. (6.134) and (6.135), we could have selected any other constant field near it. $E = 1$ is convenient because this value is always reached at the back and the front of any dipole wave.

The field profile of a detached high-field dipole wave consists of:

1. A trailing wavefront (backfront) within which the field increases from $E = E_1(J)$ to $E = E_M$.
2. A leading wavefront (forefront) within which the field decreases from $E = E_M$ to $E = E_1(J)$.

The reduced description of the dipole consists of its phase plane construction plus equations for E_M and J. Within a wavefront, we may define the moving coordinate

$\chi = x - X_\pm(\tau)$. Then, we have

$$E = E(\chi; c_\pm), \qquad \frac{dX_\pm}{d\tau} = c_\pm(J, E_M),$$

instead of Eq. (6.113). $E(\chi; c_\pm)$ is a solution of Eq. (6.114), which satisfies $E(0; c_\pm) = 1$ and the boundary conditions:

$$E(-\infty; c_+) = E_1(J), \qquad \frac{\partial E(\chi_+; c_+)}{\partial \chi} = 0, \qquad E(\chi_+; c_+) = E_M, \qquad \text{with} \quad \frac{\partial E}{\partial \chi} \geq 0, \tag{6.136}$$

$$\frac{\partial E(-\chi_-; c_-)}{\partial \chi} = 0, \qquad E(-\chi_-; c_-) = E_M, \qquad E(\infty; c_-) = E_1(J), \qquad \text{with} \quad \frac{\partial E}{\partial \chi} \leq 0. \tag{6.137}$$

On the phase plane $(E, \frac{\partial E}{\partial \chi})$, $E = E(\chi; c_+)$ corresponds to the unique separatrix which (given J and E_M fixed) joins the saddle $(E_1, 0)$ to $(E_M, 0)$ on the E-axis with $\frac{\partial E}{\partial \chi} > 0$. Similarly, $E = E(\chi; c_-)$ corresponds to the unique separatrix which (given J and E_M fixed) joins $(E_M, 0)$ on the E-axis to the saddle $(E_1, 0)$ with $\frac{\partial E}{\partial \chi} < 0$. Thus, we fix J and E_M, solve Eq. (6.114) together with Eq. (6.136) and find a unique $c_+(J, E_M)$ and the profile $E(\chi; c_+)$. $c_-(J, E_M)$ and $E(\chi; c_-)$ are found by solving Eq. (6.114) together with Eq. (6.137). Once all of these functions are determined, we can calculate the width of both the backfront and the forefront, $\chi_\pm$:

$$\frac{\partial E}{\partial \chi}(\chi_+; c_+) = 0, \qquad E(\chi_+; c_+) = E_M, \tag{6.138}$$

$$\frac{\partial E}{\partial \chi}(-\chi_-; c_-) = 0, \qquad E(-\chi_-; c_-) = E_M, \tag{6.139}$$

with $\frac{\partial^2 E}{\partial \chi^2}(\pm\chi_\pm; c_\pm) < 0$. Notice that $\chi_\pm > 0$ and

$$\frac{dX_+}{d\tau} = -\frac{d\chi_+}{d\tau} = c_+, \qquad \frac{dX_-}{d\tau} = \frac{d\chi_-}{d\tau} = c_-, \tag{6.140}$$

with these definitions. The function

$$l(J, E_M) = X_- - X_+ = \chi_- + \chi_+, \tag{6.141}$$

is the width of the dipole wave, and it can be calculated solely from the consideration of the phase plane Eqs. (6.115) and (6.116). *Specific expressions for $E(\chi; c_\pm)$, $c_\pm(J, E_M)$, and $l(J, E_M)$ can be found asymptotically in the limit $\delta \to 0$.* In the general case, they should be calculated numerically.

We can now use the bias condition Eq. (6.23) to relate E_M to J and ϕ:

$$\phi \sim E_1 + \epsilon\, \Phi(J, E_M),$$

$$\Phi(J, E_M) = \int_{-\infty}^{\chi_+} [E(\chi; c_+) - E_1]\, d\chi + \int_{-\chi_-}^{\infty} [E(\chi; c_-) - E_1]\, d\chi. \tag{6.142}$$

We shall see that $E_M = O(\epsilon^{-\frac{1}{2}})$ and $\chi_\pm = O(\epsilon^{-\frac{1}{2}})$. Then, $\epsilon\Phi(J, E_M)$ is actually of order 1 and this equation yields a function $E_M = \mathcal{E}(J) = O(\epsilon^{-\frac{1}{2}})$ for a fixed ϕ. Inserting it into Eq. (6.141), we obtain $\Lambda(J) = l(J, \mathcal{E}(J))$. $[(\Lambda(J) = O(\epsilon^{-\frac{1}{2}})]$.

To obtain an equation for $J(\tau)$, all we need to do is to differentiate Eq. (6.141) with respect to τ, considering Eq. (6.140):

$$\frac{dJ}{d\tau} = \frac{c_+(J, \mathcal{E}(J)) - c_-(J, \mathcal{E}(J))}{-\Lambda'(J)} . \tag{6.143}$$

Given that $\Lambda(J) = O(\epsilon^{-\frac{1}{2}})$, the evolution of the current occurs on a slow time scale $\sigma = \epsilon^{\frac{1}{2}}\tau$. This justifies our initial assumption that $J(\tau)$ could be taken as constant when constructing the wavefronts which the electric field pulse is made of. Equation (6.143) has a stable fixed point for which $c_+ = c_-$. This fixed point corresponds to a dipole wave moving at fixed J without changing its speed and size. The dipole speed is J, and its maximum field is given by the equal area rule. In the phase plane, this dipole wave corresponds to a homoclinic orbit of the saddle point $(E_1(J), 0)$.

Proof that a dipole wave moving without changing size has a speed J. Let $c_+ = c_- \equiv c$ at the fixed point of Eq. (6.143). The corresponding dipole wave is a homoclinic orbit in the phase plane (E, E_χ) given by Eqs. (6.115) and (6.116):

$$\frac{dE}{d\chi} = F ,$$
$$\frac{dF}{d\chi} = \frac{[v(E) - c]\,F + (c - J)}{F\delta} .$$

These equations yield

$$\frac{F}{F+1}\frac{dF}{dE} = \frac{v(E) - c}{\delta} + \frac{c - J}{(F+1)\,\delta} , \tag{6.144}$$

which may be integrated to obtain

$$F_\pm(E) - \ln[F_\pm(E) + 1] - \int_{E_M}^{E} \frac{v(s) - c}{\delta}\, ds = \frac{c - J}{\delta} \int_{E_M}^{E} \frac{ds}{F_\pm(s) + 1} . \tag{6.145}$$

Here, $F_+(E) > 0$ [resp. $F_-(E) < 0$] is the positive (resp. negative) half part of the homoclinic orbit, whose maximum is reached at $(E_M, 0)$. Clearly, $F_+(E_1) = F_-(E_1) = 0$, so that by subtracting Eq. (6.145) evaluated for F_+ from the same expression evaluated for F_-, we obtain

$$\frac{c - J}{\delta} \oint \frac{ds}{F(s) + 1} = 0 .$$

The integral in this expression is not zero because the integrand is always larger for $F = F_-$ than for $F = F_+$. Then, $c = J$, as claimed above. By inserting this result in Eq. (6.145) evaluated at $E = E_1$, we find

$$\int_{E_1}^{E_M} [v(E) - J]\, dE = 0\,. \tag{6.146}$$

This is the *equal area rule* for the homoclinic dipole wave. Notice that the present derivation holds for a general $v(E)$ having a first increasing branch for $0 < E < 1$ followed by a decreasing branch, so that a homoclinic orbit of Eqs. (6.115) and (6.116) is possible. This is also the case for an N-shaped $v(E)$. In such a case. there is a single $J = J^*$ for which a homoclinic orbit has a maximum $E_M = E_3(J)$. For this value of the current, we obtain $c_+ = c_-$ in Eq. (6.119). Then, Eq. (6.146) yields the previous equal area rule (6.121).

6.6.2
The Dipole Wave Arrives at the Anode

Let us assume that a single pulse has reached its asymptotic shape ($J = J^*$) and it advances with speed J^* until its forefront reaches $X_- = L$ at time τ_1. Afterwards, it begins leaving the sample. As time elapses, the wave *exits* through the receiving boundary and therefore, the area under the wave decreases. Since the total area has to satisfy the bias condition, this loss of area has to be compensated by a corresponding increase in E_1 so that Eq. (6.23) still holds.

Let us denote by E_L the value of $E(x, \tau)$ at the receiving boundary. Shortly after τ_1, E_L is obtained from the condition $E_L = E(L - X_-; c_-)$ (leading front of the pulse). While $0 < X_- - L < \chi_-$, the leading front is still leaving the sample and the maximum of the pulse has not yet arrived at the anode. The bias condition (6.23) becomes

$$\phi = E_1 + \epsilon\, \psi(J, E_M, X_- - L) + O(\epsilon)\,, \tag{6.147}$$

$$\psi(J, E_M, X_- - L) = \int_0^{\chi_+} [E(\chi; c_+) - E_1]\, d\chi + \int_{-\chi_-}^{L-X_-} [E(\chi; c_-) - E_1]\, d\chi\,. \tag{6.148}$$

$X_-(\tau)$ can be explicitly calculated from the equation

$$\frac{dX_-}{d\tau} = c_-(J, E_M)\,, \quad X_-(\tau_1) = L \tag{6.149}$$

(provided we know J and E_M). From Eqs. (6.147) and (6.148), we may determine E_M as a function of J and $X_- - L$: $E_M = \mathcal{E}(J, X_- - L)$. Then, Eq. (6.141) becomes

$$l(J, \mathcal{E}(J, X_- - L)) = X_- - X_+\,.$$

Time differentiation yields

$$\left(\frac{\partial l}{\partial J} + \frac{\partial l}{\partial E_M}\frac{\partial \mathcal{E}}{\partial J}\right)\frac{dJ}{d\tau} + \frac{\partial l}{\partial E_M}\frac{\partial \mathcal{E}}{\partial X_-}\, c_- = c_- - c_+\,,$$

from which we obtain

$$\frac{dJ}{d\tau} = \frac{c_+ - c_- + \frac{\partial l}{\partial E_M}\frac{\partial \mathcal{E}}{\partial X_-} c_-}{-\frac{\partial l}{\partial J} - \frac{\partial l}{\partial E_M}\frac{\partial \mathcal{E}}{\partial J}} . \tag{6.150}$$

Notice that this equation coincides with Eq. (6.143) if the third term in the numerator is omitted. This occurs provided $X_- < L$. Equations (6.149) and (6.150), together with $E_M = \mathcal{E}(J, X_- - L)$ given by Eqs. (6.147) and (6.148), determine the unknowns J, E_M and X_-. A numerical solution of these equations shows that J increases with time. Depending on the bias ϕ, one of the following two events may occur first:

i. J reaches J_c, or
ii. $X_- - L = \chi_-$. Then, the maximum of the pulse reaches the anode.

In both cases, the stage described by the previous equations ends. In case (i), a new wave is created at $x = 0$, whereas in case (ii), Eqs. (6.147) and (6.148) should be changed to

$$\phi = E_1 + \epsilon\, \Psi(J, E_M, L - X_+) + O(\epsilon), \tag{6.151}$$

$$\Psi(J, E_M, L - X_+) = \int_0^{L-X_+} [E(\chi; c_+) - E_1]\, d\chi . \tag{6.152}$$

These equations may be used to obtain $E_M = \mathcal{E}(J, L - X_+)$. The phase plane trajectory Eqs. (6.115) and (6.116) for $E(\chi; c_+)$ yields

$$L - X_+ = \int_1^{E(L-X_+;c_+)} \frac{dE}{F(E)} \Longrightarrow L - X_+ = \mathcal{L}(J, E_M) . \tag{6.153}$$

By inserting $E_M = \mathcal{E}(J, L - X_+)$ in this equation and differentiating the result with respect to time, we obtain the following equation for J:

$$\frac{dJ}{d\tau} = -\frac{1 + \frac{\partial \mathcal{L}}{\partial E_M}\frac{\partial \mathcal{E}}{\partial X_+}}{\frac{\partial \mathcal{L}}{\partial J} + \frac{\partial \mathcal{L}}{\partial E_M}\frac{\partial \mathcal{E}}{\partial J}} c_+ . \tag{6.154}$$

In this case, J also increases until J_c is reached. Then, a new dipole wave is created at the cathode.

6.6.3
Coexistence of Two Dipole Waves

Once a new dipole wave is created, there are two waves present at the sample. Let us denote the quantities belonging to the new wave by the same symbols as those for the old wave, but with an additional accent to distinguish them. Let us suppose that the new wave was created before the maximum of the pulse had time to reach the anode. The bias condition (6.23) now becomes

$$\phi = E_1 + \epsilon[\Phi(J, E'_M) + \psi(J, E_M, X_- - L)] + O(\epsilon) , \tag{6.155}$$

where the area under the new dipole, $\Phi(J, E'_M)$, is given by Eq. (6.142), and $\psi(J, E_M, X_- - L)$ is given by Eq. (6.148). Equation (6.155), together with the equations $dX_\pm/d\tau = c_\pm(J, E_M)$, and

$$l(J, E'_M) = X'_- - X'_+ , \tag{6.156}$$

$$l(J, E_M) = X_- - X_+ , \tag{6.157}$$

determine the unknowns J, $X_\pm$, $X'_\pm$, E_M and E'_M. After the maximum of the old dipole reaches $x = L$, the description of this stage changes, as indicated above. Once the old wave has completely disappeared, we are back at the initial situation (only one dipole wave moving towards the anode) and a period of the Gunn oscillation has been completed.

6.6.4 Explicit Formulation of Asymptotics for $\delta \to 0+$

In the limit $\delta \to 0+$, the description of the Gunn oscillation becomes much more explicit than previously described.

6.6.4.1 One Pulse Far from the Contacts

A single pulse moving far from the contacts is described by its backfront at $X_+(\tau)$ and its leading front at $X_-(\tau)$. The backfront is a shock wave moving at speed $c_+ = V(E_M, E_1)$ given by the equal area rule (6.128). The leading front of the pulse is a straight line of slope -1 moving at speed $c_- = J$: $E(\chi; J) = 1 - \chi$, $\chi = x - X_-(\tau) = x - \int_0^\tau J(s)\, ds - X_+(0) - E_M + 1 - X_+(0)$, that is,

$$E = E_M - x + X_+(0) + \int_0^\tau J(s)\, ds .$$

The bias condition (6.142) and the pulse width are

$$\phi \sim E_1 + \frac{\epsilon\,(E_M - E_1)^2}{2} , \tag{6.158}$$

$$X_-(\tau) - X_+(\tau) = E_M - E_1 , \tag{6.159}$$

respectively. Equation (6.158) implies that

$$(E_M - E_1) \sim \sqrt{\frac{2(\phi - E_1)}{\epsilon}}\,. \tag{6.160}$$

By inserting this into Eq. (6.159) and time-differentiating the result, we find

$$\frac{dJ}{d\tau} \sim \epsilon^{\frac{1}{2}}\, v_1' \sqrt{2(\phi - E_1)}\,[V(E_M, E_1) - J]\,. \tag{6.161}$$

This equation is an explicit form of Eq. (6.143). On the slow time scale $\sigma = \sqrt{\epsilon}\tau$, J tends to J^* given by simultaneously solving Eqs. (6.146) and (6.160).

6.6.4.2 The Pulse Reaches the Anode

This stage is described by Eqs. (6.147)–(6.150). They become much simpler in the limit of small diffusivity. The field at the anode is $E_L = E_M - (L - X_+)$. Equation (6.147) becomes

$$\phi \sim E_1 + \epsilon \left(E_M - E_1 - \frac{L - X_+}{2} \right) (L - X_+)\,. \tag{6.162}$$

Here, X_+ solves

$$\frac{dX_+}{d\tau} = V(E_M, E_1)\,, \tag{6.163}$$

with an appropriate initial condition. As the pulse width is Eq. (6.159), we find

$$\frac{dE_M}{d\tau} \sim J - V(E_M, E_1)\,, \tag{6.164}$$

for $(E_M - E_1) \sim E_1$. Equations (6.162) to (6.164) constitute a complete description of this stage. An evolution equation for J is found by differentiating Eq. (6.162) with respect to time and using the other equations. The result [to be compared to Eq. (6.150)] is

$$\frac{dJ}{d\tau} \sim \epsilon\, v_1' \left[(E_M - E_1)\, V(E_M, E_1) - (L - X_+)\, J \right]\,, \tag{6.165}$$

$$L - X_+ \sim E_M - E_1 - \sqrt{(E_M - E_1)^2 - \frac{2(\phi - E_1)}{\epsilon}}\,. \tag{6.166}$$

The unknowns evolve on the same slow time scale as before because E_M and $(L - X_+)$ are of order $\epsilon^{-\frac{1}{2}}$.

6.6.4.3 Coexistence of Two Pulses

When J reaches J_c, another wave is nucleated at the cathode on a fast time scale. After this, a stage begins during which two pulses coexist. Using accents for the variables of the new pulse, we should simultaneously solve the following equations in order to describe this stage:

$$\phi \sim E_1 + \epsilon \left[\frac{(E'_M - E_1)^2}{2} + \left(E_M - E_1 - \frac{L - X_+}{2} \right) (L - X_+) \right], \tag{6.167}$$

$$\frac{dX_+}{d\tau} = V(E_M, E_1), \quad \frac{dX'_+}{d\tau} = V(E'_M, E_1), \tag{6.168}$$

$$\frac{dE_M}{d\tau} \sim J - V(E_M, E_1), \tag{6.169}$$

$$(X'_- - X'_+) \sim (E'_M - E_1). \tag{6.170}$$

An equation for J is found as follows. We first solve Eq. (6.167) for $(E'_M - E_1)$, then insert the result in Eq. (6.170), differentiate with respect to time and use Eq. (6.168) and $dX'_-/d\tau = J$. We obtain

$$\frac{dJ}{d\tau} \sim \epsilon^{\frac{1}{2}} v'_1 \Bigg[\epsilon^{\frac{1}{2}} [(E_M - E_1) V(E_M, E_1) - (L - X_+) J] + [V(E'_M, E_1) - J] \times 2^{\frac{1}{2}} \sqrt{\phi - E_1 - \epsilon (L - X_+) \left(E_M - E_1 - \frac{L - X_+}{2} \right)} \Bigg]. \tag{6.171}$$

When $X_+ = L$, this stage ends and only one pulse remains on the sample. We are back at the first stage, having completed one period of the Gunn oscillation.

6.7 References on the 1D Gunn Effect and Closing Remarks

In this chapter, we have presented a study of the Gunn effect in n-GaAs as an oscillatory instability of the 1D Kroemer model. This model consists of a parabolic equation for the electric field and the total current density, an integral constraint (dc voltage bias), two boundary conditions at the contacts and an initial condition for the field. We have constructed the stationary solution and found that it may become unstable to oscillations for biases on a certain interval. We have derived amplitude equations for this instability (onset of the Gunn effect) and also described the fully developed oscillation of current and field far from the onset region by asymptotic methods. Pieces of this picture were obtained in the very large literature associated with the Gunn effect in the 1960s and 1970s. We would like to mention the following books and reviews where many references can be found:

- Bonch-Bruevich, V.L., Zvyagin, I.P., and Mironov, A.G. (1975) *Domain Electrical Instabilities in Semiconductors*, Consultants Bureau, New York.
- Butcher, P.N. (1967) The Gunn effect. *Rep. Prog. Phys.*, **30**, 97–148.

- Kroemer, H. (1972) Gunn effect – bulk instabilities, In: *Topics in Solid State and Quantum Electronics*, Chap. 2, (ed. W.D. Hershberger), pp. 20–98, Wiley & Sons, Inc, New York.
- Shaw, M.P., Grubin, H.L., and Solomon, P.R. (1979) *The Gunn–Hilsum effect*, Academic Press, New York.
- Volkov, A.F. and Kogan, Sh.M. (1969) Physical phenomena in semiconductors with negative differential conductivity. *Sov. Phys. Usp.*, **11**, 881–903 [U.F.N. **96**, 633–672 (1968)].
- Volume 13 of the IEEE Transactions on Electron Devices (1966) contains many valuable papers at a time where many results were being found almost simultaneously by different authors. Most relevant to our presentation, are the works by Kroemer and by McCumber and Chynoweth.

The following two works are among the best early theoretical papers on the Gunn effect. The first paper contains the outer approximations of dipole domains by shock and other waves as $\delta \to 0$. The second paper contains general phase plane analyses of dipole domains and other solutions at fixed current on infinitely long samples. Our derivation of the equal area rule in Section 6.6 is taken from this paper.

- Knight, B.W. and Peterson, G.A. (1966) Nonlinear analysis of the Gunn effect. *Phys. Rev.*, **147**, 617–621.
- Knight, B.W. and Peterson, G.A. (1967) Theory of the Gunn effect. *Phys. Rev.*, **155**, 393–404.

Sections 6.3 and 6.4 are based upon

- Bonilla, L.L., Higuera, F.J. and Venakides, S. (1994) The Gunn effect: Instability of the steady state and stability of the solitary wave in long extrinsic semiconductors. *SIAM J. Appl. Math.*, **54**, 1521–1541.
- Bonilla, L.L. and Higuera, F.J. (1995) The onset and end of the Gunn effect in extrinsic semiconductors. *SIAM J. Appl. Math.*, **55**, 1625–1649.

Section 6.5 is adapted from the two following papers:

- Higuera, F.J. and Bonilla, L.L. (1992) Gunn instability in finite samples of GaAs. II. Oscillatory states in long samples. *Physica D*, **57**, 161–184.
- Bonilla, L.L., Cantalapiedra, I.R., Gomila, G., and Rubí, J.M. (1997) Asymptotic analysis of the Gunn effect with realistic boundary conditions. *Phys. Rev. E*, **56**, 1500–1510.

Section 6.6 adapts the calculations carried out in the paper

- Bonilla, L.L., Kindelan, M., and Keller, J.B. (2007) Periodically generated propagating pulses. *SIAM J. Appl. Math.*, **65**, 1053–1079.

to the Kroemer model with saturating drift velocity.
Instabilities similar to the Gunn effect, in that they consist of oscillations of the current through a semiconductor accompanied by periodic recycling and motion of

charge dipole domains, appear in other materials different from bulk n-GaAs. For example, impurity dynamics may cause negative differential resistivity in ultrapure p-Ge even though the carrier velocity is a monotone increasing function of electric field. The frequency of the resulting oscillation is much smaller (kHz range) due to the slow dynamics of the impurities. Descriptions and further references may be found in Chapter 7 of this book and in the papers:

- Gwinn, E.G. and Westervelt, R.M. (1986) Frequency locking, quasiperiodicity, and chaos in extrinsic Ge. *Phys. Rev. Lett.*, **57**, 1060–1063.
- Kahn, A.M., Mar, D.J., and Westervelt, R.M. (1991) Spatial measurements of moving space-charge domains in p-type ultrapure germanium. *Phys. Rev. B*, **43**, 9740–9749.
- Kahn, A.M., Mar, D.J., and Westervelt, R.M. (1992) Dynamics of space-charge domains in ultrapure Ge. *Phys. Rev. Lett.*, **68**, 369–372.
- Kahn, A.M., Mar, D.J., and Westervelt, R.M. (1992) Spatial measurements near the instability threshold in ultrapure Ge. *Phys. Rev. B*, **45**, 8342–8347.
- Teitsworth, S.W., Bergmann, M.J., and Bonilla, L.L. (1995) Space charge instabilities and nonlinear waves in extrinsic semiconductors. *Nonlinear Dynamics and Pattern Formation in Semiconductors and Devices* (ed. F.-J. Niedernostheide), pp. 44–69, Springer, Berlin.
- Bergmann, M.J., Teitsworth, S.W., Bonilla, L.L., and Cantalapiedra, I.R. (1996) Solitary-wave conduction in p-type Ge under time-dependent voltage bias. *Phys. Rev. B*, **53**, 1327–1335.
- Bonilla, L.L., Hernando, P.J., Herrero, M.A., Kindelan, M., and Velázquez, J.J.L. (1997) Asymptotics of the trap-dominated Gunn effect in p-type Ge. *Physica D*, **108**, 168–190.

Similar slow-down of the oscillation due to impurity dynamics occurs in semi-insulating materials such as n-GaAs with EL2 impurities. See Chapter 9 and the papers:

- Samuilov, V.A. (1995) Nonlinear and chaotic charge transport in semi-insulating semiconductors. *Nonlinear Dynamics and Pattern Formation in Semiconductors and Devices* (ed. F.-J. Niedernostheide), pp. 220–249, Springer, Berlin.
- Piazza, F., Christianen, P.C.M., and Maan, J.C. (1997) Propagating high-electric-field domains in semi-insulating GaAs: experiment and theory. *Phys. Rev. B*, **55**, 15591–15600.
- Bonilla, L.L., Hernando, P.J., Kindelan, M., and Piazza, F. (1999) Determination of EL2 capture and emission coefficients in semi-insulating n-GaAs. *Appl. Phys. Lett.*, **74**, 988–990.

The experiments by Piazza *et al.* (1997) show evidence of chaotic oscillations due to the dynamics of the dipole domains under dc voltage bias. Undriven chaos was also reported by Kahn *et al.* (1992) from experiments in p-Ge. In this latter system, spatiotemporal chaos was observed under ac + dc voltage bias conditions by the same authors. The references and the corresponding theory can be found in the papers by Teitsworth *et al.* and Bergmann *et al.* above. Numerical simulations of driven and undriven chaos in different systems presenting Gunn-type instabilities can be found in, for example,

- Mosekilde, E., Feldberg, R., Knudsen, C., and Himdsholm, M. (1990) Mode locking and spatiotemporal chaos in periodically driven Gunn diodes. *Phys. Rev. B*, **41**, 2298–2306.
- Oshio, K. and Yahata, H. Chaotic current oscillations in the Gunn-effect device under the dc and rf bias voltages. *J. Phys. Soc. Japan*, **62**, 3639–3650 (1993).
- Oshio, K. and Yahata, H. (1995) Non-periodic current oscillations in the Gunn effect with the impact-ionization effect. *J. Phys. Soc. Japan*, **64**, 1823–1836.
- Bonilla, L.L., Kindelan, M., and Hernando, P.J. (1998) Photorefractive Gunn effect. *Phys. Rev. B*, **58**, 7046–7050.

For significantly different models, the asymptotic methods presented here can be applied and yield a good description of the Gunn effect. Further discussion can be found in

- Bonilla, L.L. and Cantalapiedra, I.R. (1997) Universality of the Gunn effect: self-sustained oscillations mediated by solitary waves. *Phys. Rev. E*, **56**, 3628–3632.

References

1 Gunn, J.B. (1965) Instabilities of current and potential distribution in GaAs and InP, from *Proceedings of the Symposium on Plasma Effects in Solids*, (ed. J. Bok), pp. 199–207, Dunod, Paris.

2 Kroemer, H. (1972) Gunn effect – bulk instabilities, in: *Topics in Solid State and Quantum Electronics*, Chapt. 2, (ed. W.D. Hershberger), pp. 20–98, Wiley & Sons, New York.

3 Ridley, B.K. and Watkins, T.B. (1961) Possibility of negative resistance effects in semiconductors. *Proc. Phys. Soc. Lond.*, **78**, 293; Hilsum, C. (1962) Transferred electron amplifiers and oscillators. *Proc. IRE*, **50**, 185–189.

4 Gunn, J.B. (1963) Microwave oscillations of current in III–V semiconductors. *Solid State Commun.*, **1**, 88–91; Gunn, J.B. (1964) Instabilities of current in III–V semiconductors. *IBM J. Res. Dev.*, **8**, 141.

5 Butcher, P.N. (1967) The Gunn effect. *Rep. Prog. Phys.*, **30**, 97–148.

6 Gomila, G., Rubí, J.M., Cantalapiedra, I.R., and Bonilla, L.L. (1997) Stationary states and phase diagram for a model of the Gunn effect under realistic boundary conditions. *Phys. Rev. E*, **56**, 1490–1499.

7 Bonilla, L.L., Cantalapiedra, I.R., Gomila, G., and Rubí, J.M. (1997) Asymptotic analysis of the Gunn effect with realistic boundary conditions. *Phys. Rev. E*, **56**, 1500–1510.

8 Rosner, D.E. (2000) *Transport Processes in Chemically Reacting Flow Systems*. Dover, New York.

9 Kroemer, H. (1966) Nonlinear space-charge domain dynamics in a semiconductor with negative differential mobility. *IEEE Trans. Electron Devices*, **ED-13**, 27–40.

10 Bonilla, L.L. and Higuera, F.J. (1991) Gunn instability in finite samples of GaAs I. Stationary states, stability and boundary conditions, *Physica D*, **52**, 458–476.

11 Bonilla, L.L., Higuera, F.J., and Venakides, S. (1994) The Gunn effect: Instability of the steady state and stability of the solitary wave in long extrinsic semiconductors. *SIAM J. Appl. Math.*, **54**, 1521–1541. Copyright (1994) by the Society for Industrial and Applied Mathematics.

12 Bonilla, L.L. and Higuera, F.J. (1995) The onset and end of the Gunn effect in extrinsic semiconductors. *SIAM J. Appl. Math.*, **55**, 1625–1649. Copyright (1995) by the Society for Industrial and Applied Mathematics.

13 Kindelan, M., Higuera, F.J., and Bonilla, L.L. (1996) Onset of the Gunn effect in semiconductors: bifurcation analysis and numerical simulations. *Z. Angew. Math. Mech. (ZAMM)*, **76**(suppl. 2), 575–576.

14 Higuera, F.J. and Bonilla, L.L. (1992) Gunn instability in finite samples of GaAs. II Oscillatory states in long samples. *Physica D*, **57**, 161–184.

15 Bonilla, L.L. (1991) Solitary waves in semiconductors with finite geometry and the Gunn effect. *SIAM J. Appl. Math.*, **51**, 727–747.

16 Bonilla, L.L., Kindelan, M., and Keller, J.B. (2007) Periodically generated propagating pulses. *SIAM J. Appl. Math.*, **65**, 1053–1079.

7
Electric Field Domains in Bulk Semiconductors II: Trap-mediated Instabilities

7.1
Introduction

In this chapter, we examine the situation in which the negative differential conductivity is due to *negative differential charge carrier* concentration. In other words, there is a range of electric fields F for which the concentration of charge carriers n decreases and the field increases, that is, $dn/dF < 0$. This is in contrast with instabilities that result from negative differential drift velocity, such as the Gunn effect. Qualitatively speaking, because the dynamics of trapping and de-trapping of charge carriers relative to impurities is generally much slower than the k-space transfer mechanisms that underlie negative differential velocity, the dynamics of space charge instability due to trapping generally manifests itself on relatively slow time scales.

In this discussion, we focus on trap-controlled instabilities that occur in extrinsic semiconductors, particularly ultrapure p-type Ge at liquid Helium temperature. This latter system is of particular interest largely due to a series of precise experiments carried out in the 1980s and early 1990s, and also because of the success in developing accurate theoretical models. From a technical perspective, p-Ge is of interest as a very sensitive detector material for far-infrared radiation [1]. The methods developed in this chapter are also applicable to other material systems exhibiting trap-controlled instabilities, for example, semi-insulating GaAs; some of these are discussed briefly in Chapter 9. Firstly, we review the status of the theory of nonlinear charge waves in the p-Ge system and discuss the extent of agreement with the key experiments. As in previous chapters, an essential element of our approach is the use of singular perturbation and bifurcation methods. Singular perturbation methods allow for the tremendous simplification of the full set of drift-diffusion transport equations to a reduced form that, nonetheless, captures most of the experimental phenomena. Bifurcation methods are used to study the transitions which occur in the reduced models. In particular, transitions from time-independent to nonstationary behavior.

Before turning to the transport model, we present a brief review of some of the key experiments to elucidate the nonlinear transport properties of p-Ge.

Nonlinear Wave Methods for Charge Transport. Luis L. Bonilla and Stephen W. Teitsworth

ISBN: 978-3-527-40695-1

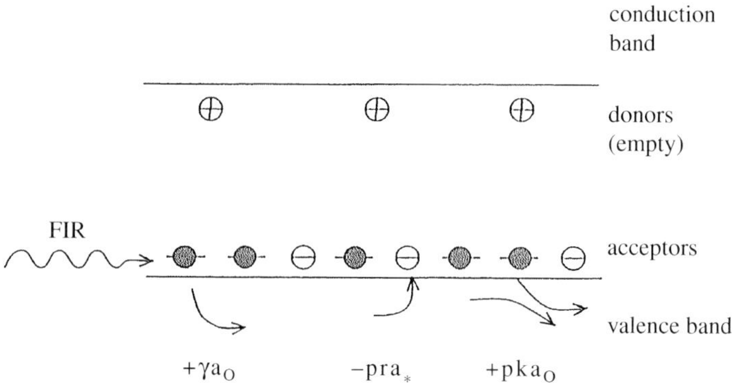

Fig. 7.1 Energy band diagram of p-type Ge also illustrating the key processes of free hole generation and capture: FIR illumination, capture, and impact ionization (from [8]).

The energy band schematic of p-Ge is shown in Figure 7.1. Impurity breakdown refers to a field strength where impact ionization leads to near complete liberation of all charge carriers (in this case holes). Here, by impact ionization, we refer to a process by which a sufficiently energetic free hole collides with a bound hole and imparts a sufficient amount of energy to liberate it. Below the field strength of breakdown, the density of free holes is highly sensitive to the far-infrared (FIR) illumination which explains applicability of this material as a photodetector.

In 1983, period doubling and temporal chaos associated with noisy spontaneous current oscillations were reported in a regime of applied dc voltage just below the electric field for which a phenomena known as *impurity breakdown* occurs [2]. Experiments were later performed for conditions of combined dc and ac voltage bias in the pre-breakdown region and these revealed period-doubling bifurcations and temporal chaos as the amplitude of the ac bias was varied [3]. The pre-breakdown results were mostly explained in terms of low-dimensional dynamical models which included the nonlinear dependencies of quantities such as impurity capture rate and impact ionization rate on local electric field strength. However, these early models generally did not not include spatial degrees of freedom, nor did they include boundary conditions associated with the electrical contacts that are responsible for charge injection into the material. Later experiments shifted to the regime of behavior above the threshold for impurity breakdown, which revealed negative differential resistance (NDR) and large amplitude spontaneous oscillations of the electrical current flowing through the structures [4]. This regime of post-breakdown space-charge instability was exploited in high precision tests of predictions from the theory of the frequency-locking route to chaos [5]. This series of experiments on p-Ge then culminated with measurements of the spatial profile of electric field using movable capacitive probing. These results verified the presence of traveling solitary waves of electric field, that is, high field domains, which were found to move periodically across the samples [6]. Figure 7.2 shows

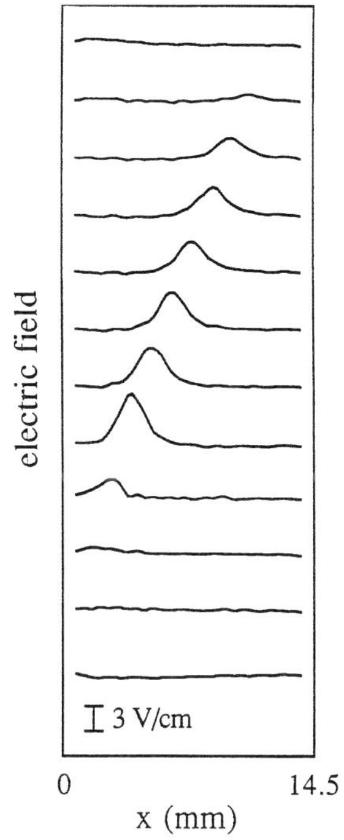

Fig. 7.2 Experimental data showing the local electric field vs. position along the sample length for p-Ge. The successive traces are taken at times separated by 98 μs (from [6]).

an example of the experimentally measured electric field profiles associated with a moving high-field domain.

In this chapter, we first present the nonlinear drift-diffusion transport model that is appropriate for describing charge transport in p-Ge. Then, we introduce nondimensional variables which allows us to identify the crucial dimensionless small parameters of this problem. Taking these small parameters to zero provides us with a maximally "reduced" model. We first consider the properties of this reduced model on an infinite line under current bias. We identify the heteroclinic and homoclinic orbits in a co-moving frame and associate these with traveling nonlinear waves. For the case of current bias, we must analytically determine the dynamical stability of these waves. Then, we turn to the analytically more challenging case of voltage bias in samples of finite length. We use bifurcation theory to identify the processes by which nonlinear waves come into existence as the external bias parameters are tuned. Finally, we review the key role of these nonlinear waves in the asymptotic limit of long samples under voltage bias.

7.2 Drift-Diffusion Transport Model for Trap-Mediated System

In the drift-diffusion approximation, the dynamical equations that describe carrier (either electrons or holes) transport in an extrinsic semiconductor such as p-Ge can be written in the form [2]:

$$\frac{\partial a_*}{\partial T} = \gamma\, a_0 + p\left[k(\mathcal{E})a_0 - r(\mathcal{E})a_*\right], \tag{7.1}$$

$$\epsilon\frac{\partial \mathcal{E}}{\partial T} = j(T) - e\left[p\,v(\mathcal{E}) - D\frac{\partial p}{\partial X}\right], \quad \text{and} \tag{7.2}$$

$$\epsilon\frac{\partial \mathcal{E}}{\partial X} = e(p + d - a_*)\,. \tag{7.3}$$

Here, $\mathcal{E}$ denotes the electric field, a_*, $a_0 = a - a_*$ and p denote the ionized acceptor, neutral acceptor, and free hole concentrations, respectively. The total acceptor concentration is given by a and is determined during the crystal growth process [1]. Here, X and T are dimensional space and time coordinates. Equation (7.1) is a rate equation which describes the local dynamics of trapping and de-trapping of free holes vis-à-vis the shallow acceptor impurities; in the case of p-Ge, the impurity is typically Boron [1]. The physical significance of the various terms on the right-hand side is depicted in Figure 7.1. The term $\gamma\, a_0$ describes the generation of ionized acceptors corresponding to the liberation of holes from neutral acceptors which is due to far-infrared illumination; the total photon flux is proportional to γ. The rate of impact ionization of neutral acceptors due to collisions with energetic free holes is given by the second term, and the rate of hole recombination, wherein a free hole falls into a vacant ionized acceptor site is represented by the third term. Equation (7.2) is simply Ampère's Law with the drift-diffusion approximation taken for the electrical current. Here, $v(\mathcal{E})$ is the field-dependent drift velocity of the free holes and D is the hole diffusivity, which is assumed field-independent in this treatment. The quantity $j(T)$ represents the total current that flows through a hypothetical external circuit and includes the contribution of displacement current. The third equation, Eq. (7.3), is a form of Poisson's law for the electric field. Here, d denotes the residual donor concentration in the material which partially "compensates" the acceptors. In fact, at the low temperatures for which most the p-Ge experiments were performed, the donors are completely ionized. Finally, ϵ denotes the semiconductor permittivity.

The presence of NDR and the resultant space-charge instabilities are largely determined by the forms of electric field-dependence in the kinetic coefficients and drift velocity. We have used the following functional forms which are expressed in terms of a nondimensionalized electric field, $E \equiv (\mu_0/v_s)\mathcal{E}$, where $\mu_0 \simeq 1 \times 10^6\ \mathrm{cm^2/V\,s}$ denotes the low field mobility and $v_s \simeq 10^7\ \mathrm{cm/s}$ denotes the saturation drift velocity [7, 8]:

$$v(E) = v_s\{9(E + 0.33)/10 + (18/10\pi)(E - 0.5)\tan^{-1}[5 - 10E] + (9/100\pi)\ln[1 + (5 - 10E)^2]\}, \tag{7.4}$$

$$r(E) = r_0\{0.05 + [1.04 + 100.0E^2]^{-1.5}\}, \tag{7.5}$$

$$k(E) = k_0\left\{\left(1.0 + \exp\left[\frac{0.55 - E}{0.015}\right]\right)^{-1}(0.25 + 2.0\exp[-E/0.34]) + 0.1(E/1.15)^4\right\}. \tag{7.6}$$

These equations are based on both experimental data and heuristic considerations. The values of the prefactors in each expression are, respectively, $r_0 = k_0 = 6.0 \times 10^{-6}\,\mathrm{cm^3/s}$ and $v_s = 1.0 \times 10^7\,\mathrm{cm/s}$. When the electric field is below the breakdown value, of order 5 V/cm, the drift velocity increases linearly according to the well-known mobility approximation. However, at larger fields, the velocity approaches a saturation value which is due, in p-Ge, to the emission of optical phonons once holes have acquired sufficient kinetic energy. The capture coefficient r_0 is approximately constant for small fields, and hence for small hole velocity with respect to the stationary impurities. At larger fields, however, the capture coefficient decreases as E^{-3} since the atomic cross section for capture also decreases [4]. The shape of the impact ionization coefficient can also be understood in terms of an atomic cross section process. For field strength below a threshold value, the average kinetic energy of the holes is below the level needed to liberate a bound hole. When the average kinetic energy matches the binding energy of the bound hole to the acceptor, the cross section is maximal. Then, as the kinetic energy increases beyond this threshold, the cross section falls off, an effect that can be qualitatively understood by the fact that the fast hole spends less time in the neighborhood of the acceptor. At the highest field strengths, the impact ionization coefficient again increases due to secondary ionization processes mediated, for example, by optical phonons. The negative differential field-dependence in the k–E curve, as shown in Figure 7.3a, is essential for both NDR and space-charge instabilities in this system.

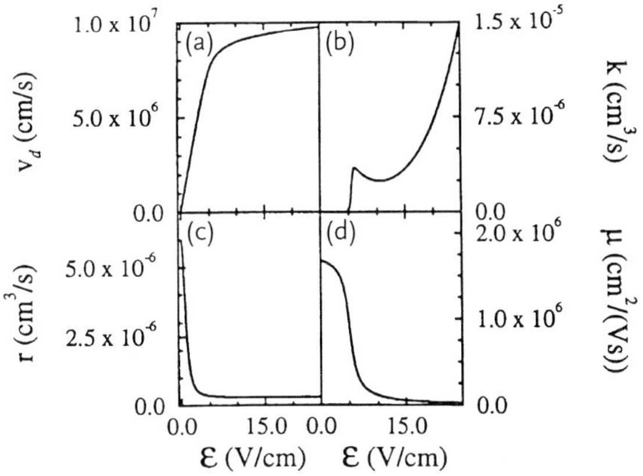

Fig. 7.3 Electric field dependencies of the key processes: (a) hole drift velocity, (b) impact ionization, (c) capture coefficient, and (d) mobility (from [8]).

When the sample of interest has finite length L, we must also append boundary conditions to this model. The boundary conditions depend on the nature of metal-semiconductor contact and how it is created. The precise microscopic boundary conditions can be accurately modeled by macroscopic boundary conditions of the following form:

$$\epsilon \frac{\partial \mathcal{E}}{\partial T} + j_{\text{con}}(\mathcal{E}) = j(T) \quad \text{at} \quad X = 0 \quad \text{and} \quad X = L\,, \tag{7.7}$$

where j_{con} is a particular function that models the field-dependence transport across the contacts. For contacts with low resistivity, one may use a linear (i. e., Ohmic) approximation for the contact characteristic:

$$j_{\text{con}}(\mathcal{E}) = \frac{\mathcal{E}}{\rho_{\text{con}}}\,, \tag{7.8}$$

where ρ_{con} is the contact resistivity. This is similar to the boundary condition assumed for the Gunn effect in the preceding chapter.

The total external current density $j(T)$ is determined by the nature of the bias condition. For current bias, $j(T)$ is a specified function of time that is appended to the dynamical equations above. However, as with the Gunn effect, most experiments use voltage bias, which means that we must append the following constraint to the model:

$$\int_0^L \mathcal{E}(X, T)\,dX = \Phi\,, \tag{7.9}$$

where Φ is the total applied voltage. This is a nonlocal constraint on the manner in which the space- and time-dependent fields can evolve. We should also note here that, in this discussion – as with the Gunn effect, we are neglecting spatial dynamics in the direction perpendicular to the applied voltage. For the p-Ge system, this is reasonable since dielectric relaxation occurs rapidly and stably in the transverse direction. Equivalently, the transverse field strengths do not reach the levels for NDR to occur in these transverse directions.

7.3
Nondimensional Form and the Reduced Model

Following the procedure developed in previous chapters, we first identify convenient scales in which to measure time, position, electric field strength, and concentration or density. However, due to the presence of trapping dynamics, we now find an additional time scale relative to our previous treatment of the Gunn effect. Thus, there are two characteristic time scales: $T_1 \equiv \epsilon/(ed\mu_0) \sim 10^{-10}$ s is the dielectric relaxation time and $T_2 \equiv (k_0 d)^{-1} \sim 10^{-6}$ s is the characteristic time for trapping and de-trapping of holes relative to the acceptors. Then, the nondimensional variables are expressed as follows:

$$E = (\mu_0/v_s)\mathcal{E}\,, \quad P = (p/d)A\,, \quad A = a_*/d - 1\,, \quad \tau = T/T_2\,,$$
$$J(\tau) = j(T)/(edv_s)\,,$$
$$x = X(\mu_0 ed/\epsilon v_s)V(E) = v(\mathcal{E})/v_s,\, K(E) = k(\mathcal{E})/k_0\,,$$
$$R(E) = r(\mathcal{E})/k_0\,, \quad \rho_0 = \rho_{\text{con}}e\mu_0 d\,. \tag{7.10}$$

Another important dimensionless quantity to introduce is the compensation ratio $\alpha \equiv a/d$, which is of order 1 for the experimental samples with which to compare. With these definitions, the full dynamical equations may be recast as

$$\frac{\partial A}{\partial \tau} = \Gamma\,(\alpha - 1 - A) + P[K(\alpha - 1) - R - (K + R)A]\,, \tag{7.11}$$
$$\beta\frac{\partial E}{\partial t} = J(t) - VP + \delta\frac{\partial P}{\partial x}\,, \tag{7.12}$$
$$\frac{\partial E}{\partial x} = P - A\,, \tag{7.13}$$

where we have introduced the following dimensionless small parameters:

$$\beta = T_1/T_2 = 5.2 \times 10^{-5}\,,$$
$$\delta = \mu_0 ed D/\left(\epsilon v_s^2\right) = 0.04\,, \tag{7.14}$$
$$\Gamma = \epsilon^2\gamma/\left(\mu_0^2 e^2 d^3 k_0\right) \sim 1.9 \times 10^{-8}\,.$$

The existence of these small parameters strongly suggests the applicability of singular perturbation methods, and we begin by setting β, δ, and Γ equal to zero which gives the *outer* approximation to the full set of equations, Eqs. (7.11)–(7.13). Physically, this limit corresponds to neglecting the displacement current, the diffusion current, and the external illumination term. As noted in previous chapters, when we study the properties of the outer problem, we may well encounter unphysical behavior such as the development of shocks. When effects such as these arise, we must go back to the full equations and solve an inner problem for the shock region. Then, to achieve a full solution, we should match the inner and outer solutions, for example, using asymptotic matching as described in Chapter 2, in order to construct a complete solution to the desired accuracy. Let us also observe that the neglect of the Γ term is reasonable only for field strengths that are above impurity breakdown. Below the breakdown threshold (in fact, the regime where the semiconductor behaves as a good photodetector), the generation term plays an important role.

Setting β and δ to zero allows us to eliminate variables A and P in favor of E and $\partial E/\partial x$ using Eqs. (7.12) and (7.13). If we also take $\Gamma = 0$, we arrive at a second order nonlinear partial differential equation of hyberbolic type which has previously been called the *reduced model* for charge waves in extrinsic semiconductors such as p-Ge [9, 10]

$$\frac{\partial^2 E}{\partial x \partial \tau} + J\frac{K+R}{V^2}\left(\frac{V'}{K+R}\frac{\partial E}{\partial \tau} + V\frac{\partial E}{\partial x} + j(E) - J\right) = \frac{1}{V}\frac{dJ}{d\tau}, \tag{7.15}$$

$$\frac{1}{L}\int_0^L E(x,\tau)\,dx = \phi\,, \tag{7.16}$$

$$E(0,\tau) = \rho_0 J(\tau)\,, \tag{7.17}$$

where $V' \equiv dV/dE$ is the dimensionless differential mobility, $j(E)$ denotes the steady-state current-field characteristic defined in the next section, and ϕ is the nondimensional applied voltage. Note that we have only written the boundary condition at the injecting contact $x = 0$. Of course, a similar boundary condition must also hold at the receiving contact at $x = L$ in the full model. It has been shown [9] that this boundary condition can be recovered by inserting a diffusive boundary layer at the receiving contact, in other words, by solving and matching the appropriate inner problem in the region near $x = L$ to the solutions of the reduced model.

7.4
Steady States, *J–E* Curves, and Steady Wave Solutions on the Infinite Line under Current Bias

We now briefly turn to the case of dc current bias in a sample of effectively infinite length, that is, dimensionless length much greater than unity. Because we are looking for traveling wave solutions, it makes sense to transform the reduced model into a co-moving frame. Therefore, we introduce the following change of variable, $\xi \equiv x - C\tau$ where C is the expected wavespeed which should be of order 1. Then, we can rewrite Eq. (7.15) as

$$\begin{aligned}\frac{\partial^2 E}{\partial \xi \partial \tau} + \frac{JV'}{V^2}\frac{\partial E}{\partial \tau} &= C\frac{\partial^2 E}{\partial^2 \xi} + J\left[\frac{CV'}{V^2} - \frac{K+R}{V}\right]\frac{\partial E}{\partial \xi} \\ &\quad + \frac{J}{V^2}(K+R)(J-\rho V)\,,\end{aligned} \tag{7.18}$$

$$\text{where} \quad \rho \equiv \frac{\alpha K}{K+R} - 1\,. \tag{7.19}$$

In attempting to understand the properties of this equation, we begin by examining the spatially uniform steady state solutions. This is equivalent to finding a particular class of fixed points as discussed in Chapter 2. In order to find the fixed points in both space and time, we set all of the derivatives to zero which gives a simple equation for the current–voltage curves under current bias in long samples:

$$j(E) = \rho(E)V(E)\,. \tag{7.20}$$

Numerical evaluations of Eq. (7.20) for several different values of compensation ratio α are shown in Figure 7.4a. For compensation just above unity, a region of

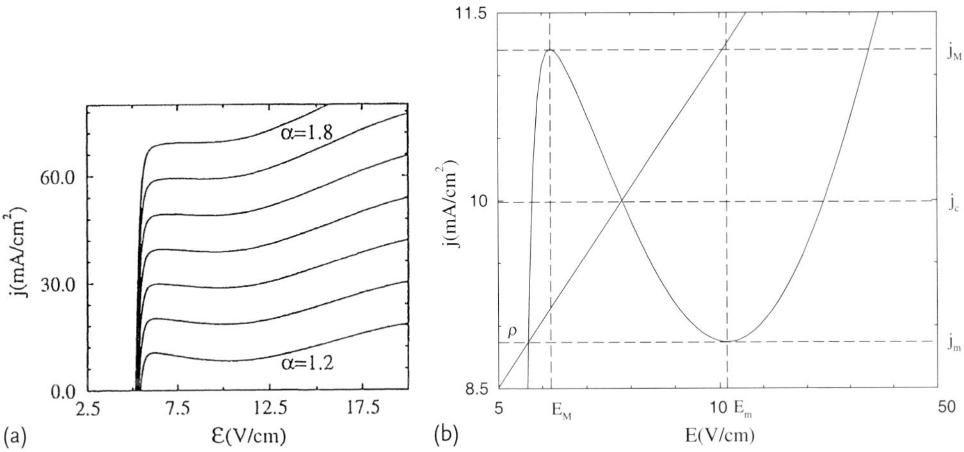

Fig. 7.4 (a) Current density vs. electric field for equally spaced values of compensation ratio, $\alpha \equiv a/d = 1.2$–1.8. (b) Blow-up of the NDR region for $\alpha = 1.21$ (from [8]).

NDR is evident for fields just above the breakdown value. Note that impurity breakdown occurs for a field strength of approximately 5 V/cm and is marked by a sharp increase in the current by several orders of magnitude. It should also be noted that the curves shown here are not expected to be valid for smaller fields since photogeneration has been neglected in the reduced model.

The presence of NDR in this system is a direct consequence of the behavior of the impact ionization coefficient $K(E)$. This is easily seen if we compute the differential conductivity from Eq. (7.20) and make the approximations that $R/K \ll 1$ and $R' \simeq 0$, which are reasonable for fields above breakdown. Then, we have

$$\frac{dj}{dE} \simeq \alpha \frac{RVK'}{K^2} + (\alpha - 1)V' .$$

The first term on the right-hand side is negative in regions where $K' < 0$ while the second term, which is always positive, can be made to have magnitude smaller than the first by choosing suitably small values of compensation α. However, as α increases substantially from unity, the second term must eventually dominate, and for larger compensation there should be no NDR in this system. This behavior is confirmed in Figure 7.4a for the largest values of α that are displayed. It is interesting to note that all p-Ge experiments showing nonlinear wave phenomena were performed on closely compensated samples. Some studied samples did not show any NDR effects, and it seems plausible that the compensation may have been too high to observe in these cases.

Figure 7.4b shows a more detailed J–E curve in the NDR range for $\alpha = 1.21$. This figure also clearly shows that there is a range of currents $J_m < J < J_M$ for which there are three possible values of field corresponding the given current value. The two outermost field values are locally stable while the middle value in the NDR region is not. As in previous discussions, we label the field values as $E_1 < E_2 < E_3$.

Let us now search for nontrivial traveling wave solutions to the reduced model Eq. (7.18). Such solutions should appear as spatially-nonuniform steady states of this equation and therefore we can equate the time derivatives to zero to write

$$C\frac{\partial^2 E}{\partial^2 \xi} + J\left[\frac{CV'}{V^2} - \frac{K+R}{V}\right]\frac{\partial E}{\partial \xi} + \frac{J}{V^2}(K+R)(J-\rho V) = 0\,. \tag{7.21}$$

This is a two-dimensional nonlinear *ordinary* differential equation and we can analyze it using simple phase plane methods. In fact, this equation has a good deal of qualitative similarity with the FHN system treated in Chapter 3. As we did there, we treat the wavespeed C as a bifurcation parameter which is to be fine-tuned in order to achieve heteroclinic or homoclinic connections. These orbits then correspond to a traveling wave of monotone front or pulse type. As previously noted, these wave solutions are only possible in the NDR region, that is, $J_m < J < J_M$. In order to find the closed trajectories, we adjust C and find that closed orbits are produced by a Hopf bifurcation about E_2. Performing a linear stability analysis about this point, we find that there is a purely imaginary pair of eigenvalues of a linearized system for a value C^* such that

$$C^* = \left.\frac{(K+R)V}{V'}\right|_{E_2}, \tag{7.22}$$

equivalent to setting the damping coefficient of the appropriate eigenvalues obtained from linearization of Eq. (7.21) equal to zero for $E = E_2$. This holds provided that the following auxiliary condition is satisfied:

$$\rho'|_{E_2} < \left.\frac{JV'}{V^2}\right|_{E_2} < 0\,, \tag{7.23}$$

which causes the first derivative of the source term in Eq. (7.21) to be positive, implying that the linearization of Eq. (7.21) about E_2 becomes the equation of the linear harmonic oscillator.

This Hopf bifurcation can be subcritical or supercritical. The stability of the limit cycles within the two-dimensional system is not particularly important for constructing traveling wave solutions. The key issue is that such solutions exist. In this case, the sub- or super-critical nature of the bifurcation simply determines whether the limit cycles bifurcate for $C > C^*$ or $C < C^*$. Normal form calculations for this system were reported in [11]. It was found in the simulations of that paper that bifurcations for J near to J_m tend to subcritical, while those for J near to J_M tend to be supercritical. In both cases, the amplitude of the limit cycle grows until it collides with one of the saddle points (E_1 or E_3) at a unique value of wavespeed $\overline{C}$. At this point, the limit cycle becomes a homoclinic orbit in the phase plane, departing one of the saddle points along the unstable manifold and returning along the stable manifold of the same point. By plotting these closed orbits in the E–ξ plane, we can see in Figure 7.5a that the homoclinic orbit emanating from $E_1(E_3)$ corresponds to the a high-(low-)field solitary wave. Similarly, we note that there is a special value of J where the growing limit cycles will simultaneously collide

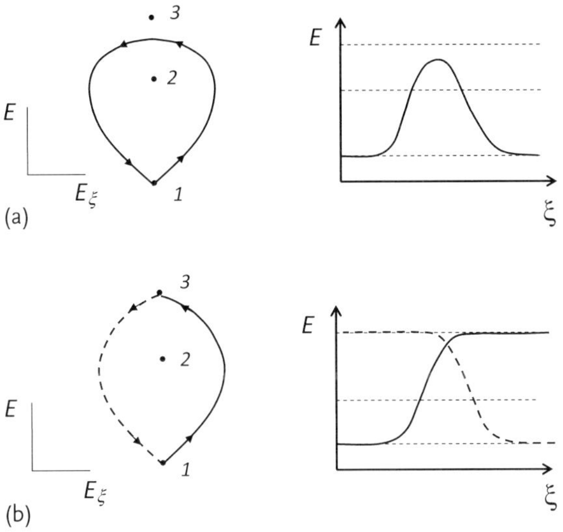

Fig. 7.5 Co-moving phase plane diagrams for (a) solitary waves and (b) monotone front waves.

with both saddle points. This results in a pair of heteroclinic connections which are interpreted as traveling fronts that are either monotonically decreasing or increasing, as in Figure 7.5b. Finally, we note that while the values of C for which the homoclinic or heteroclinic bifurcations occur must be determined numerically, the analytical expression of Eq. (7.22) provides a close estimate to these values.

Next we turn to the question of the dynamical stability of the traveling wave solutions on the infinite line under current bias. The basic approach is to linearize the reduced model about the spatially non-uniform solutions and then to ascertain the key properties of the eigenvalues. The stability analysis was first performed for this problem in [13]. One begins by assuming a separation of variables

$$E(\xi, \tau) = E^0(\xi) + \eta\varphi(\xi)e^{\lambda\tau}, \quad \text{with } \eta \ll 1 . \tag{7.24}$$

Here, $E^0(\xi)$ is the traveling wave whose linear stability we want to ascertain. Upon substitution of this ansatz into Eq. (7.18) and retaining only terms linear in the small parameter η, we have the following eigenvalue problem:

$$-C\varphi_{\xi\xi} + [\lambda + f_1(\xi)]\varphi_\xi + [\lambda f_2(\xi) + f_3(\xi)]\varphi = 0 , \tag{7.25}$$

where the functions f_i are defined as follows:

$$f_1(\xi) \equiv J\left(\frac{K+R}{V} - \frac{CV'}{V^2}\right)\bigg|_{E^0(\xi)} , \tag{7.26}$$

$$f_2(\xi) \equiv J\left[\frac{V'}{V^2}\right]\bigg|_{E^0(\xi)} , \tag{7.27}$$

$$f_3(\xi) \equiv \frac{d f_1}{d E^0}\frac{d E^0}{d\xi} - J\frac{d}{d E^0}\left[\frac{K+R}{V^2}(J-\rho V)\right]\bigg|_{E^0(\xi)} . \tag{7.28}$$

Now, we can see that if any eigenvalue λ of Eq. (7.26) has positive real part, then the traveling wave must be *unstable*. The analysis follows somewhat similar lines as in the discussion in Chapter 3 (see Eq. (3.21) in Chapter 3) for stability of waves in the FHN, but the arguments are inherently more complex due to the functions f_i and a more complex structure of the equation.

Let us now sketch the analysis of this eigenvalue problem. It is straightforward to show, by direct substitution, that $\lambda = 0$ is always an eigenvalue of this problem with corresponding eigenfunction given by $dE^0/d\xi$. Note that this eigenfunction decays to zero exponentially as $\xi \to \pm\infty$. Following [13], we introduce the following change of variable in order to eliminate the first order terms from the eigenvalue problem:

$$\varphi(\xi) \equiv \psi(\xi)e^F, \quad \text{where} \quad F \equiv \frac{1}{2C}\left[\lambda\xi + \int f_1(E^0(\xi))d\xi\right]. \tag{7.29}$$

This transforms the original eigenvalue problem into the following form:

$$\psi_{\xi\xi} + \left[\Lambda^2 + w(\xi)\Lambda + \Omega(\xi)\right]\psi = 0, \tag{7.30}$$

where we have introduced the following definitions:

$$\Lambda \equiv \frac{\lambda}{2C}, \tag{7.31}$$

$$w(\xi) \equiv J\left[\frac{CV' + (K+R)V}{CV^2}\right]\bigg|_{E^0(\xi)} > \Delta > 0, \tag{7.32}$$

$$\Omega(\xi) \equiv \left[\frac{f_1(E^0)}{2C}\right]^2 + \frac{1}{2C}\frac{df_1}{dE^0}\frac{dE^0}{d\xi} - \frac{J}{C}\frac{d}{dE^0}\left[\frac{K+R}{V^2}(J-\rho V)\right]. \tag{7.33}$$

Here, a small positive parameter Δ has been introduced to illustrate that w is always non-zero and positive. It is interesting to note that Eq. (7.30) is in an analogous form as the one-dimensional time-independent Schrödinger equation. However, the eigenvalue Λ appears quadratically rather than linearly as in the usual Schrödinger equation. In fact, one can use techniques of spectral theory to prove a one-one correspondence between the eigenvalues of Eq. (7.27) and those of Eq. (7.30) [13]. Therefore, we can conclude that a particular traveling is unstable provided that at least one of the Λ such that $\mathrm{Re}(\Lambda) > 0$ exists.

Based on our previous observation that the function $dE^0/d\xi$ solves Eq. (7.27) with eigenvalue $\lambda = 0$, we see that $\Lambda = 0$ is an eigenvalue of Eq. (7.30) with a corresponding eigenfunction given by

$$\psi_0(\xi) = \frac{dE^0}{d\xi}\exp\left[-\frac{1}{2C}\int f_1(E^0(\xi))d\xi\right]. \tag{7.34}$$

Thus, if E^0 denotes a solitary wave, then both $dE^0/d\xi$ and ψ_0 possess one zero, corresponding to the position of the local minimum (or maximum) of the wave.

In the context of a one-dimensional Schrödinger equation, the presence of one zero suggests an immediate analogy with the *first-excited state* of the corresponding quantum problem. In contrast, both the monotone front wave $dE^0/d\xi$ and also ψ_0 possess no zeros and they correspond to the *ground state* of the related quantum problem. We now multiply Eq. (7.30) by the complex conjugate ψ^* and the complex conjugate of Eq. (7.30) by ψ, subtract the resulting equations and integrate from $\xi = -\infty$ to ∞. The result is

$$(\Lambda - \Lambda^*)\left[\Lambda + \Lambda^* + \langle w(\xi)\rangle\right] = 0\,, \tag{7.35}$$

$$\text{where} \quad \langle w(\xi)\rangle = \frac{\int_{-\infty}^{\infty} w\,|\psi|^2 d\xi}{\int_{-\infty}^{\infty} |\psi|^2 d\xi} = 0\,. \tag{7.36}$$

Since w is positive, $\langle w\rangle$ is larger than the minimum value of $w(\xi)$, which we denote by 2Δ. Then, Eq. (7.35) implies that all eigenvalues whose real part is larger than $-\Delta$ are real. Any eigenvalues with a positive real part are therefore real and we can restrict our study of stability to real Λ. We now study the following auxiliary problem [13]:

$$\psi_{\xi\xi} + \left[\Lambda^2 + w(\xi)\Lambda + \Omega(\xi)\right]\psi = \mu(\Lambda)\psi\,, \tag{7.37}$$

in which the effective energy parameter $\mu(\Lambda)$ is the lowest eigenvalue of the Schrödinger equation (7.37) for a real parameter Λ. Bonilla and Vega were able to prove that $d\mu/d\Lambda = 2\Lambda + \langle w(\xi)\rangle$, so that $\mu(\Lambda)$ is a real monotonic increasing function of Λ, at least for $\Lambda > -\Delta$. From the preceding discussion, it is clear that one of its eigenvalue–eigenfunction pairs is just $\psi_0(\xi)$ and $\mu(0) = 0$. In fact, it is generally true that those values of Λ for which the solutions of Eq. (7.37) satisfy $\mu(\Lambda) = 0$ will automatically be eigenvalues of Eq. (7.30). First, consider the case of monotone fronts: $\mu(0) = 0$ corresponds to the ground state with $\Lambda = 0$. However, $\mu(\Lambda)$ is an increasing real-valued function and this means that $\mu(\Lambda) > 0$ for all $\Lambda > 0$. In other words, all of the zeroes of $\mu(\Lambda)$ occur for $\Lambda < 0$ which immediately tells us that monotone fronts are *dynamically stable*. On the other hand, the solitary wave corresponds to the first excited state of Eq. (7.37), which implies that there must be a ground state with lower effective energy, that is, $\mu(0) < 0$. Therefore, this means that there exists $\Lambda^* > 0$ such that $\mu(\Lambda^*) = 0$ and we conclude that the solitary wave is *dynamically unstable*.

In summary, we have found that monotone waves (which only occur for a special value of J) are dynamically stable, while the solitary waves are dynamically unstable. One implication of this is that it would seem unlikely that one would ever observe stable solitary waves in experiments under current bias since the slightest variation in J – due, for example, to noise – would cause the solitary wave to shrink or to grow until its top reaches $E = E_3$ and then expand as its sides become monotone fronts moving in opposite directions. In experiments under voltage bias, the voltage constraint implies that the area under the electric field is kept constant. This constraint may stabilize the solitary wave.

7.5
Nonlinear Wave Solutions in Finite Samples under Voltage Bias

Now, we turn to the case of voltage bias applied to a sample of finite length L. First of all, we discuss the steady state solution under voltage bias. These have time-independent but spatially varying solutions. The problem to solve is obtained from Eq. (7.15) by setting all time derivatives to zero:

$$V(E_{ss})\frac{\partial E_{ss}}{\partial x} + j(E_{ss}) - J = 0\,, \tag{7.38}$$

$$E_{ss}(0) = \rho_0 J\,, \tag{7.39}$$

$$\int_0^L E_{ss}(x)dx = \phi\,, \tag{7.40}$$

where E_{ss} denotes the steady state electric field profile. An analytical solution to these equations is obtained by integration as follows:

$$x = -\int_{\rho_0 J}^{E_{ss}(x)} \frac{V(E)}{J - j(E)} dE\,, \tag{7.41}$$

where it is understood that the external current J needs to be adjusted to satisfy the voltage constraint. The analysis of this implicit form of the solution and its dependence on voltage ϕ and contact resistivity ρ_0 has been studied in detail in [9]. We provide only a brief description of the key qualitative features. As the voltage ϕ moves into the NDR region, the steady state profile takes the form of a downward step function. Near the injecting contact, it adopts a larger value which persists to a position $0 < \bar{x} < L$ where it transitions to a lower value appropriate to the first branch of the j–E characteristic. As ϕ is further increased, this position $\bar{x}$ moves across the sample in order to maintain the voltage constraint until it hits the receiving contact at L. A typical example of this profile obtained from carrying out the numerical integration of Eq. (7.41) is presented in Figure 7.6. The cases of very small and very large contact resistivity have been treated in [9].

As soon as the average electric field reaches a threshold level corresponding approximately to the onset of NDR in the j–E curves, the steady states become dynamically unstable. It turns out that the bifurcation leading to this behavior can be understood analytically and we summarize the key results of that analysis at the end of this section. However, first, let us review the key numerical results showing the presence of large and small amplitude current oscillations and the associated spatiotemporal profiles in the electric fields. In the panels of Figure 7.7, we show numerically obtained current–time traces and space–time profiles of electric field $E(x,t)$ for two values of applied voltage that are above the oscillatory onset value. Panel (a) shows the small amplitude current oscillations that develop for an applied average field of 6.055 V/cm just above the onset value ($\mathcal{E}_{th} \simeq 6.051$ V/cm). The amplitude of these oscillations are found to smoothly transition to zero as the field

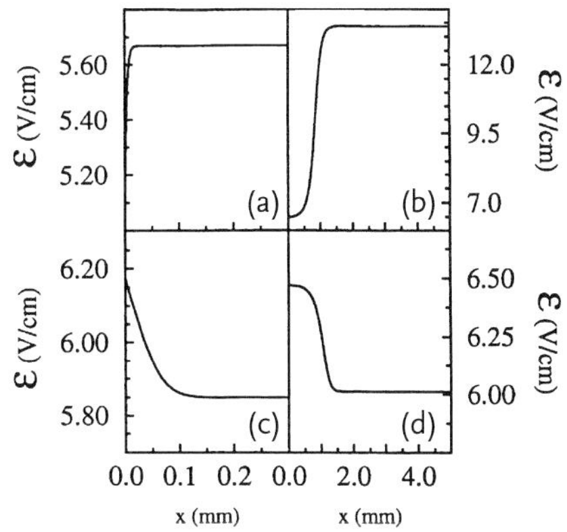

Fig. 7.6 Spatial profile of electric field vs. position x, for different applied voltages: (a) 5.67 V/cm; (b) 12.58 V/cm; (c) 5.85 V/cm; and (d) $\Phi/L = 6.048$ V/cm slightly below the onset value for oscillatory behavior, that is, $\Phi_c/L \simeq 6.054$ V/cm (from [8]).

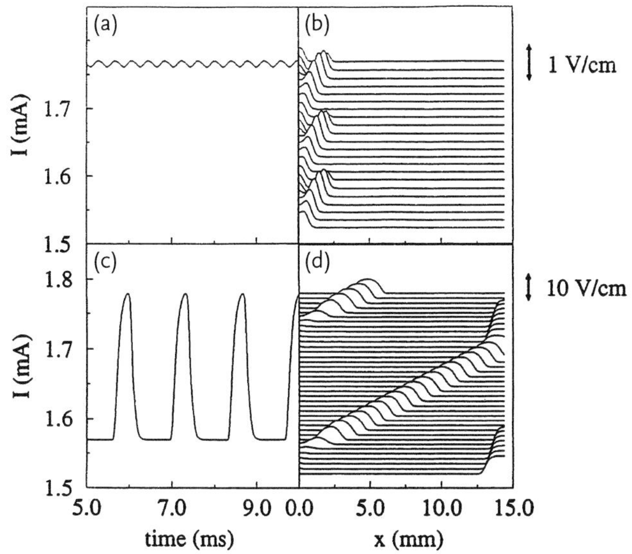

Fig. 7.7 Panels (a) and (b) show current vs. time and several snapshots of the electric field profile at time intervals of 0.05 ns, respectively, for $\Phi/L = 6.055$ V/cm, and exhibits spatially-decaying waves. Panels (c) and (d) show the same thing for mature solitary waves at $\Phi/L = 6.50$ V/cm (from [8]).

approaches the threshold value, suggesting that the initial bifurcation is of a supercritical Hopf type. Panel (b) shows the corresponding field profiles and we can see that these oscillations are due to small amplitude solitary waves that nucleate at the injecting contact and then move only a short distance into the sample before decaying to zero. Panel (c) shows large amplitude current oscillations that occur for an applied average field of 6.500 V/cm, well above the onset value ($\mathcal{E}_{th} \simeq 6.051$ V/cm). One can say that, in addition to having large amplitude, the current trace is very non-sinusoidal. This is reflected in Panel (d) which shows the corresponding field profiles. The large amplitude solitary waves (i. e., high field domains) that nucleate at the injecting contact and then move steadily across the sample until they arrive at the receiving contact are clearly evident. It is also clear from this figure that as an old wave disappears, the current has a spike upward which in turn, promotes the nucleation of the next wave. On the other hand, during the periods when the wave is far from the contacts, the current is nearly constant in time. The traveling solitary wave shown in Figure 7.7b obtained from the reduced model agrees rather nicely with the experimental data shown in Figure 7.2.

In Figure 7.8, we show the numerically obtained time-average current (with amplitude of the oscillatory part of the current indicated by the vertical bars) and the frequency of the current oscillations vs. the average applied electric field. Numerically, one also identifies that the solitary wave branches are born in a bifurcation as the current increases through the instability value. The numerically obtained bifurcation diagram is shown in Figure 7.8b. Note that the inset shows a hysteretic transition between small amplitude decaying oscillations and the solitary waves. This is, in part, understandable as a manifestation of the quasicontinuum of successive Hopf bifurcations.

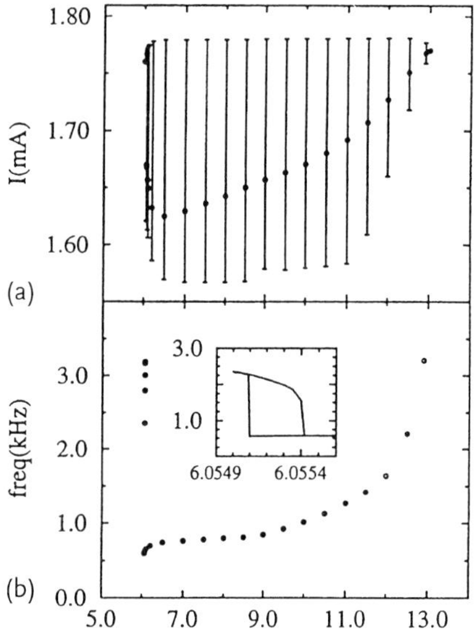

Fig. 7.8 (a)Time-averaged current vs. average electric field where vertical bars indicate the amplitude extremes of current oscillations. (b) The fundamental frequency vs. average electric field where the inset illustrates the hysteretic nature of the transition from the decaying wave to solitary wave branches (from [8]).

What can one say about the stability of the solitary wave state or the loss of stability of the steady state? Regarding the former question, we note that Bonilla, Higuera and Venakides were able to obtain rigorous stability results in a highly simplified model of the Gunn effect [12]. On the other hand, we can say quite a lot about the loss of stability of the spatially non-uniform steady state. We begin by assuming the usual separation of variables ansatz:

$$E(x,\tau) = E_{ss}(x) + \eta\,\hat{E}(x)e^{\lambda\tau}, \quad \text{and} \quad J(\tau) = J_{ss} + \eta\,\hat{J}(x))e^{\lambda\tau}, \quad \text{with} \quad \eta \ll 1. \tag{7.42}$$

This form is then substituted into the full reduced model Eqs. (7.15)–(7.17) to yield an eigenvalue problem. Rather than repeating steps here, let us note that this eigenvalue problem and the solution method are entirely analogous to what was shown for the onset of the Gunn effect under voltage bias in samples of finite length in the preceding chapter. The result of the eigenvalue calculation in the p-Ge case is [14]

$$\lambda_n \sim i\,(2n+1)\,\pi \frac{(K+R)\,|J'|}{\left(\frac{|J'|}{J} + \frac{V'}{V}\right)\ln L}, \quad \text{where} \quad n = 0, \pm 1, \pm 2, \ldots, o(L)\,. \tag{7.43}$$

Here, the $n = \{0, -1\}$ complex conjugate eigenvalue pair corresponds to the first mode to go unstable in the Hopf bifurcation which is evidently supercritical. This explains the origin of the small amplitude oscillations that signal the onset of time-dependent behavior. Note also that, in the limit of long samples $L \to \infty$, the higher order modes go unstable for values of bias ϕ that are only infinitesimally greater than for the first mode. In this case, the steady state loses stability through a quasicontinuum of Hopf bifurcation in analogy with the Gunn effect discussed in the previous chapter.

7.6 Multiple Shedding of Wavefronts in Extrinsic Material

In this section, we summarize numerical simulation results and an asymptotic theory which shows chaos under dc voltage bias associated with multiple shedding of wavefronts [15]. The treatment in this section follows closely the discussion in [15]. In particular, we introduce a model in the form of a finite-dimensional dynamical system which provides a simplified description of space charge wave motion in long samples. This model uses relevant information from the asymptotics, although we do not rigorously derive it from such asymptotic calculations. Nevertheless, solutions of the simplified model are in good agreement with the results of direct numerical simulations.

7.6.1 Numerical Results of Wavefront Shedding

In order to numerically solve the system of Eqs. (7.15)–(7.17) for E and J, one discretizes the equations using finite difference approximations to the derivatives and employs an implicit method to generate the solution. The initial condition for the electric field is taken to be spatially uniform with a value that is consistent with the global constraint, Eq. (7.17). In Figure 7.9, we show a space–time plot of the electric field and associated current density for $\Phi = 6.25\,\mathrm{V/cm}$, just above the threshold voltage value for which propagating domain behavior occurs. The gray scale ranges from 5.3 V/cm (black) to 14.2 V/cm (white), and this scale is used in all similar plots that follow. The dimensionless sample length is 3800 corresponding to a real p-Ge sample of length 3.87 cm, and the contact resistivity ρ_{con} is 780 Ω cm corresponding to a value of 10.0 in dimensionless units. This case corresponds well to experimental data, but published data in p-Ge were only presented for a few samples and relatively low bias values and contact resistivity. We clearly see that a single domain moves across the sample at constant speed until it reaches the receiving contact. As it disappears, a new wave is created at the injecting contact and the process periodically repeats. The current versus time plot indicates that the current is steady when the domain moves in the sample interior, while there is an increase when the domain reaches the receiving contact. It is important to note that the fronts of changing electric field (or equivalently, the regions of non-zero charge density) are sharp in space relative to other physical length scales for this

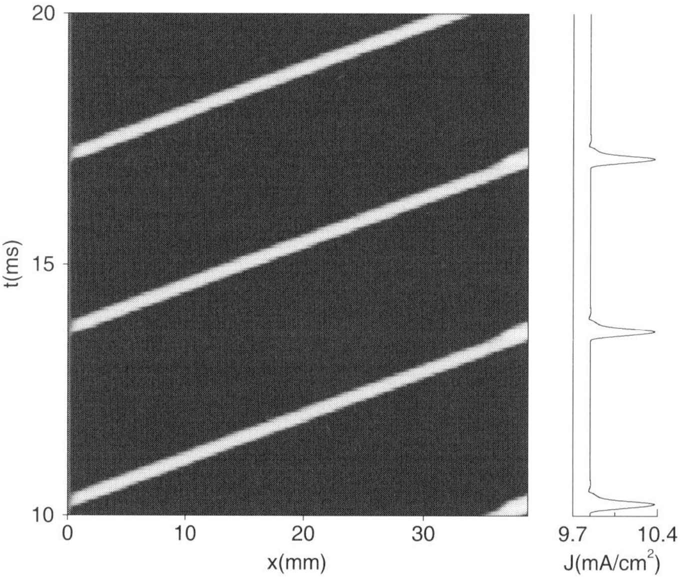

Fig. 7.9 Spatiotemporal evolution of the full reduced model showing periodic solitary wave motion for $\Phi = 6.25\,\mathrm{V/cm}$ in a long sample (from [15]).

problem, that is, the extent of the flat top domains or the sample length. It has been found that this separation of length scales increases with sample length. It is only for voltages near the onset point that one tends to observe rounded solitary waves rather than well-separated pairs of fronts.

For larger bias, the propagating domain becomes fatter and eventually, a second small domain nucleates and propagates part way into the sample; but it dissipates before reaching the receiving contact or merging with the larger domain. At even larger bias values, the second domain merges with the primary domain near the receiving contact and this situation is shown in Figure 7.10 which corresponds to $\Phi = 7.25\,\text{V/cm}$. Again, the current is plotted on the far right of the figure. When the first domain reaches the receiving contact, the current increases. Instead of immediately starting the nucleation of a new wave, the area lost by the dying wave is gained by the trailing wave. Note that the width of the trailing wave increases after the leading wave starts to disappear. The current increases, reaching a local maximum just before the trailing domain touches the leading domain, that is, the fronts collide. The current abruptly increases after the front collision and it rises to a global maximum at which point a new domain begins to nucleate at the injecting contact. As the domain forms, the current decreases and reaches a minimum at which point a new domain detaches and begins to propagate. Then, the current increases until a second smaller domain is nucleated. Finally, the current settles to a rather low constant level as the two domains move steadily and in unison across the interior region of the sample. Current behavior is apparently dominated by the major events involving the fronts: collisions with the contacts or with each other.

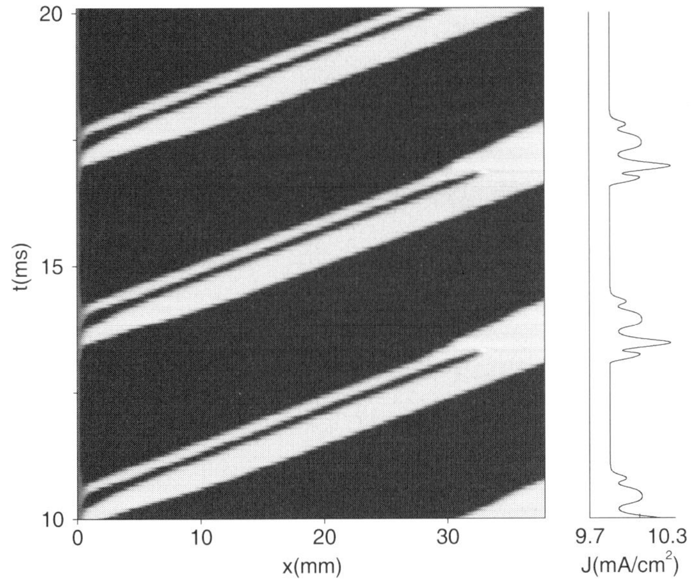

Fig. 7.10 Spatiotemporal evolution of a more complex traveling wave with clear merging $\Phi = 7.25\,\text{V/cm}$ (from [15]).

This suggests the viability of a dynamical model that focuses on discrete front motions and the current $J(t)$.

At even larger biases, the portion of the sample occupied by the high-field value E_3 is larger, reducing the separation between domains and giving more complicated $E(x, t)$ structure and J–t behavior. Figure 7.11 shows a space–time plot and current–time plot (on the side bar) for what appears to be a chaotic state for an applied bias of $\Phi = 10.0\,\text{V/cm}$. The spatiotemporal dynamics possess a great deal of structure and complexity. The process of multiple domain shedding is similar to that for the previous case. The large current peaks correspond to nucleation of leading domains. The leading domains cross the sample without catching up or forward-colliding with any other high-field regions, and are indicated by dark regions that extend all the way across the space–time diagram. Note that the spatial extent of the leading domains is larger than in Figure 7.7. The aperiodicity of the current is reflected in the irregular appearance of the maximum current peaks or, equivalently, of the dark strips that extend across the entire sample. A number of local maxima corresponding to the shedding of trailing domains are in between them.

Figure 7.12a shows a bifurcation diagram in which all values of successive current maxima as a function of Φ are plotted. An important feature in this diagram is the apparent presence of windows of chaotic behavior with a large number of points being visited. In contrast, for periodic states we see a small number of points corresponding to perfectly repeating current maxima. Also, in the periodic regimes there are points where various branches merge or disappear, corresponding to the

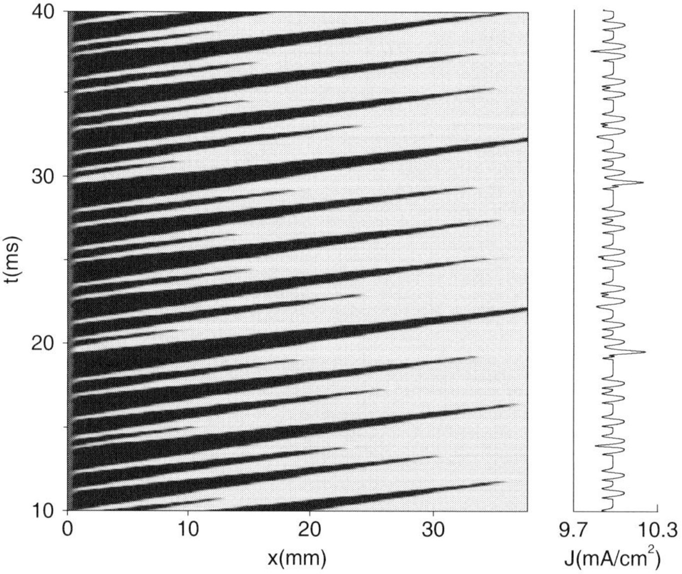

Fig. 7.11 Spatiotemporal evolution of the full reduced model showing chaotic wave motion for $\Phi = 10.0\,\text{V/cm}$ (from [15]).

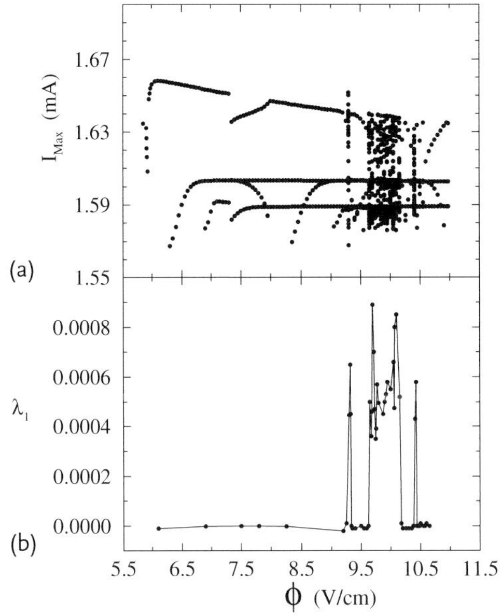

Fig. 7.12 (a) Bifurcation diagram constructed from the local current maximum, and (b) the largest Lyapunov exponent (from [15]).

development or destruction of trailing domains. For the parameter values selected here, we do not observe period doubling. It seems likely that the route to chaos is of boundary crisis type in which the attractor collides with a periodic orbit on its basin boundary [18].

In Figure 7.12b, we show the largest Lyapunov exponent λ_1 versus Φ for the reduced drift-diffusion model. This unambiguously confirms the presence of chaos in the "chaotic" windows. The next two exponents have been calculated and are never positive, so that the chaos is of a low-dimensional variety. To compute the exponents in this case, one can use an algorithm adapted for use with partial differential equations that uses adaptive control of the integration time step [8]. The values of λ_1 are zero in the periodic regimes as they should be for periodic behavior. The smallness of λ_1 in the chaotic regime is easily understood by recalling the period of the system, about 1000 nondimensional time units. This indicates that the chaos originates in processes that occur on time scales on the order of the front transit time across the sample. It is interesting to note that this type of chaotic behavior was never reported for experiments on p-Ge with time-independent voltage bias, most likely because experimentally studied samples were too short, biases were not sufficiently large, or contact resistivity was too low. However, in early experimental studies of the Gunn effect in GaAs, Gunn [16] observed that for long samples, current oscillations were almost completely random, resembling white noise. It is plausible that Gunn may have observed a similar form of chaos to the one described here.

7.6.2
Asymptotic Model for Wavefront Shedding

In order to describe the evolution of the current when all wavefronts are detached from the injecting contact, we introduce an asymptotic model that tracks the position of the wavefronts and also incorporates a simplified mechanism for creation and destruction of wavefronts. We begin by assuming that $E_M < \phi < E_m$ in Figure 7.4b and that $E_M/j_M < \rho < E_m/j_m$. For appropriate parameter values, there appears a regular oscillation of the current caused by repeated creation, motion and destruction of high-field domains in the sample. High-field domains are formed by two wavefronts separating a region where the electric field is uniform and large from regions of uniform low-field. Clearly, there are positively and negatively charged wavefronts, having $\partial E/\partial x > 0$ or $\partial E/\partial x < 0$, respectively. Near the contacts, there are narrow boundary layer regions where the electric field abruptly changes. Creation of high-field domains occurs at the injecting contact via an instability of the boundary layer which expels a high-field domain from the injecting contact to the bulk of the sample. Typically, the total current mostly changes during wavefront creation and destruction events. In the limit as $L \to \infty$, space and time scales are x/L and t/L, where we now use t to denote the dimensionless time. Then, $j(E) = J$, except in wavefronts and boundary layers at the contacts. If the field profile consists of a single high-field domain detached from the contacts, we have $E = E_3(J)$ inside the domain and $E = E_1(J)$ outside, where $E_1 < E_2 < E_3$ are the three zeros of $j(E) - J$ for $j_m < J < j_M$. High and low-field regions are joined by wavefronts which are the unique solution of Eq. (7.15) (with zero right-hand side) in the moving coordinate $\chi = x - X_\pm(t)$, $dX_\pm/dt = c_\pm(J)$ (the signs $+$ or $-$ refer to the charge inside the wavefront) and appropriate boundary conditions. For example, at a positively charged wavefront, $E \to E_1(J)$ as $\chi \to -\infty$ and $E \to E_3(J)$ as $\chi \to +\infty$. The numerically determined values of $c_\pm(J)$ are shown in Figure 7.13. Boundary layers obey (most of the time) a quasistationary version of Eq. (7.15) with appropriate boundary conditions on a semi-infinite spatial support. The instantaneous value of the current $J(t/L)$ determines the field profile in the low and high uniform-field regions and the velocity of the wavefronts.

Next, assume that we have an initial field profile consisting of N high-field domains (solitary waves), each formed by two wavefronts located at $X_+^{(i)}(t) < X_-^{(i)}(t)$. We shall number the wavefronts so that $X_\pm^{(i)}(t) > X_\pm^{(i+1)}(t)$, and if necessary, we shall consider $X_-^{(1)} = L$ and $X_+^{(N)} = 0$. Then, the positions $X_\pm^{(i)}(t)$ are given by

$$X_\pm^{(i)}(t) = \int_{t_{b,\pm}^{(i)}}^{t} c_\pm(J(s))\, ds\,, \tag{7.44}$$

where $t_{b,\pm}^{(i)}$ denotes the time at which the i-th monopole (with positive or negative charge) was born at $x = 0$.

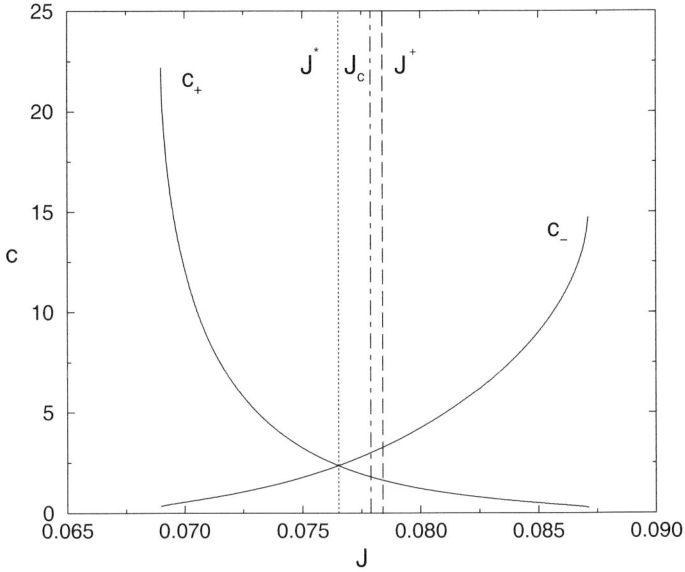

Fig. 7.13 Velocities of heteroclinic orbits between E_1 and E_3 in dimensionless units (from [15]).

The evolution of the total current density is determined by the bias condition (7.17), which may be approximately evaluated as

$$\phi = E_1(J) + [E_3(J) - E_1(J)] \sum_{i=1}^{N} \frac{X_-^{(i)} - X_+^{(i)}}{L} \tag{7.45}$$

(terms of order $1/L$ and smaller have been ignored here; note that the $X_\pm^{(i)}/L$ are of order unity). We can get an ordinary differential equation for J by differentiating Eq. (7.45) with respect to time and then substituting $dX_\pm^{(i)}/dt = c_\pm(J)$ in the result. We obtain

$$\frac{dJ}{dt} = \frac{1}{L} \frac{(E_3 - E_1)^2}{\frac{\phi - E_1}{j_3'} + \frac{E_3 - \phi}{j_1'}} (n_+ c_+ - n_- c_-) , \tag{7.46}$$

$$\frac{dX_+^{(i)}}{dt} = c_+(J) , \quad \frac{dX_-^{(i)}}{dt} = c_-(J) , \tag{7.47}$$

where i goes from 1 to N. The quantities n_+ and n_- are, respectively, the number of positive and negative monopoles *detached* from the contacts (i. e., excluding possible monopoles at $x = 0$ and $x = L$), while j_1' and j_3' denote the derivative of the static $j(E)$ characteristic with respect to electric field, evaluated at E_1 and E_3, respectively. Notice that the system of Eqs. (7.44)–(7.47) completely specifies the behavior of current and field profile on the scales x/L and t/L, except that we do not have conditions for determining when new fronts are emitted from the injecting contact.

We start with the simple case of Figure 7.9: the motion of a single high-field domain far from the contacts. J satisfies Eq. (7.46) with $n_+ = n_- = 1$, that is,

$$dJ/ds = A(J)[c_+(J) - c_-(J)],$$

where $s = t/L$ and $A(J) > 0$. As shown in Figure 7.4b, $c_+(J)$ [resp. $c_-(J)$] is a decreasing (resp. increasing) function of the current. Therefore, J evolves exponentially fast toward the zero of the right-hand side of this equation, $J = J^*$. When the leading wavefront at $x = X_-$ arrives at $x = L$, it disappears almost instantaneously in the scale s, and we obtain $dJ/ds = A(J)\, c_+(J) > 0$, so that the current increases. The injecting boundary layer near $x = 0$ ceases to be quasi-stationary when J surpasses the value J_c at which the line $J = E/\rho$ intersects the second (decreasing) branch of $J = j(E)$. This results in the expulsion of a narrow high-field domain from $x = 0$ to the interior of the sample. It is enough to say that a certain semi-infinite problem has to be numerically solved and matched to the resulting situation with a narrow high-field domain (consisting of a region of $E = E_3(J)$ bounded by positively and negatively charged wavefronts) near $x = 0$ and a high-field region from a positively charged wavefront to $x = L$. In the new situation, we have $dJ/ds = A(J)\,[2\,c_+(J) - c_-(J)]$, and J tries to go toward the zero $J = J^+$ of $2\,c_+(J) - c_-(J)$. Depending on the resistivity ρ, J_c can be larger than J^+ (and then J decreases toward J^+), or $J_c \in (J^*, J^+)$ (and then J starts increasing, and a second high-field domain may be expelled from $x = 0$). The simplest case, $J_c > J^+$, was described asymptotically in [17]. Provided ϕ is large enough, J evolves exponentially fast towards J^+. When the old domain leaves the sample, only two wavefronts (bounding the new high-field domain) remain, and J evolves exponentially fast toward J^* so that a period of the oscillation is completed; see Figure 7.9. The second case, $J_c \in (J^*, J^+)$ is more complicated: numerical simulations show that multiple high-field domains may coexist in the sample at the same time as in Figure 7.10.

To achieve a simplified description of the current oscillation, valid for any positive value of J_c, we proceed to further examine the shedding process. The simplest rule to determine when a new wavefront is shed from the injecting contact would be as follows: a positive (negative) front is emitted at the instant that J passes through J_c with positive (negative) time derivative. However, this rule neglects the time needed for a sufficient charge to be injected at the contact to form a propagating front. We may estimate the effective delay time by considering the time evolution of Eq. (7.15) evaluated at the injecting contact,

$$\dot{u} + J\,\frac{K+R}{V}\,u = \left(\frac{1}{V} - \rho J\frac{V'}{V^2}\right)\dot{J} + J\,\frac{K+R}{V^2}\,[J - j(E)], \tag{7.48}$$

where $u(t) = \partial E(0,t)/\partial x$ and the argument of V, K and R is $E(0,t) = \rho J(t)$, that is, the value of electric field at the injecting contact. In this equation, we can think of $J(t)$ as the driving charge injection processes which determine front launching. Based on extensive simulations of the reduced model, one finds that $u(t)$ must attain a sufficiently large positive or negative value for the front to detach and begin to propagate. This value is generally found to lie between 50% and 90% of the steady state value that $|u|$ would have in the case of no propagating fronts, that is, where E near the injecting contact rapidly rises to the E_3 value or rapidly falls to the E_1 value. The asymptotic model system is fully defined once the threshold is set and consists in Eqs. (7.44)–(7.47). Thus, in the limit of an infinitely long sample, terms of order $\epsilon = 1/L$ drop and we arrive at a low-dimensional dynamical system which essentially consists of: (i) propagating negative and positive charge points that move according to the Eq. (7.44), (ii) subject to the conservation law Eq. (7.47), (iii) which produce a measurable current according to Eq. (7.46), and are created according to Eq. (7.48). Note that in the $\epsilon \to 0$ limit, the current will exhibit slope discontinuities at the formation times $t_{b,\pm}^{(i)}$ and destruction times $t_d^{(i)}$, but will be otherwise continuous and governed by Eq. (7.46). We note that the rigorous foundations of this and similar asymptotic models have been explored recently using singular perturbation methods [17].

We estimate the order-of-magnitude of the time delay associated with wavefront formation by evaluating Eq. (7.48) for $J \approx J_c$. This implies a relaxation time of

$$\tau^* \approx \frac{V}{\frac{dJ_c}{dt} - (\alpha - 1)\, KV}, \tag{7.49}$$

that is, the approximate time for $u(t)$ to go from a value of $u(0) = -1$ to $u(\tau) = 0$. By dJ_c/dt, we refer to the value of dJ/dt when J crosses J_c. Then, we adopt the criterion that a new front is born at $x = 0$ at the time $t + a\tau$ where t is the time at which $J = J_c$, and $a\tau$ is a delay time. Here, a is a number of order one which is determined from simulation of the reduced model for a particular bias voltage and then assumed to apply over the complete range. Wavefront destruction is assumed to occur instantaneously at times $t_d^{(i)}$ when $X_-^{(i)} = X_+^{(i-1)}$ or when $X_+^{(i)} = X_-^{(i)} = L$. We ignore the finite duration of (fast) monopole destruction stages which is equivalent to the well-justified approximation of neglecting the diffusive boundary layer at the receiving contact [17]. It should also be kept in mind that the index instantaneously decreases by one when wavefronts downstream collide with one another.

Let us now use the above asymptotic model to interpret the simulation results for Eqs. (7.15)–(7.17). The case of contact resistivity such that $J_c > J^+$ has been explained already: we obtain the usual Gunn effect with, at most, one solitary wave detached from the contacts for any time; see Figure 7.9. Let us now assume that the contact resistivity is such that $J_c \in (J^*, J^+)$. Then, the current will increase after creation of a solitary wave because $2c_+ - c_- > 0$, and multiple wave shedding is possible [17]. This situation is shown in Figure 7.14, which shows simulation results of our simplified asymptotic model for similar parameters to those of Figure 7.10. The latter is depicted using data from direct numerical simulation of the reduced partial differential equation model Eqs. (7.15)–(7.17). To obtain Figure 7.14, the values of a were set to 13.75 for positive front emission and 7.82 for negative front emission. We use these same values in the data of Figure 7.15, which has the same bias values as Figure 7.11. The chaos appears to be closely tied to the

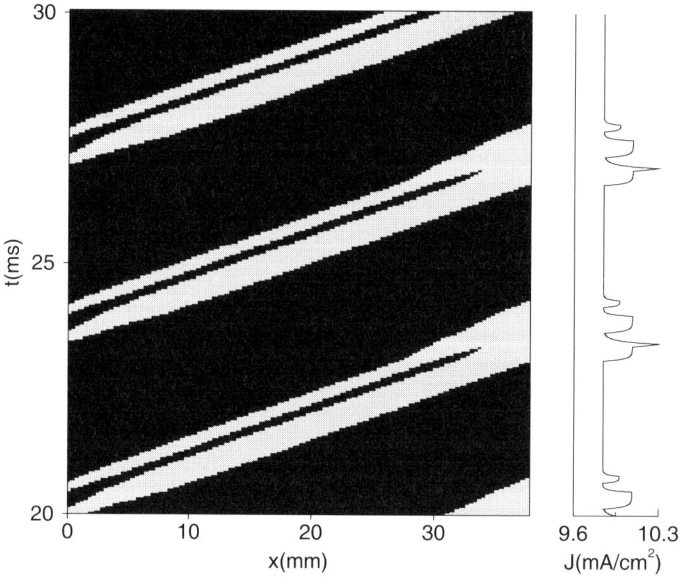

Fig. 7.14 Spatiotemporal evolution of the asymptotic model showing merging motion for $\Phi = 7.25\,\mathrm{V/cm}$. Compare with Figure 7.10 (from [15]).

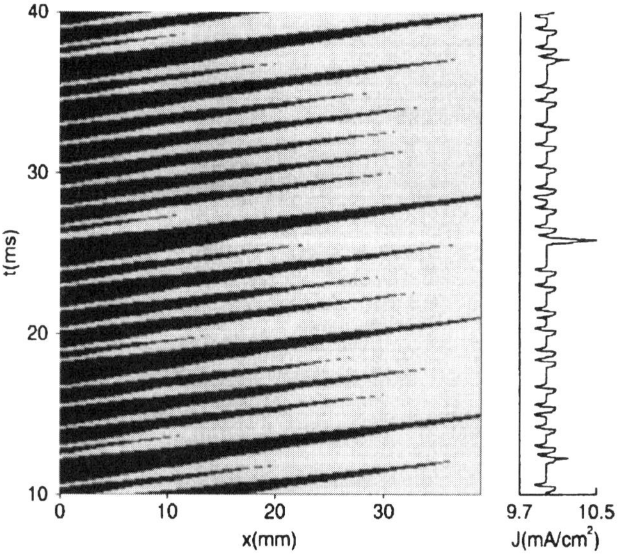

Fig. 7.15 Spatiotemporal evolution of the asymptotic model chaotic motion for $\Phi = 10.0\,\mathrm{V/cm}$. Compare with Figure 7.11 (from [15]).

asynchronous emission of fronts. This explains why the chaos observed for this partial differential equation system is low-dimensional. It is still an open question as to how the maximal number of domains possible might scale with system size and contact resistivity.

References

1 Haller, E.E., Hansen, W.L., and Goulding, F.S. (1981) Review of photoconductor properties in p-Ge, *Adv. Phys.*, **30**, 93.

2 Teitsworth, S.W., Westervelt, R.M., and Haller, E.E. (1983) Non-linear oscillations and chaos in electrical breakdown in Ge. *Phys. Rev. Lett.*, **51**, 825–828.

3 Teitsworth, S.W. and Westervelt, R.M. (1984) Chaos and broadband noise in extrinsic semiconductors. *Phys. Rev. Lett.*, **53**, 2587–2590.

4 Teitsworth, S.W. and Westervelt, R.M. (1986) Subharmonic and chaotic response of periodically driven extrinsic Ge photoconductors. *Phys. Rev. Lett.*, **56**, 516–519.

5 Gwinn, E.G. and Westervelt, R.M. (1986) Frequency locking, quasiperiodicity, and chaos in extrinsic Ge. *Phys. Rev. Lett.*, **57**, 1060–1063.

6 Kahn, A.M., Mar, D.J., and Westervelt, R.M. (1991) Spatial measurements of moving space-charge domains in p-type ultrapure germanium. *Phys. Rev. B*, **43**, 9740–9749.

7 Teitsworth, S.W., Bergmann, M.J., and Bonilla, L.L. (1995) Space charge instabilities and nonlinear waves in extrinsic semiconductors. *Nonlinear Dynamics and Pattern Formation in Semiconductors and Devices* (ed. F.-J. Niedernostheide), Proc. Phys. Vol. 79, Springer, Berlin.

8 Bergmann, M.J. (1996) Numerical investigations of space-charge instabilities and nonlinear waves in extrinsic semiconductors, Ph.D. thesis, Duke University.

9 Bonilla, L.L. (1992) Small signal analysis of spontaneous current instabilities in extrinsic semiconductors with trapping: application to p-type ultrapure germanium. *Phys. Rev. B*, **45**, 11642–11654.

10 Cantalapiedra, I.R., Bonilla, L.L., Bergmann, M.J., and Teitsworth, S.W. (1993) Solitary wave dynamics in extrinsic semiconductors under dc voltage bias. *Phys. Rev. B*, **48**, 12278–12281.

11 Bonilla, L.L. and Teitsworth, S.W. (1991) Theory of periodic and solitary space charge waves in extrinsic semiconductors. *Physica D*, **50**, 545–559.

12 Bonilla, L.L., Higuera, F., and Venakides, S. (1994) The Gunn effect: Instability of the steady state and stability of the solitary wave in long extrinsic semiconductors, *SIAM J. Appl. Math.*, **54**, 1521–1541.

13 Bonilla, L.L. and Vega, J.M. (1991) On the stability of wavefronts and solitary space charge waves in extrinsic semiconductors under current bias. *Phys. Lett. A*, **156**, 179–182.

14 Bonilla, L.L., Cantalapiedra, I.R., Bergmann, M.J., and Teitsworth, S.W. (1994) *Semicond. Sci. Technol.*, **9**, 599.

15 Cantalapiedra, I.R., Bonilla, L.L., Bergmann, M.J., and Teitsworth, S.W. (2001) *Phys. Rev. E*, **63**, 056216. Copyright (2001) by the American Physical Society.

16 Gunn, J.B. (1965) Instabilities of current and potential distribution in GaAs and InP, *Proceedings of the Symposium on Plasma Effects in Solids* (ed. J. Bok), pp 199 207, Dunod, Paris.

17 Bonilla, L.L., Hernando, P.J., Herrero, M.A., Kindelan, M., and Velázquez, J.L. (1997) Asymptotics of the trap-dominated Gunn effect in p-type Ge. *Physica D*, **108**, 168–190.

18 Ott, E. (2002) *Chaos in Dynamical Systems*, 2nd edn., Cambridge University Press.

8
Nonlinear Dynamics in Semiconductor Superlattices

8.1
Introduction

In this chapter, we discuss some salient issues of nonlinear electronic transport in semiconductor superlattices. We will have a chance to see how concepts and techniques introduced in previous chapters on waves in excitable media can be applied to study the rich nonlinear dynamics in these materials. In Chapter 5, we discussed how to use singular perturbation methods in order to obtain a reduced description of a strongly coupled superlattice (SL) with a single miniband in terms of electron density and electric field. Here, we shall analyze those reduced equations and related ones to describe nonlinear phenomena in weakly coupled SL, in which the barrier width is much larger than the reciprocal of the typical electron wave number inside the barrier. In the opposite limit, nonlinear transport in strongly coupled SL can be often described by continuum balance equations (e. g., drift-diffusion equations), semiclassical Boltzmann equations inside minibands or quantum Wigner equations, as explained in Chapter 5. The continuum limit of discrete drift-diffusion equations used to model a weakly coupled SL is a partial differential equation of drift-diffusion type, and therefore our analysis of such an equation in this chapter is similar to the analysis of strongly coupled SL. Thus, we also describe some aspects of transport in strongly coupled SL in this chapter. The discussion in this chapter is based on the treatment presented in [22].

Semiconductor superlattices (SL) are unique nonlinear systems with regards to electron transport properties. To form a simple type of SL, several identical periods comprising two layers of compatible semiconductor materials (e. g., GaAs and AlAs, with similar lattice constants) are grown in a vertical direction, [1–3]. The conduction band of an infinitely long SL looks like a one-dimensional crystal formed by a succession of quantum barriers and wells (AlAs and GaAs in our previous example) [2]. In typical experiments of vertical transport, a finitely long, doped or undoped SL is placed in the central part of a diode (forming a n^+-n-n^+ or a n-i-p structure) and contacts are attached at the two ends thereof. Depending on bias conditions, SL configuration, doping, temperature and other control parameters, the current across the SL and the electric potential inside display a great wealth of patterns and dynamical behavior, some of which we shall describe in this

Nonlinear Wave Methods for Charge Transport. Luis L. Bonilla and Stephen W. Teitsworth

ISBN: 978-3-527-40695-1

chapter. For more information, see the review [4]. What makes the SLs unique nonlinear systems is that, in certain regimes, they present features of spatially discrete systems, whereas their behavior is more typical of continuous systems in other regimes. These different properties may be described by discrete or continuous balance equation models. In this chapter, we will describe the more relevant features of nonlinear dynamics in SLs, particularly emphasizing wave fronts and wave propagation, which are the keystone of our present understanding.

The current–voltage (I–V) characteristic curve of a weakly coupled SL presents the distinctive multibranch aspect shown in Figure 8.1a when there is enough electric charge inside the SL. This charge may be produced by doping or irradiating the SL with appropriate laser intensity (thereby generating electron–hole pairs). The multiplicity of branches in the I–V curve is due to the formation of electric field domains (EFDs), as explained below. When the electron density inside the SL is sufficiently low, the electric field is almost spatially uniform. The I–V curve is smooth and it has a number of peaks as depicted in the dotted photocurrent curve of Figure 8.1b corresponding to an undoped reference sample with the same configuration as the doped sample. (The photocurrent is proportional to the number of electrons that arrive at the collector per unit time after a pulse excitation and can be measured by the varying current through a resistor in series with the SL [5].) These peaks of the current occur at field values at which the subbands of two adjacent wells are aligned. Take, for instance, the second peak in Figure 8.1b. If the applied field is $(\mathcal{E}_{C2} - \mathcal{E}_{C1})/(el)$ (e is minus the electron charge and l is the SL period), electrons in the lowest subband of a given well, $C1$ with energy $\mathcal{E}_{C1}$ in the absence of bias, tunnel resonantly across the barrier to the $C2$ subband of the adjacent well (with energy $\mathcal{E}_{C1}$ in the absence of bias). They then undergo scattering processes (with LO phonons, interface roughness, ...) and fall to the $C1$ subband

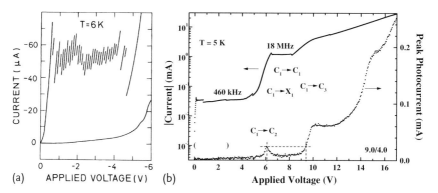

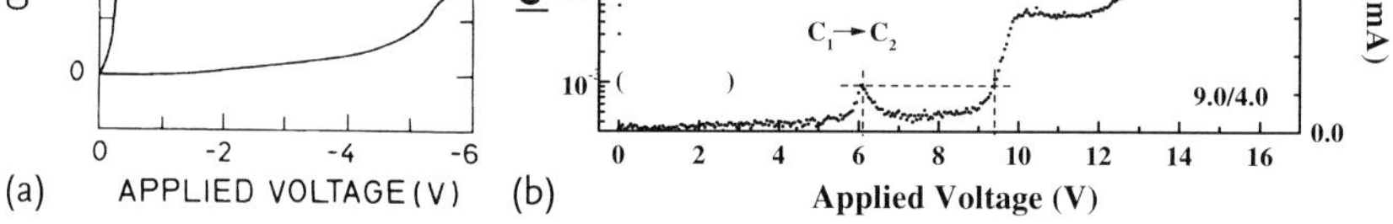

Fig. 8.1 Current–voltage characteristic curves of GaAs/AlAs SL. (a) Multibranched characteristic for the first plateau of a negatively biased sample with $w = 9$ nm, $d = 4$ nm, doping $N_D^w = 1.5 \times 10^{11}$ cm^{-2}. The smooth curve with lower current corresponds to an undoped SL with the same configuration (reference sample). Reprinted from [6]. (b) Characteristic with flat plateaus for the same sample, positively biased over a larger voltage range. The dotted characteristic curve underneath is the peak photocurrent vs. applied voltage obtained by time-of-flight experiments in the undoped reference sample. Reprinted from [7].

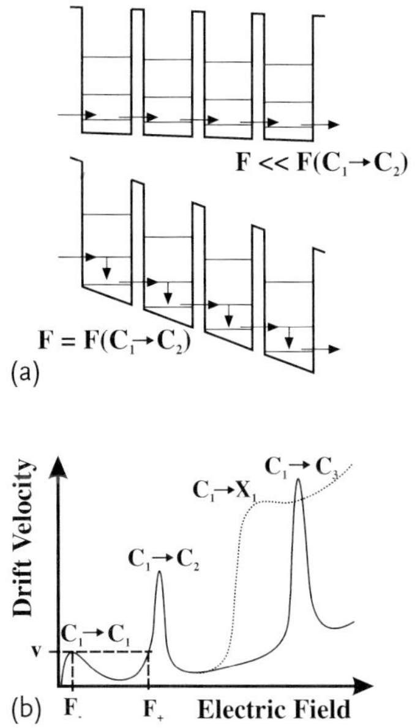

Fig. 8.2 (a) Conduction band profile of a superlattice with three subbands in an applied electric field. (b) Drift velocity vs. field characteristics of this superlattice consisting of the nonresonant background, the low-field maximum ($C_1 \rightarrow C_1$), and the first two resonant tunneling maxima. The dotted line indicates a possible shape of the $v(F)$ curve in the case of a $\Gamma \rightarrow X$ resonance located between the $C_1 \rightarrow C_2$ and $C_1 \rightarrow C_3$ resonances. The dashed line is explained in the text. Reprinted from [7].

of that well. This process is called sequential tunneling; see Figure 8.2a. Outside resonances, charge transport is less efficient, resulting in lower values of the current. Whatever the value of the charge in the SL, the bias regions between peaks of the current are called plateaus. The first peak is approximately located at a field equal to the scattering energy divided by el, while the second peak is approximately given by $(\mathcal{E}_{C2} - \mathcal{E}_{C1})/(el)$, and so on.

Provided the charge inside the SL is large enough, a spatially uniform field configuration is unstable and a stationary field configuration having two EFDs appears. A rough description of these domains is as follows. Assume the bias is on the first plateau of the I–V curve and that both current and voltage are independent of time. The electron drift velocity is proportional to the photocurrent of the undoped reference sample in Figure 8.1b, which is schematically indicated in Figure 8.2b. Part of the SL (the first m SL periods) is at an almost zero field F_- and resonant tunneling between $C1$ subbands of adjacent wells dominates. The rest of the SL ($N - m$ SL periods, where N is the total number of periods) is at the field F_+ having the same velocity (proportional to the photocurrent) as the $C1$–$C1$ resonant field F_-; see Figure 8.2b. F_+ is slightly smaller than that corresponding to the $C1$–$C2$ resonant tunneling process described above, $F = (\mathcal{E}_{C2} - \mathcal{E}_{C1})/(el)$. The extension of the lower field domain is given by the condition that $mF_- + (N - m)F_+ \approx (N - m)(\mathcal{E}_{C2} - \mathcal{E}_{C1})/(el)$ equal the volt-

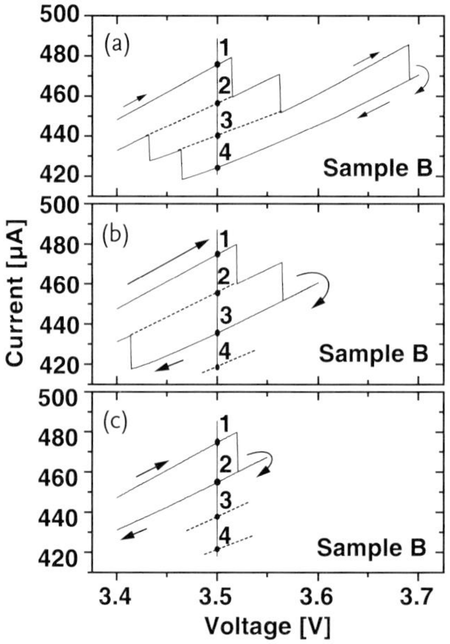

Fig. 8.3 Details of multistability of stationary field configurations in Figure 8.1a obtained by successive adiabatic up and down sweeping of the current–voltage characteristic curve. (a–c) correspond to different up and down sweeping processes designed to reveal different stable branches. Reprinted from [9].

age across the SL. As shown in Figure 8.3, we may have coexistence of different domain branches for a given value of the voltage. The field configuration at each branch differs in the extension of low and high- field domains. The domain wall separating low and high-field domains is at the last SL period for the first branch after the $C1$–$C1$ resonance in Figure 8.1a. Then, $m = N - 1$. For the following branch, $m = N - 2$, and so on. The extension of the high-field domain is smaller for a branch having higher current value than another one at the same voltage. Adiabatic up and down sweeping of the I–V curve show hysteresis cycles and experimentally demonstrates multistability of different branches, as shown in Figure 8.3. The coexistence of two EFDs inside the SL is confirmed by photoluminescence measurements [8].

At intermediate values of the charge inside the SL, the I–V curve is flat between peaks as depicted in the curve with larger current of Figure 8.1b. When time-resolved measurements of the current are carried out in these flat regions, self-sustained oscillations are observed, as in Figure 8.4. In certain cases, experimental evidence shows that self-oscillations appear as recycling and motion of domain walls separating two EFDs. This can be experimentally verified by indirect means because the current time trace is directly measurable but the electric potential or field inside the SL is not. Typical indirect measurements are the time-resolved photoluminescence spectra. The photoluminescence intensity (shown in grey scale in Figure 8.4c so that darker regions correspond to larger intensity) is

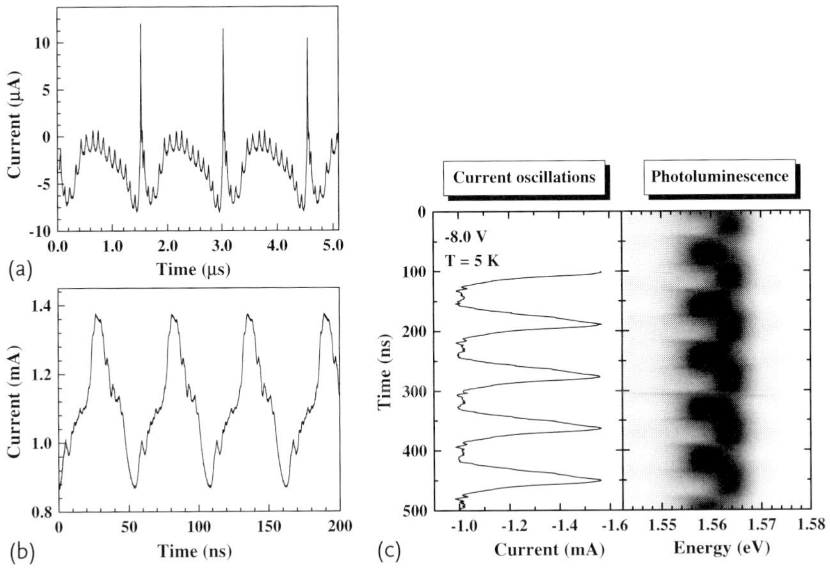

Fig. 8.4 Time-resolved current oscillations for a GaAs/AlAs SL with $w = 9$ nm, $d = 4$ nm, $N_D^w = 10^{11}$ cm^{-2}. (a) Time-resolved signal showing current self-oscillations for a bias on the first plateau (2.77 V) at 6 K. Notice the high-frequency spikes superimposed on the self-oscillations. (b) The same for a voltage on the second plateau (7.3 V at 5 K): current spikes are much harder to observe. Reprinted from [10]. (c) Comparison of time-resolved current oscillations and the photoluminescence spectrum for a bias on the second plateau. Reprinted from [11].

recorded for each time as a function of the wavelength or energy of light emitted due to electron–hole recombination. Let us assume that there are two EFDs in the SL, one at zero field so that electrons tunnel from the $C1$ subband of one well to the $C1$ subband of the next well, and the other domain at a field corresponding to $C1$–$C2$ resonant tunneling. Recombination light emitted by wells in the high-field domain has less energy than light emitted at zero field due to the Quantum Confined Stark Effect [12]. If most of the SL is at zero field, the photoluminescence spectrum will present a peak at a higher energy than that corresponding to a SL, most of which is at the $C1$–$C2$ resonant field. Thus, the strength of the photoluminescence peaks gives an idea of the spatial extension of the corresponding EFDs. Figure 8.4c shows that photoluminescence peaks corresponding to two domains (the values of the field at these two domains correspond to the $C1$–$C2$ and the $C1$–$C3$ resonant fields) alternate in strength, in phase with current self-oscillations. This suggests that the extension of these domains is periodically changing with time, as indicated by numerical simulations of the models later presented. An interesting feature of the self-oscillations in Figure 8.4a is the presence of current spikes superimposed on the signal. Experimental and numerical evidence shows that each spike occurs as the domain wall moves from one SL well to the next one. Then, the number of spikes during each self-oscillation period indicates the number of wells traversed by the domain wall during its motion. In Figure 8.4a, the number of spikes in one period of the current time trace is significantly lower (17) than the number of SL periods (40). This suggests that recycling and motion of domain walls is confined to part of the SL [7]. Theory and numerical simulations of discrete models [13, 14] (see below) show that current self-oscillations may be due to recycling and motion of two different type of waves: accumulation wave fronts (which are moving charge monopoles) and pulses of the electric field (which are moving charge dipoles). The current time traces and the corresponding electric field profiles of these two oscillatory states are different, and the main difference is the following. Dipoles recycle at the injecting region and move until they arrive at the receiving region, whereas monopoles become first appreciable well inside the SL and then move until they arrive at the receiving contact. Correspondingly, the current time trace of dipole-type self-oscillations should exhibit a number of spikes close to the number of SL periods (40 for Figure 8.4), whereas the current time trace of monopole-type self-oscillations should exhibit less current spikes [13]. Together with the evidence provided by photoluminescence measurements, the current time traces in Figure 8.4a support the conclusion that self-sustained current oscillations for that SL at the indicated bias conditions are due to recycling and motion of monopole wave fronts.

8.2
Spatially Discrete Model for the Doped Weakly Coupled SL

In weakly coupled SLs, the main transport mechanism is sequential resonant tunneling. Nonlinear phenomena seen in experiments can be described by discrete balance equations. Early models were motivated by static domain formation, and we may cite Likharev *et al.* [15], Laikhtman [16] and Laikhtman and Miller [17] among the first authors who recognized the importance of using discrete equations. In this chapter, we will describe a model of SL transport introduced in 1994 by our group [18], aspects of which have been steadily improved in collaboration with other groups, particularly Platero's and Schöll's. Models of the same type (involving two electron populations corresponding to the lowest two subbands instead of only one electron density) were put forth by Schöll and his group, also beginning in 1994 [14, 19].

For a weakly coupled doped SL, there are certain important time scales that are well separated. The scattering time is the time that an electron originally in an excited subband takes to lose energy and fall to the first subband. The escape time or tunneling time is the average time that an electron needs to escape from one well to the adjacent one. Finally, the dielectric relaxation time is the longer time scale over which the local electric field at a SL period evolves. The latter yields the order of magnitude of the period of self-sustained oscillations, and it is of the order of the SL length divided by an average electron velocity. In a weakly coupled SL, the scattering time is much smaller than the tunneling time, and the latter is much smaller than the macroscopic dielectric relaxation time. Then, the dominant mechanism of vertical transport is sequential tunneling, only the first subband of each well is appreciably occupied and the tunneling current is quasistationary. The latter statement means that we can calculate the well-to-well tunneling current density across a barrier assuming a constant value of the applied electric field and a constant electron density at the wells adjacent to the barrier. The resulting expression is then considered to be the *constitutive relation* between the tunneling current and the local electric field(s) and electron densities at the wells adjacent to the barrier.

Consider a SL with $N + 1$ barriers and N wells whose widths are d_B and d_W, respectively. Then, the SL period is $l = d_B + d_W$. The zeroth barrier separates the injecting region from the first SL well, whereas the N-th barrier separates the N-th well from the collecting region. At zero field, the barriers have energies V_b. Assume that F_i is the average field across one SL period consisting of the i-th well and the $(i-1)$-th barrier. Here, we use F_i instead of the actual electric field which is $-F_i$ in the direction perpendicular to the interfaces between the two different semiconductors conforming the SL. Similarly, n_i is the 2D electron density at the i-th well, concentrated in a plane perpendicular to the growth direction inside the i-th well. Then, the Poisson equation (averaged over the i-th period) and the charge continuity equation are

$$F_i - F_{i-1} = \frac{e}{\varepsilon}(n_i - N_D)\,. \tag{8.1}$$

$$e\frac{dn_i}{dt} = J_{i-1\to i} - J_{i\to i+1}\,. \tag{8.2}$$

Here, $J_{i\to i+1}$ is the tunneling current density across the i-th barrier, that is, from well i to well $i+1$ with $i = 1, \ldots, N$.

By differentiating Eq. (8.1) with respect to time and inserting the result into Eq. (8.2), we obtain

$$\varepsilon\frac{dF_i}{dt} + J_{i\to i+1} = J(t)\,, \tag{8.3}$$

in which the total current density $J(t)$ is the same function for all i. Equation (8.3) is a discrete version of Ampère's equation. We shall now assume that $J_{i\to i+1}$ depends on F_i, n_i, and n_{i+1}. The tunneling current across the i-th barrier clearly depends on the electrochemical potentials of the adjacent wells, thereby depending on n_i and n_{i+1}. That $J_{i\to i+1}$ depends on F_i, but not on F_{i+1}, is a simplifying assumption. We complete our description by adding the voltage bias condition

$$\frac{1}{(N+1)}\sum_{i=0}^{N} F_i = \frac{V(t)}{(N+1)l}\,, \tag{8.4}$$

for the known voltage $V(t)$.

Our discrete system of equations consists of Eq. (8.1) for $i = 1, \ldots, N$, Eq. (8.3) for $i = 0, \ldots, N$, and Eq. (8.4). In total, we have $(2N + 2)$ equations for the unknowns $n_1, \ldots, n_N, F_0, \ldots, F_N$, and $J(t)$. We need to derive the constitutive relations $J_{0\to 1}$ (tunneling from the injecting region to the SL), $J_{i\to i+1}$ (tunneling across an inner barrier of the SL), and $J_{N\to N+1}$ (tunneling from the SL to the collecting region). Equation (8.3) evaluated for $i = 0$ and $i = N$ determines the boundary conditions.

8.2.1 Tunneling Current Density

The functions $J_{i\to i+1} = \mathcal{J}(F_i, n_i, n_{i+1})$ should be calculated from quantum mechanical considerations. However, it is important to point out that a consistent calculation from first principles should also provide the reduced equations for F_i and n_i in the form of Eqs. (8.1)–(8.4). This has not been done by anyone to this date. However, on the evidence that many predictions based on discrete equations have been verified in experiments, we adopt them and use a reasonable form of constitutive equations for $J_{i\to i+1} = \mathcal{J}(F_i, n_i, n_{i+1})$. See [20] for a more general discrete model. Following ideas put forth in [21], Appendix A of [22] shows how to derive from this more general model the following analytical expressions for the tunneling current density:

$$n_i = \frac{m^* k_B T}{\pi\hbar^2} \ln\left[1 + e^{\frac{\mu_i - \mathcal{E}_1}{k_B T}}\right], \tag{8.5}$$

$$J_{i\to i+1} = \frac{e\, v^{(f)}(F_i)}{l} \left\{ n_i - \frac{m^* k_B T}{\pi\hbar^2} \ln\left[1 + e^{-\frac{eF_i l}{k_B T}} \left(e^{\frac{\pi\hbar^2 n_{i+1}}{m^* k_B T}} - 1\right)\right]\right\} \tag{8.6}$$

$$v^{(f)}(F_i) = \sum_{\nu=1}^{n_{\max}} \frac{\hbar^3 l}{2m^{*2}} \frac{(\gamma_1 + \gamma_\nu)\mathcal{T}_i(\mathcal{E}_1)}{(\mathcal{E}_1 - \mathcal{E}_\nu + eF_i l)^2 + (\gamma_1 + \gamma_\nu)^2}, \tag{8.7}$$

for $i \neq 0, N$. The electron density Eq. (8.5) is the same as Eqs. (5.109) and (5.109) of Chapter 5 in the limits of zero collision broadening and zero miniband widths which are appropriate for a weakly coupled SL. Then, μ_i and $\mathcal{E}_1$ are the chemical potential and the energy state (measured from the bottom of the well) at the i-th QW, respectively. In Eq. (8.6), $n_{\max}$ is the number of states with energies $\mathcal{E}_\nu$ in the wells. Note that the *forward drift velocity* $v^{(f)}(F_i)$ is a sum of Lorentzians centered at the resonant field values $F_\nu = (\mathcal{E}_\nu - \mathcal{E}_1)/(el)$. These Lorentzians have widths $\gamma_1 + \gamma_\nu$, where γ_ν is the broadening of the energy state due to scattering. The other functions appearing in Eq. (8.6) are

$$\mathcal{T}_i(\epsilon) = T_i(\epsilon)\, B_{i-1,i}(\epsilon)\, B_{i,i+1}(\epsilon), \tag{8.8}$$

$$B_{i,i+1} = \frac{k_{i+1}}{d_W + \alpha_i^{-1} + \alpha_{i+1}^{-1}}, \tag{8.9}$$

$$T_i = \frac{16 k_i k_{i+1} \alpha_i^2 e^{-2\alpha_i d_B}}{\left(k_i^2 + \alpha_i^2\right)\left(k_{i+1}^2 + \alpha_i^2\right)}. \tag{8.10}$$

Here, T_i is the transmission coefficient through the i-th barrier while k_i and α_i are the wave vectors at the i-th well and barrier, respectively, given by

$$\hbar k_i = \sqrt{2m^*\epsilon}\,, \quad \hbar k_{i+1} = \sqrt{2m^*[\epsilon + e(d_W + d_B)F_i]}\,,$$

$$\hbar\alpha_{i-1} = \sqrt{2m^*\left[eV_b + e\left(d_B + \frac{d_W}{2}\right)F_i - \epsilon\right]}\,,$$

$$\hbar\alpha_i = \sqrt{2m^*\left(eV_b - \frac{ed_W F_i}{2} - \epsilon\right)}\,,$$

$$\hbar\alpha_{i+1} = \sqrt{2m^*\left[eV_b - e\left(d_B + \frac{3d_W}{2}\right)F_i - \epsilon\right]}\,. \tag{8.11}$$

Equations (8.5)–(8.7) have been obtained by the following procedure:

- Using the energy states (subbands) and wave functions of isolated QW as a basis, calculate the tunneling rate through the i-th barrier (from the μ subband of the i-th well to the ν subband of the $(i+1)th$ well) by means of Fermi's golden rule (which is first order time dependent perturbation theory). The tunneling rate multiplied by a factor $2e$ (due to spin degeneracy) and by the difference between the Fermi–Dirac distribution functions of both wells is the tunneling current between the subbands. The tunneling current is then found by summing over all energies and states in both wells.
- When transforming the sums over the lateral momenta and energies into integrals, broadened densities of states for both wells should be included because of scattering.
- Assuming that the broadening functions do not depend on lateral momentum, the integrals over lateral momenta only affect the Fermi–Dirac distribution functions and can be explicitly calculated as in Eq. (5.101) of Chapter 5.
- The integrals over energy can be approximately calculated in the limit of a scattering broadening which is small compared to subband energies and chemical potentials. Then, Eqs. (8.5)–(8.7) are obtained.

The tunneling current is a linear function of n_i, but it is a strongly nonlinear function of n_{i+1}. Moreover, $J_{i\to i+1} \sim ev^{(f)}(F_i)n_i/l$, for F_i of the order of the first resonant value or larger. For such values, the resulting tunneling current density has the same shape as assumed in the original *discrete drift model* [18, 23]. In the limit $k_B T \gg \pi\hbar^2 n_{i+1}/m^* \approx \pi\hbar^2 N_D/m^*$, we can approximate Eq. (8.6) by

$$J_{i\to i+1} = \frac{v^{(f)}(F_i)}{l}\left(n_i - n_{i+1}\, e^{-\frac{eF_i l}{k_B T}}\right)\,. \tag{8.12}$$

This equation can be rewritten as a discrete drift-diffusion electronic current density

$$J_{i\to i+1} = \frac{e\, n_i\, v(F_i)}{l} - \frac{e\, D(F_i)}{l^2}\,(n_{i+1} - n_i)\,, \tag{8.13}$$

$$v(F_i) = v^{(f)}(F_i)\left(1 - e^{-\frac{eF_i l}{k_B T}}\right)\,, \tag{8.14}$$

$$D(F_i) = v^{(f)}(F_i) l e^{-\frac{eF_i l}{k_B T}}\,, \tag{8.15}$$

in which $v^{(f)}(F_i)$ is given by Eq. (8.7). Note that the drift velocity in Eq. (8.14) is somewhat similar to the Lorentzian dependence given by the simpler theory by Kazarinov and Suris [24]. The additional prefactor in Eq. (8.6) yields a more complicated field dependence in our formulas.

In the opposite limit $k_B T \ll \pi\hbar^2 n_{i+1}/m^* \approx \pi\hbar^2 N_D/m^*$, Eq. (8.6) becomes

$$J_{i\to i+1} = \frac{e\, v^{(f)}(F_i)}{l}\left[n_i - \left(n_{i+1} - \frac{m^* e F_i l}{\pi\hbar^2}\right)\theta\left(n_{i+1} - \frac{m^* e F_i l}{\pi\hbar^2}\right)\right]. \tag{8.16}$$

Here, $\theta(x)$ denotes the Heaviside unit step function, and we have ignored an exponentially small term for $n_{i+1} < m^* e F_i l/(\pi\hbar^2)$.

8.2.2
Boundary Conditions

We also need constitutive relations for tunneling currents from the emitter and to the collector. Following similar arguments as before, it is possible to show that the tunneling currents in the contact regions are

$$J_{0\to 1} = j_e(F_0)\,, \quad J_{N\to N+1} = \frac{n_N}{N_{Dc}}\, j_c(F_N)\,, \quad i = 1,\ldots,N-1\,, \tag{8.17}$$

where the emitter and collector current densities are functions of the field. Linear approximations of these currents are often used:

$$j_e(F) = \sigma F, \quad j_c(F) = \sigma_c F\,, \tag{8.18}$$

where σ and σ_c are contact conductivities. Using the constitutive relations (8.6) and (8.17), the discrete tunneling model consists of solving the N Poisson equations (8.1), the $N+1$ Ampère's equations (8.3), and the bias condition (8.4) for the $2N+2$ unknowns, F_i ($i = 0, 1, \ldots, N$), n_i ($i = 1, \ldots, N$) and $J(t)$. The same equations with the high field approximation $J_{i\to i+1} \approx e n_i v^{(f)}(F_i)/l$ is called *the discrete drift model*, and with the high temperature approximation Eq. (8.13) is called *the discrete drift-diffusion model.*

8.2.3
Photoexcitation in an Undoped SL

In the case of an undoped photoexcited weakly coupled SL, the discrete equations describing nonlinear charge transport are

$$\varepsilon\,(F_i - F_{i-1}) = e\,(n_i - p_i)\,, \qquad (8.19)$$

$$\varepsilon\,\frac{dF_i}{dt} + J_{i\to i+1} = J(t)\,, \qquad (8.20)$$

$$\frac{dp_i}{dt} = \gamma - r(F_i, I)\,n_i p_i\,, \qquad (8.21)$$

where p_i is the hole 2D density at the i-th well. Tunneling of holes is neglected so that only photogeneration and recombination with electrons enter Eq. (8.21). Photogeneration γ is a function of the intensity I of laser light with frequency ω_{exc} given by $\gamma = I\,\alpha_{3D}(\hbar\omega_{\text{exc}})\,d_W/(\hbar\omega_{\text{exc}})$, where α_{3D} is the three-dimensional absorption coefficient. The recombination coefficient is

$$r(F, I) = \left(\frac{n_{\text{ref}}}{n_{\text{in}}\pi c}\right)^2 \int_0^\infty \frac{\omega^2 \alpha_{2D}(\hbar\omega, F)}{\exp(\frac{\hbar\omega}{k_B T}) - 1}\,d\omega\,, \qquad (8.22)$$

where n_{ref} and $n_{\text{in}} \approx \gamma\,n_{\text{ref}} d_W/c$ are the refraction index and the density of intrinsic charge carriers, respectively. The 2D absorption coefficient α_{2D} is proportional to the square of the modulus of the electron–hole overlap integral for a constant electric field F. For fixed I, the recombination coefficient decreases as the field $F > 0$ increases [25].

8.2.4
Continuous Drift-Diffusion Model for a Strongly Coupled SL

The simplest model for a strongly coupled SL is the drift-diffusion Eq. (5.155) of Chapter 5:

$$\varepsilon\,\frac{\partial F}{\partial t} + F\mu_{ES}(F)\left(\varepsilon\,\frac{\partial F}{\partial x} + \frac{eN_D}{l}\right) - D_{ES}(F)\,\varepsilon\,\frac{\partial^2 F}{\partial x^2} = J(t)\,, \qquad (8.23)$$

$$\mu_{ES}(F) = \frac{e\Delta^2 l^2}{8k_B T\hbar^2 \nu_e}\,\frac{1}{1 + \left(\frac{eFl}{\hbar\nu_e}\right)^2}\,, \qquad (8.24)$$

$$D_{ES}(F) = \frac{k_B T}{e}\,\mu_{ES}(F)\,. \qquad (8.25)$$

Equation (8.25) is a nonlinear Einstein relation linking the diffusion coefficient $D_{ES}(F)$ and the Esaki–Tsu electron mobility $\mu_{ES}(F)$. The total current density is determined from the voltage bias condition:

$$\int_0^L F(x, t)\,dx = V\,, \qquad (8.26)$$

where $L = (N+1)l$. We should impose boundary conditions at the end of the SL. Simple ones are Ohm's law:

$$\sigma F = J(t) - \varepsilon \frac{\partial F}{\partial t}, \qquad (8.27)$$

at $x = 0$ and at $x = L$, where σ is the conductivity of contacts. Mathematically speaking, Eqs. (8.23)–(8.27) are just a variant of the Kroemer model for the Gunn effect in bulk GaAs.

8.3
Nondimensionalization of the Discrete Drift-Diffusion Model

To account for most of the observed phenomena in weakly coupled SLs, it is sufficient to study the simpler discrete drift-diffusion (DDD) model, comprising Eqs. (8.1), (8.3), (8.4), (8.13), (8.17) and (8.18). To analyze the DDD model, it is convenient to render all equations dimensionless. Let $v(F)$ reach its first positive maximum at (F_M, v_M). We adopt F_M, N_D, v_M, $v_M\, l$, eN_Dv_M/l and $\varepsilon F_M l/(eN_Dv_M)$ as the units of F_i, n_i, $v(F)$, $D(F)$, J and t, respectively. For the first plateau of the 9/4 SL of [7], we find $F_M = 6.92$ kV/cm, $N_D = 1.5 \times 10^{11}$ cm^{-2}, $v_M = 156$ cm/s, $v_M\, l = 2.03 \times 10^{-4}$ cm^2/s and $eN_Dv_M/l = 2.88$ A/cm^2. For a circular sample with a diameter of 120 μm, the units of current and time are 0.326 mA and 2.76 ns, respectively. Then, Eqs. (8.1), (8.3), (8.4), (8.13), (8.17) and (8.18) become

$$\frac{dF_i}{dt} + v(F_i)\, n_i - D(F_i)\,(n_{i+1} - n_i) = J, \qquad (8.28)$$

$$F_i - F_{i-1} = \nu\,(n_i - 1), \qquad (8.29)$$

$$\frac{1}{N+1}\sum_{i=0}^{N} F_i = \phi, \qquad (8.30)$$

$$\frac{dF_0}{dt} + j_e(F_0) = J, \qquad (8.31)$$

$$\frac{dF_N}{dt} + j_c(F_N)\, n_N = J. \qquad (8.32)$$

Here, we have used the same symbols for dimensional and dimensionless quantities. The parameters $\nu = eN_D/(\varepsilon F_M)$ and $\phi = V/[(N+1)F_M l]$ are dimensionless doping density and average electric field (bias), respectively. In the simplest case, we may use Ohm's law, $j_e = \sigma F = j_c(F)$ for the emitter and collector current densities. For the 9/4 SL, $\nu \approx 3$. We recall that $i = 1, \ldots, N-1$ in Eq. (8.28) and $i = 1, \ldots, N$ in Eq. (8.29). As a handy reference, a table is provided with the definitions of the units we have used to nondimensionalize the model equations and their numerical value for the first plateau of the 9/4 SL. Given a dimensionless magnitude, we should multiply it by its corresponding unit in the table to obtain its value in physical units. In addition to the numerical values included in the table

Table 8.1 Definitions of the units used to nondimensionalize the DDD model and their numerical values for the first plateau of the 9/4 SL.

F	n	v	D	J	t
F_M	N_D	v_M	$v_M l$	$\frac{eN_D v_M}{l}$	$\frac{\varepsilon F_M l}{eN_D v_M}$
kV/cm	10^{11} cm^{-2}	cm/s	10^{-4} cm^2/s	A/cm^2	ns
6.92	1.5	156	2.03	2.88	2.76

(corresponding to the first plateau of the 9/4 SL of [7]), we can calculate F_M, v_M, and so on for any SL or plateau we may be interested in. Then, we can use the corresponding numerical values to compute the units listed in the table, and translate the dimensionless values of the magnitudes given by the theory into physical dimensional units. Using dimensionless units has the following advantages:

i. redundant parameters in the equations are eliminated, and
ii. different terms in the equations can be properly compared.

Thus, the task of neglecting small terms can be appropriately carried out.

8.4
Wave Fronts and Stationary States under Current Bias

Nonlinear phenomena in weakly coupled SLs are dominated by the formation and motion of EFDs. Domain walls separating two EFDs may be pinned because of the spatially discrete structure of a weakly coupled SL, thereby producing multistable branches of static EFDs under voltage bias conditions. By switching the voltage, we can probe the motion of domain walls and observe the complex nonlinear behavior of the EFDs.

This type of phenomena does not occur in strongly coupled SLs, where the nonlinear phenomena due to motion of EFDs are undriven and driven self-oscillations of the current, which are akin to the Gunn effect in bulk semiconductors. In strongly coupled SLs, the current self-oscillations are due to the periodic motion of charge dipoles (which are smooth pulses of the electric field) on an EFD. The multistable static domains studied in the remainder of this section occur only in weakly coupled SLs.

Typically, EFDs are regions of a weakly coupled SL having a uniform (or almost uniform) value of the electric field. An EFD is bounded by either a domain wall that connects it to a different domain or by the SL boundaries. To characterize the behavior of EFDs, we should investigate the behavior of their domain walls. Mathematically speaking, these domain walls are stable traveling wave front solutions for the chosen model equations describing the SL in the absence of any boundaries and under dc current bias conditions. Wave fronts are transition regions in which the electric field varies monotonically from its value in one EFD to its value in the other EFD. An increasing field profile corresponds to a charge accumulation layer (CAL), while a decreasing field profile corresponds to a charge depletion layer (CDL). Even pulses of the electric field can be described by the motion of two wave fronts, CAL and CDL, which form their leading and trailing edge.

To be precise, let us consider an infinite SL under constant current bias J described by Eqs. (8.28) and (8.29), or equivalently,

$$\frac{dF_i}{dt} + v(F_i)\,\frac{F_i - F_{i-1}}{\nu} = D(F_i)\,\frac{F_{i+1} + F_{i-1} - 2F_i}{\nu} + J - v(F_i)\,. \tag{8.33}$$

Clearly, there are two stable spatially homogeneous stationary solutions, namely, $F^{(1)}(J)$ and $F^{(3)}(J)$, where $v(F^{(k)}) = J$, $F^{(1)}(J) < F^{(2)}(J) < F^{(3)}(J)$. We are interested in nonuniform front states of the DDD model which satisfy $F_i \to F^{(1)}(J)$ as $i \to -\infty$ and $F_i \to F^{(3)}(J)$ as $i \to \infty$. These states correspond to CALs: $n_i - 1 = (F_i - F_{i-1})/\nu > 0$. It is also possible to have CDLs such that $F_i \to F^{(3)}(J)$ as $i \to -\infty$ and $F_i \to F^{(1)}(J)$ as $i \to \infty$, with $n_i - 1 < 0$. CALs are either stationary or time-dependent. In the second case, they are wave fronts moving with constant velocity $c = c(J, \nu)$, such that $F_i(t) = F(i - ct)$, $i = 0, \pm 1, \ldots$ $F(\tau)$ is a smooth profile which solves the following nonlinear eigenvalue problem for c (measured in wells traversed per unit time) and $F(\tau)$:

$$c\,\frac{dF}{d\tau} = v(F) - J + v(F)\,\frac{F - F(\tau - 1)}{\nu} - D(F)\,\frac{F(\tau + 1) + F(\tau - 1) - 2F}{\nu}\,, \tag{8.34}$$

$$F(-\infty) = F^{(1)}(J)\,, \quad F(\infty) = F^{(3)}(J)\,. \tag{8.35}$$

CDLs obey the same equation with the obvious change in boundary conditions. Numerical simulations have always shown CDLs moving with positive velocity. As we will see later, wave fronts are the key to understanding more complex phenomena related to the dynamics of EFDs. From now until the end of this section, we shall consider CALs only. By using a comparison principle, the existence of stationary fronts has been rigorously proved [26]. Outside the interval of current values in which there are stationary fronts, we can only prove that there are fronts moving to the right or the left [26]. Moving and stationary fronts cannot exist simultaneously at the same value of the current [27].

What is the role of these idealized wave fronts in finite, voltage-biased SLs? In finite SLs, boundary conditions modify the extent and shape of EFDs, but wave fronts are still useful to describe their field profile. The reason for this is that static EFDs in a voltage-biased SL occur naturally with constant current, and boundary conditions affect very small regions near the ends of a long SL. If the current is changing with time in a long voltage-biased SL, its envelope changes on time scales that are much longer than the dielectric relaxation time. Then, the electric field profile adjusts adiabatically to the instantaneous value of the average current. Such a profile can then be approximated by EFDs and by CALs or CDLs defined as wave fronts of an infinite SL with a constant value of the current. This behavior will be explained in more detail later.

8.4.1
Pinning

Numerical simulations of Eq. (8.33) show that, after a short transient, a variety of initial conditions such that $F_i \to F^{(1)}(J)$ as $i \to -\infty$ and $F_i \to F^{(3)}(J)$ as $i \to \infty$ evolve towards either a stationary or moving monopole. For systematic numerical studies, we have therefore adopted an initial step-like profile, with $F_i = F^{(1)}(J)$ for $i < 0$, $F_i = F^{(3)}(J)$ for $i > 0$ and $F_0 = F^{(2)}(J)$. The boundary data are taken to be $F_{-N} = F^{(1)}(J)$, $F_N = F^{(3)}(J)$ with N large. Figure 8.5 is a phase diagram showing the regions in the plane (J, ν) where different wave fronts are stable. There are two important values of ν, $\nu_1 < \nu_2$. The critical value ν_1 defines the minimum doping value needed for a stationary monopole to exist, whereas ν_2 is the minimum doping needed for a monopole to move upstream with $c < 0$:

- For $0 < \nu < \nu_1$ and each fixed J in the interval $(v_m, 1)$, only traveling monopole fronts moving downstream (to the right) are observed. This implies that there will be no experimental observation of static current branches if the 2D doping obeys $N_D < \nu_1 \epsilon F_M/e$. For $\nu > \nu_1$, stationary monopoles are found (and static current branches in the SL I–V characteristic can be experimentally observed); see Figure 8.5. Upper bounds for ν_1 are [26]

$$\nu_{1b}(J) = v_m \, \frac{F_m - F^{(1)}(J)}{1 - v_m} \, , \tag{8.36}$$

$$\nu_c = \min \, \nu_{1b}(J) = v_m \, \frac{F_m - 1}{1 - v_m} \, . \tag{8.37}$$

  The smallest possible bound $\nu_{1b}(J)$ is $\nu_c = 0.198$, for our numerical example. We have found that the smallest value of $\nu_1(J)$ is 0.16.
- For $\nu_1 < \nu < \nu_2$, traveling fronts moving downstream exist only if $v_m < J < J_1(\nu)$, where $J_1(\nu) < 1$ is a critical value of the current; see Figure 8.5. If $J_1(\nu) < J < 1$, the stable solutions are steady fronts (stationary monopoles). We have found that $\nu_2 = 0.33$.
- New solutions are observed for $\nu > \nu_2$. As before, there are traveling fronts moving downstream if $v_m < J < J_1(\nu)$, and stationary monopoles if $J_1(\nu) < J < J_2(\nu)$. $J_2(\nu) < 1$ is a new critical current; see Figure 8.5. For $J_2(\nu) < J < 1$, the stable solutions of Eq. (8.33) are monopoles traveling upstream (to the left). As ν increases, $J_1(\nu)$ and $J_2(\nu)$ approach v_m and 1, respectively. Thus, stationary solutions are found for most values of J if ν is large enough.

Figure 8.5 depicts $J_1(\nu)$ and $J_2(\nu)$ as functions of ν. The dashed lines are analytical expressions for $J_1(\nu)$ and $J_2(\nu)$ that are calculated by using a comparison principle and the numerical forms for $v(E)$ and $D(E)$, while the solid curves result from a full simulation of Eq. (8.33); see [26]. Notice that J_1 decreases from $J_1 = 1$ to $J_1 = v_m$ as ν increases from ν_1. Similarly, J_2 decreases from $J_2 = 1$ to a minimum value $J_2 \approx 0.53$ and then increases back to $J_2 = 1$ as ν increases. Thus, we see that the pinning interval (J_1, J_2) at which fronts are stationary is narrower for doping such

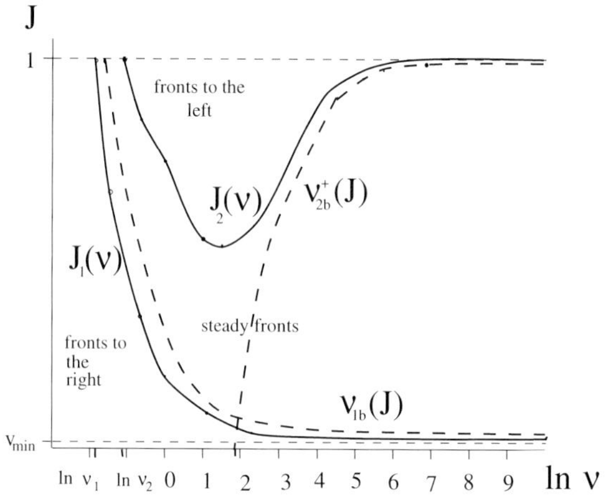

Fig. 8.5 Critical currents J_1 and J_2 as functions of the dimensionless doping ν. Monopoles move downstream for $\nu_m < J < J_1(\nu)$, are stationary for $J_1(\nu) < J < J_2(\nu)$, and move upstream for $J_2(\nu) < J < 1$. Dashed lines in this figure represent theoretical bounds $\nu_{1b}(J)$ and $\nu_{2b}^{+}(J)$ derived in [26]. Reprinted from [26].

that $J_2(\nu)$ reaches its minimum. For larger ν, the interval of J for which stationary solutions exist becomes wider again, trying to span the whole interval $(\nu_m, 1)$ as $\nu \to \infty$. For very large ν, the velocities of downstream and upstream moving monopoles become extremely small in absolute value.

Monopole velocity as a function of current has been depicted in Figure 8.6 for the first plateau of the 9/4 SL in [7]. Notice that, for the DDD model, the interval of current values at which wave fronts are stationary is shifted to lower values of

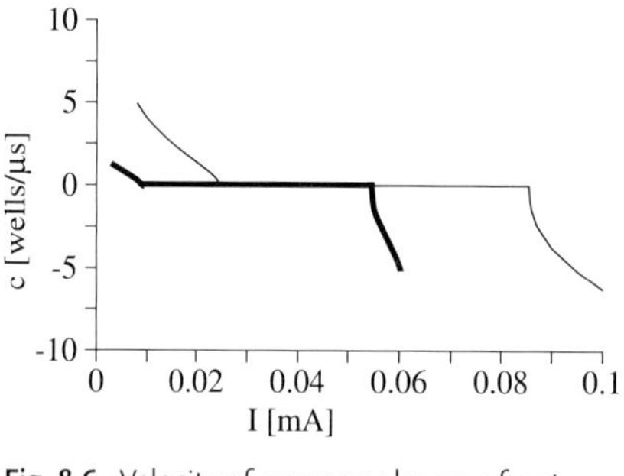

Fig. 8.6 Velocity of a monopole wave front as a function of the current for the first plateau of the 9/4 SL in [7]. Results for the discrete drift-diffusion model with $\nu = 3$ (thick line) have been compared to those obtained by using the exact tunneling current $J_{i\to i+1} = \mathcal{J}(n_i, n_{i+1}, F_i; T)$ (thin line) instead of the approximation given by Eq. (8.6). Notice that both the DDD model and the model with the more general tunneling current exhibit a pinning interval for the current at which $c = 0$, and current intervals for which the monopoles move upstream or downstream the electron flux. Reprinted from [26].

the current with respect to the more general tunneling current $e\mathcal{J}(n_i, n_{i+1}, F_i; T)$ given by the THM or Green function methods; see [26].

There are two limits in which analytical understanding of wave front motion has been achieved: the continuum limit $\nu \to 0$ (with finite values of νN) and the strongly discrete limit of sufficiently large ν. The transition from moving to stationary wave fronts is best understood in the strongly discrete limit [29], while we have a theory of wave front recycling in the continuum limit.

8.4.1.1 Pinning of Wave Fronts with a Single Active Well

At the critical currents, $J_1(\nu)$ and $J_2(\nu)$, wave fronts moving downstream (to the right, following the electron flow, $c > 0$) for smaller J or upstream (to the left, against the electron flow, $c < 0$) for larger J fail to propagate. What happens is that the wave front field profile $F(\tau)$ becomes sharper as J approaches the critical currents; see Figure 8.7. Exactly at J_k, $k = 1, 2$, gaps open up in the wave front profile which therefore loses continuity. The resulting field profile is a stationary front $F_i = F_i(J, \nu)$: the wave front is pinned for $J_1 < J < J_2$. The depinning transition (from stationary fronts to moving wave fronts) is technically speaking, a global saddle-node bifurcation. We shall study it first in the simplest case of large dimensionless doping ν, and then indicate what happens in the general case.

For sufficiently large doping and J close to a critical current (either J_1 or J_2), the moving front is led by the behavior of a single well which we will call the *active well*.

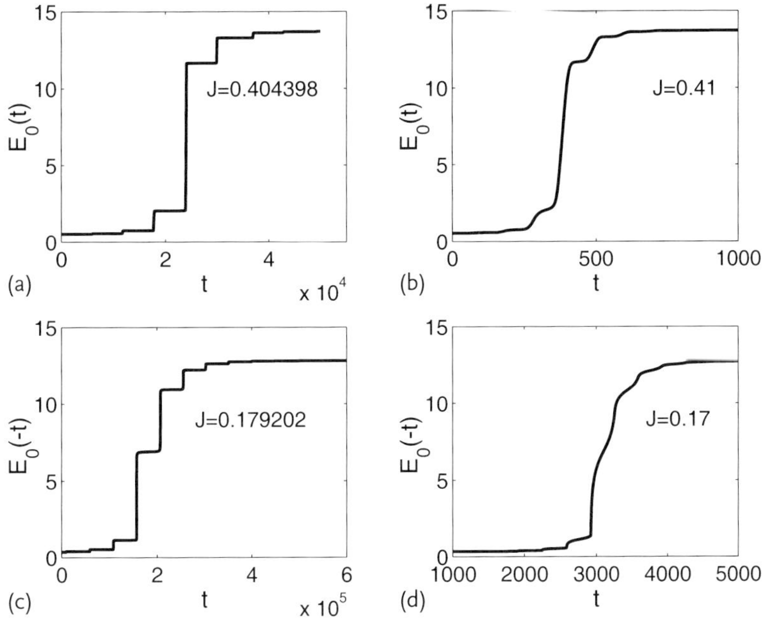

Fig. 8.7 Sharpening of wave front profiles as the dimensionless current J approaches its critical values for $\nu = 3$. (a) $J \approx J_2$, (b) $J > J_2$, (c) $J \approx J_1$, (d) $J < J_1$. Reprinted from [28].

If we examine the shape of a stationary front near the critical current, we observe that all wells are close to either $F^{(1)}(J)$ or $F^{(3)}(J)$, except for one well which drifts slowly and eventually jumps: the active well. Let us call E_0 the electric field at the active well. Since all wells in the front perform the same motion, we can reconstruct the profile $F(i-ct)$ from the time evolution of $F_0(t) = F(-ct)$. Before the active well jumps, $F_i \approx F^{(1)}(J)$ for $i < 0$ and $F_i \approx F^{(3)}(J)$ for $i > 0$. Thus, Eq. (8.31) becomes

$$\frac{dF_0}{dt} \approx J - v(F_0) - v(F_0)\,\frac{F_0 - F^{(1)}}{\nu} + D(F_0)\,\frac{F^{(1)} + F^{(3)} - 2F_0}{\nu}\,. \tag{8.38}$$

This equation has three stationary solutions for $J_1 < J < J_2$, two stable and one unstable, and only one stable stationary solution otherwise. At the critical currents, two of these solutions coalesce, forming a saddle-node. At low values of the current, the two coalescing solutions are a double zero (multiplicity two) of the right side of Eq. (8.38), corresponding to a local maximum thereof. For high currents, the two coalescing solutions are a double zero of the right side of Eq. (8.38), corresponding to a local minimum thereof. The critical currents are such that the expansion of the right-hand side of Eq. (8.38) about the two coalescing stationary solutions,

$$J - v(F_0) - v(F_0)\,\frac{F_0 - F^{(1)}}{\nu} + D(F_0)\,\frac{F^{(1)} + F^{(3)} - 2F_0}{\nu} = 0\,, \tag{8.39}$$

has a zero linear term,

$$D_0'(F^{(1)} + F^{(3)} - 2F_0) - 2D_0 - v_0'(F_0 - F^{(1)}) - v_0 - \nu\, v_0' = 0\,. \tag{8.40}$$

Here, D_0' means $D'(F_0) = \frac{dD}{dF}(F_0)$, and so on. Equations (8.39) and (8.40) yield approximations to F_0 and the critical current J_c (which is either J_1 or J_2). The results show excellent agreement with those of numerical solutions of the model for $\nu > 2$. Our approximation performs less well for smaller ν, which indicates that more active wells are needed to improve it.

Let us now construct the profile of the traveling wave fronts after depinning for J slightly below J_1 or slightly above J_2. We shall use the method of matched asymptotic expansions [30]. The idea is to describe the jump of an active well from $F^{(1)}(J)$ to $F^{(3)}(J)$ if $J > J_2$ [or from $F^{(3)}(J)$ to $F^{(1)}(J)$ if $J < J_1$] by means of two separate stages. During the first stage, $F_0(t)$ stays very close to its stationary value at the critical current, $F_0(J_c)$, for a very long time. After reaching a certain blow-up time, the active well jumps to $F^{(3)}(J)$ [or it falls to $F^{(1)}(J)$] on a faster time scale, as suggested by the numerical results depicted in Figures 8.7a and c. The process is then repeated until the active well reaches either $F^{(3)}(J)$ or $F^{(1)}(J)$. Let us start with the slow stage. Up to terms of order $|J - J_c|$, Eq. (8.38) becomes

$$\frac{d\varphi}{dt} \approx \alpha\,(J - J_c) + \beta\,\varphi^2\,, \tag{8.41}$$

for $F_0(t) = F_0(J_c) + \varphi(t)$, as $J \to J_c$. $F_0(J_c)$ is the stationary solution of Eq. (8.38) at $J = J_c$. The coefficients α and β are given by

$$\alpha = 1 + \frac{\nu_0 + D_0}{\nu\,\nu_1'} + \frac{D_0}{\nu\,\nu_3'}\,, \tag{8.42}$$

$$2\nu\beta = D_0''(F^{(1)} + F^{(3)} - 2F_0) - 4D_0' - 2\nu_0' + \nu_0''(F^{(1)} - F_0 - 2\nu)\,. \tag{8.43}$$

β is negative if $J_c = J_1$, and positive if $J_c = J_2$. Equation (8.41) yields the outer approximation to the depinning transition [30] and has the solution

$$\varphi(t) \sim (-1)^k \sqrt{\frac{\alpha(J - J_k)}{\beta}}\,\tan\left(\sqrt{\alpha\beta\,(J - J_k)}\,(t - t_0)\right) \tag{8.44}$$

($k = 1, 2$), for J such that $\mathrm{sign}(J - J_k) = \mathrm{sign}\beta$. The amplitude in Eq. (8.44) is very small most of the time, but it blows up when the argument of the tangent function approaches $\pm\pi/2$. Thus, the outer approximation holds over a time interval $(t - t_0) \sim \pi/\sqrt{\alpha\beta\,(J - J_k)}$. The reciprocal of this time interval yields an approximation for the wave front velocity,

$$|c(J, \nu)| \sim \frac{\sqrt{\alpha\beta\,(J - J_k)}}{\pi}\,. \tag{8.45}$$

In Figures 8.8 and 8.9, we compare this approximation with the numerically computed velocity for $\nu = 3$ and $\nu = 20$, respectively. The agreement is excellent. Notice that for the smaller value, $\nu = 3$, we need to improve our approximations considering that there are several *active wells*, $F_{-L}(t), \ldots, F_0(t), \ldots, F_M(t)$ that differ from either $F^{(1)}(J)$ or $F^{(3)}(J)$ during the front motion: $F_{-i} = F^{(1)}(J)$ if $i > L$ and $F_i = F^{(3)}(J)$ if $i > M$. The corresponding theory is similar to the one that we have described except that we consider a system of finitely many differential equations (for the active wells) instead of the single Eq. (8.38) [28].

When the solution begins to blow-up, the outer solution Eq. (8.44) is no longer a good approximation, for $F_0(t)$ departs from the stationary value $F_0(J_c)$. We must go back to Eq. (8.38) and obtain an inner approximation to this equation [30]. As J is close to J_c and $F_0(t) - F_0(J_c)$ is of order 1, we numerically solve Eq. (8.38) at $J = J_c$ with the matching condition that $F_0(t) - F_0(J_c) \sim (-1)^k 2/[\pi\sqrt{\beta/[\alpha\,(J - J_c)]} - 2|\beta|\,(t - t_0)]$, as $(t - t_0) \to -\infty$. This inner solution describes the jump of F_0 to values close to $F^{(1)}$ if $J_c = J_1$, or to values close to $F^{(3)}$ if $J_c = J_2$. During this jump, the motion of F_0 forces the other points to move. Thus, for $J_c = J_1$, $F_1(t)$ can be calculated by using the inner solution in Eq. (8.33) for F_0, with $J = J_c$ and $F_2 \approx F^{(3)}$. Similarly, for $J_c = J_2$, $F_{-1}(t)$ can be calculated by using the inner solution in Eq. (8.33) for F_0, with $J = J_c$ and $F_{-2} \approx F^{(1)}$. A composite expansion [30, 31] constructed with these inner and outer solutions is compared to the numerical solution of the model in Figure 8.10. Notice that we can reconstruct the traveling wave profiles $F(i - ct)$ from the identity $F_0(t) = F(-ct)$ by rescaling the horizontal axis in Figure 8.10.

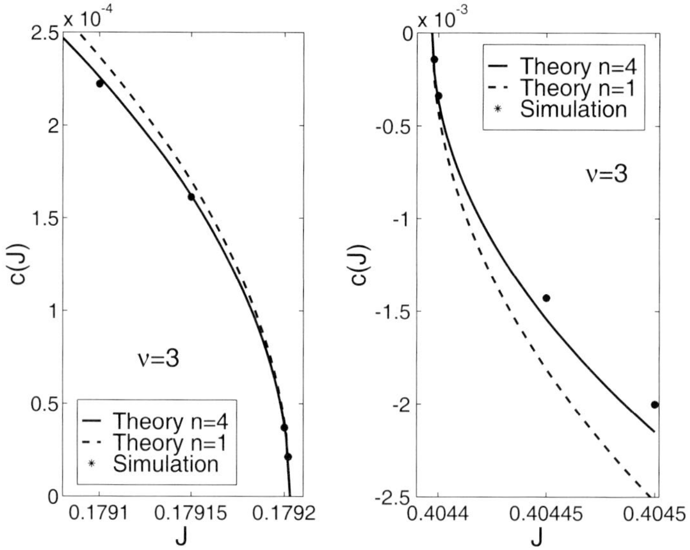

Fig. 8.8 Wave front velocity as a function of current density for $\nu = 3$. We have compared the numerically measured velocity to the results of our theory with one or four active wells. Reprinted from [28].

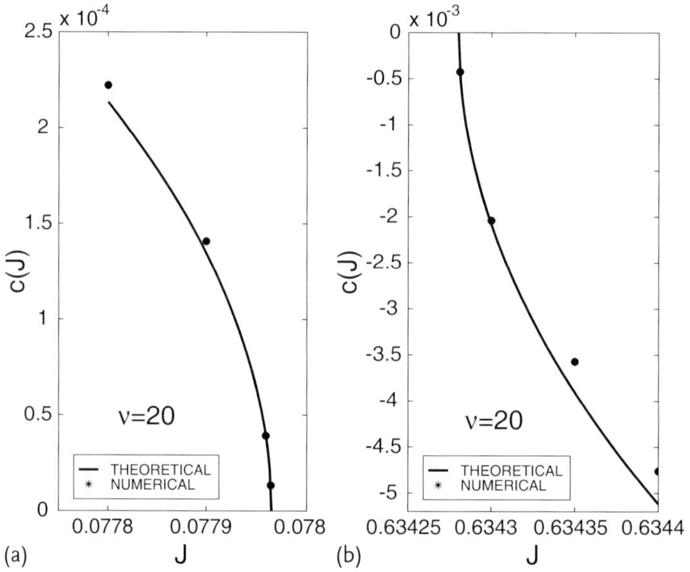

Fig. 8.9 Wave front velocity as a function of current density for $\nu = 20$. We have compared the numerically measured velocity to the results of our theory with one active well. Reprinted from [28].

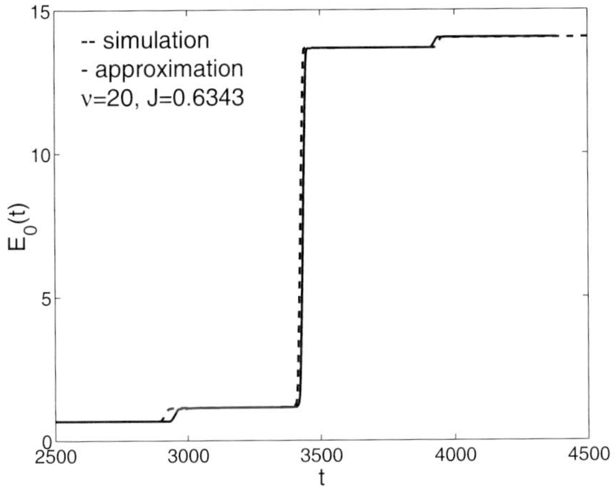

Fig. 8.10 Wave front profiles near $J = J_2$ for $\nu = 20$. The results of matched asymptotic expansions with one active well and the numerical solution of the model are compared. Reprinted from [28].

8.4.1.2 **Continuum Limit**

The continuum limit of the DDD model is useful to understand self-sustained oscillations of the current and wave front motion [23]. It consists of $\nu \to 0$, $i \to \infty$, with $\nu i = x \in [0, N\nu]$, $N\nu \gg 1$. In this limit, Eq. (8.33) yields

$$\frac{\partial F}{\partial t} + v(F)\,\frac{\partial F}{\partial x} = J - v(F)\,, \tag{8.46}$$

up to terms of order ν. Equation (8.46) corresponds to the hyperbolic limit of the well-known Kroemer model of the Gunn effect [32]. Notice that a wave front joining $F^{(1)}(J)$ to $F^{(3)}(J)$ with $\partial F/\partial x > 0$ (or with $\partial F/\partial x < 0$) and zero velocity cannot exist because $J - v(F)$ in Eq. (8.46) changes sign at $F = F^{(2)}(J)$. Thus, wave fronts cannot be pinned in the continuum limit. As we show now, wave fronts can be constructed by using shock waves (i. e., moving discontinuous electric field profiles). With constant J, shock waves are solutions of these equations and their speed can be explicitly calculated [33, 34]. Let $V(F_+, F_-)$ denote the speed of a shock wave such that F becomes F_- (resp. F_+) to the left (resp. right) of the shock wave. Inside the shock wave, we should use the discrete model. The wave front velocity should be rescaled so that $V = c\nu$ is the correct velocity for the (rescaled) continuum profile $F(x - Vt) = F(\nu\,(i - ct)) = F_i(t)$. Let $F'(\xi) = dF/d\xi$ and $\dot{F}_i(t) = dF_i/dt$ with $\xi = x - Vt$. Then, $F_i - F_{i-1} = F(\xi) - F(\xi - \nu) = \nu F'(\xi - \nu_0)$ with $0 < \nu_0 < \nu$ by the mean value theorem. However, we have $-VF'(\xi - \nu_0) = \dot{F}_i(t + \nu_0/V) =$

$\dot{F}_i(t) + O(\nu)$, and therefore, $dF_i/dt \sim -V\,(F_i - F_{i-1})/\nu$. Then,

$$\begin{aligned} F_+ - F_- &= \sum (F_i - F_{i-1}) = \nu \sum (n_i - 1) \\ &\sim V \sum \frac{F_i - F_{i-1}}{v(F_i) + D(F_i)} + \sum \frac{D(F_i)}{v(F_i) + D(F_i)} \nu n_{i+1} \\ &+ \nu \sum \left(\frac{J}{v(F_i) + D(F_i)} - 1 \right) = V \sum \frac{F_i - F_{i-1}}{v(F_i) + D(F_i)} \\ &+ \sum \frac{D(F_i)\,(F_{i+1} - F_i)}{v(F_i) + D(F_i)} + \nu \sum \left(\frac{J + D(F_i)}{v(F_i) + D(F_i)} - 1 \right) . \end{aligned}$$

This expression yields

$$V \sim \frac{\sum \frac{v(F_i)\,(F_{i+1} - F_i)}{v(F_i) + D(F_i)} - \nu \sum \frac{J - v(F_i)}{v(F_i) + D(F_i)}}{\sum \frac{F_i - F_{i-1}}{v(F_i) + D(F_i)}} .$$

Its numerator contains a term multiplied by ν which is bounded ($J = v(F_i)$ outside the wave front) and vanishes in the continuum limit. By approximating Riemann sums by integrals in the remaining formula, the result is

$$V(F_+, F_-) = \frac{\int_{F_-}^{F_+} \frac{v(E)}{v(E) + D(E)}\, dE}{\int_{F_-}^{F_+} \frac{dE}{v(E) + D(E)}} , \tag{8.47}$$

or equivalently, the following weighted equal-area rule

$$\int_{F_-}^{F_+} \frac{v(E) - V(F_+, F_-)}{v(E) + D(E)}\, dE = 0 . \tag{8.48}$$

For $D = 0$, this formula reduces to that derived for the discrete drift model in [23]. This formula can be corrected by using the trapezoid rule to evaluate integrals; see [28]. There is only one value of J, J^*, such that $V = J$ with $F_- = F^{(1)}(J)$ and $F_+ = F^{(3)}(J)$. For $J \in (v_m, J^*)$, a wave front joining $F^{(1)}(J)$ to $F^{(3)}(J)$ consists of a shock wave having $F_+ = F^{(3)}(J)$, and F_- such that $V(F^{(3)}(J), F_-) = v(F_-)$. Furthermore, to the left of the shock wave, there is a *tail* region moving rigidly with the shock wave such that

$$[v(F) - V] \frac{\partial F}{\partial \xi} = J - v(F) , \tag{8.49}$$

for negative $\xi = x - Vt$, and $F(-\infty) = F^{(1)}(J)$, $F(0) = F_-$. This whole structure (shock wave and tail region) is called a *monopole with left tail* [35]. Similarly, for $J \in (J^*, 1)$, a wave front joining $F^{(1)}(J)$ to $F^{(3)}(J)$ becomes a *monopole with right tail*. This monopole consists of a shock wave having $F_- = F^{(1)}(J)$, and F_+ such

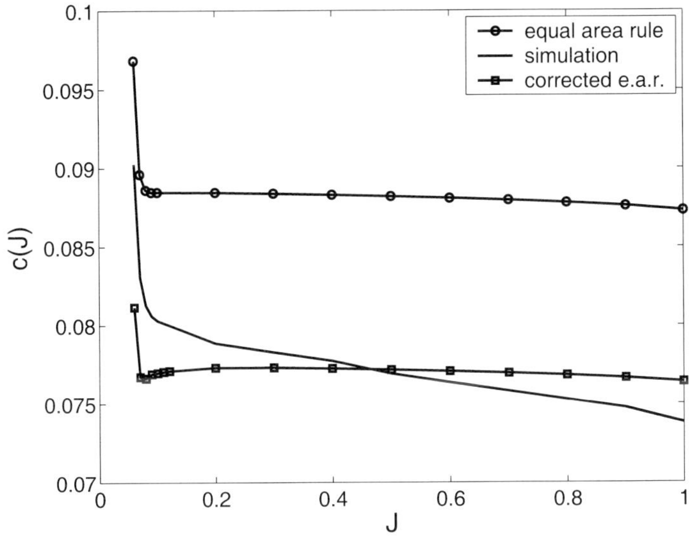

Fig. 8.11 Comparison of the equal-area rule (leading order) and corrected equal-area rule (including first order corrections) approximations to the wavefront velocity with the numerical solution of the model for $\nu = 0.01$. Reprinted from [28].

that $V(F_+, F^{(1)}(J)) = \nu(F_+)$, and a tail region satisfying Eq. (8.49) for positive ξ with the boundary conditions $F(0) = F_+$ and $F(\infty) = F^{(3)}(J)$ [35]. In conclusion, the wave front velocity as a function of J is determined by the following equations:

$$C(J) = V(F^{(3)}(J), F_-)\,, \quad \text{with} \quad \nu(F_-) = V(F^{(3)}(J), F_-)\,, \text{ if } \quad \nu_m < J < J^*, \tag{8.50}$$

$$C(J) = V(F_+, F^{(1)}(J))\,, \quad \text{with} \quad \nu(F_+) = V(F_+, F^{(1)}(J))\,, \text{ if } \quad J^* < J < 1. \tag{8.51}$$

We have compared the continuum approximation of the wave front velocity (in wells traversed per unit time, that is $c(J, \nu) = C(J)/\nu$, not rescaled) with the numerical solution of the model for $\nu = 0.01$ in Figure 8.11. The equal-area rule result corresponds to Eqs. (8.47), (8.50) and (8.51) and its maximum difference with the numerical solution is about 17.6%. This result can be significantly improved if the integrals in Eq. (8.47) are approximated by the trapezoidal rule. Then, the corrected equal-area result differs at most by 3% from the numerical solution [28].

8.4.1.3 Role of Diffusivity

Discrete diffusivity or, more generally, back tunneling as in the term containing n_{i+1} in the tunneling current of Eq. (8.6), is responsible for the existence of upstream moving monopoles [26]. Thus, for field values corresponding to the second and higher plateaus in the I–V diagram, monopoles can only move downstream because the diffusivity vanishes at high fields. For large doping, the active well $F_0(t)$ takes on a large field value if J is close to the low value J_1, and it takes on a low field value if J is close to the large value J_2. We could ignore diffusivity in certain expressions in our active well theory corresponding to the first case (monopoles moving downstream with $c > 0$), and still obtain reasonable numerical approximations for large enough doping values. If the dimensionless doping ν is small enough, monopoles can only move downstream with $c > 0$. In this case, the active well theory yields worse approximations and we can use the continuum limit theory in which diffusivity plays a minor role (it corrects the expression for monopole velocity).

8.5
Static Field Domains in Voltage-Biased SLs

The simplest solutions of discrete models for doped, weakly coupled SLs under dc voltage bias are time-independent solutions. Their shape depends on the voltage range, the SL configuration (d_W, d_B), and the doping density. If the tunneling current of the uniform field profile $F_i = F$, $n_i = N_D$, and $J_{i\to i+1} = \mathcal{J}(F, N_D, N_D)$ has several peaks corresponding to resonances between different subbands, the current–voltage characteristic of the SL exhibits different regions between peaks called plateaus. Restricting ourselves to a voltage range in one plateau, we need to consider a region with only a single maximum and a single minimum of the forward drift velocity in Eq. (8.16) (general case) or the drift velocity in Eq. (8.33) (DDD model). In the second and successive plateaus, the tunneling current is approximately proportional to the forward drift velocity times the electron density. Therefore, we have the discrete drift model which is first order in the differences and easy to graphically solve [18, 23]. Equation (8.33) with $D \equiv 0$ and $dF_i/dt = 0$ yields $J = v(F_i)[1 + (F_i - F_{i-1})/\nu]$, from which

$$F_{i-1} = F_i + \nu - \frac{J\nu}{v(F_i)} \equiv f(F_i; J)\,, \tag{8.52}$$

for $i = 1, \ldots, N$. In order to determine the field F_0 at the injecting contact, we solve the stationary version of Eq. (8.31), $J = j_e(F_0)$, for F_0 as a function of J. Many calculations have assumed Ohm's law: $F_0 = J/\sigma$ [36], or the simple boundary condition $n_1 = 1 + c$ (constant electron density) [18, 23, 37].

The first order difference Eq. (8.52) is easy to graphically solve for fixed J and the appropriate boundary condition, as shown in Figure 8.12. Once all the field profiles corresponding to J are found, we can calculate their average field,

$$\Phi(J) = \frac{1}{N}\sum_{i=0}^{N} F_i\,, \tag{8.53}$$

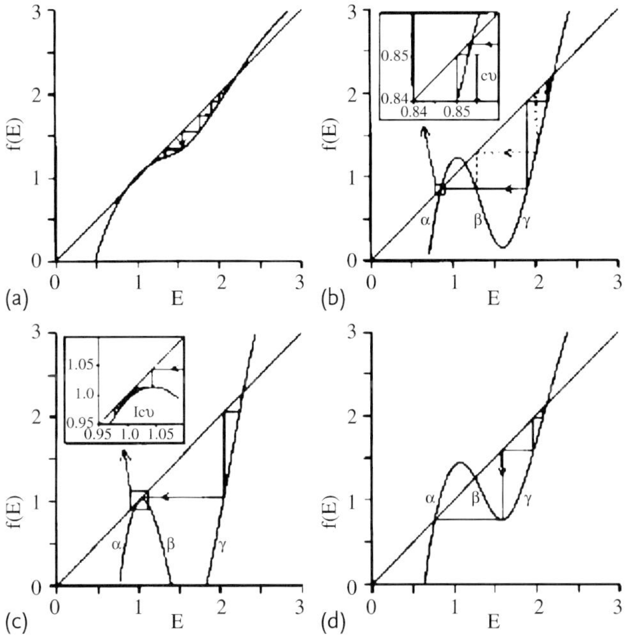

Fig. 8.12 The function $f(F, J)$ for fixed J and a trajectory F_i indicating decreasing i for various doping densities ν and total current densities J. (a) $\nu = 0.15$, $J = 0.8$, (b) $\nu = 1.0$, $J = 0.8$, (c) $\nu = 1.0$, $J = 1.008$, (d) $\nu = 1.0$, $J = 0.599 = J_c$. Reprinted from [37].

and we graphically solve the equation $\Phi(J) = \phi$. Thus, the values of the current J (and their associated field profiles) corresponding to a given ϕ are calculated. Please refer to Figure 8.13.

Let us observe a few specific features of the solution process. Note that uniform field distributions solve $v(F) = J$. Therefore, there are three such solutions for $v_m < J < v_M$, $F^{(1)}(J) < F^{(2)}(J) < F^{(3)}(J)$, which are also the fixed points of the mapping in Eq. (8.52). This equation provides F_i as a multivalued function of F_{i-1}. Only $F^{(1)}(J)$ and $F^{(3)}(J)$ are stable solutions under the discrete dynamics generated by the mapping. Let us now see how to construct solutions of Eq. (8.52) that start from F_0 given by the boundary condition. If we are looking for solutions with an increasing field profile, we locate F_0, for example, on the first branch of $f(F_i; J)$ defined in Eq. (8.52) and graphically solve the mapping. Rapidly, F_i tends toward the stable fixed point of the mapping $F^{(1)}(J)$ as i increases. Therefore, one possible solution is a single EFD with $F_i = F^{(1)}(J)$, except in a small region near the cathode needed to accommodate the boundary condition. This is the solution having an almost uniform field profile, and it corresponds to a LFD covering the whole SL. The corresponding voltage is small. Are there other solutions having the same voltage?

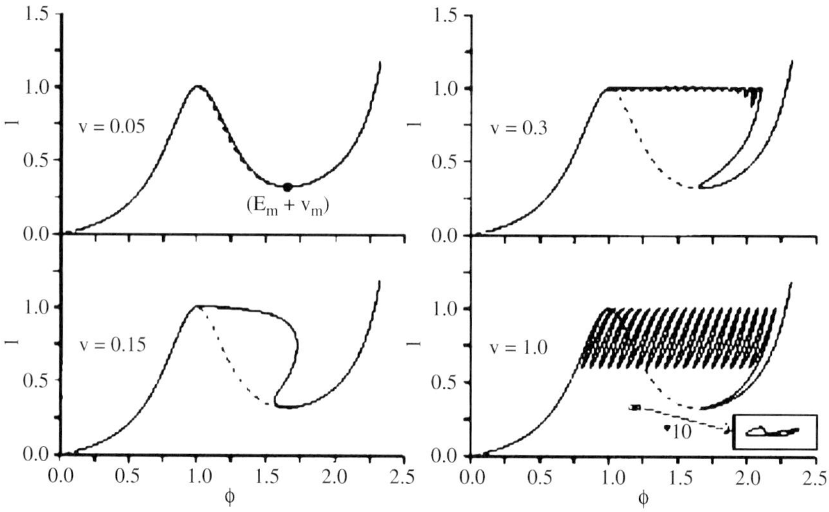

Fig. 8.13 Current–voltage characteristics for $c = 0.001$, $N = 20$ and different values of ν. The full line denotes the states where the electric field E_i is strictly monotone increasing in i. For $\nu = 1.0$, additional branches with non-monotonic field profiles E_i appear. They are isolated from the stationary branches having increasing field profiles. We have shown one such branch which has also been blown up for the sake of clarity. The dotted line is the $\nu(E)$ curve used to calculate Figures 8.12 and 8.13. Reprinted from [37].

Since F_i is a multivalued function of F_{i-1} according to Eq. (8.52), we find other solutions for a fixed value of J by allowing a jump from $F^{(1)}(J)$ (or its neighborhood) toward a different branch of the mapping at any given SL period, $i = i_0$. Such jumps are possible only if the local minimum of the mapping $f(F_i; J)$ is smaller than $f(F^{(1)}(J); J)$, which occurs for sufficiently large doping densities. Below a critical doping density, which can be estimated by Eq. (8.37), the only stable static solution is the almost uniform one [23, 37]. For larger values of the doping density, jumps between branches of the discrete mapping in Eq. (8.52) are possible and give rise to different static solutions having the same J, as in Figure 8.12b. Their field profile consists of two EFDs separated by a CAL. The second EFD has a field value $F_i = F^{(3)}(J)$, provided that the SL has a sufficient number of periods for the mapping to reach its second stable fixed point. Note that, for a weakly coupled SL with small broadening due to scattering, the value $F^{(1)}(J)$ in the LFD is close to a resonance while the value $F^{(3)}(J)$ in the HFD can be quite far from resonance, that is, if the current considered at the second resonant peak of the plateau is much larger than the current at the first peak.

It turns out that only solutions with no jumps (corresponding to a single EFD) or with jumps from the first to the third branch of the mapping are stable solutions of the time-dependent problem. Solutions with one jump have one LFD and one HFD separated by a CAL that can be located at any $i = i_0$. The corresponding field profiles have the same J, but different voltages, as indicated by Eq. (8.30). In a diagram of J versus voltage, these stable solutions of the time-dependent problem form N disconnected curves that can be observed in the current–voltage characteristic of the SL. The branches of stable solutions are connected (at $J_M = v_M$ and at some $J_{min} = v_{min}$ depending on the doping density) to branches of unstable solutions, whose field profile displays one intermediate jump from the first to the second branch of the mapping [23, 37]. It is also clear that, by including more than two peaks of the drift velocity in our construction and choosing the appropriate SL doping density, we can construct solutions with more than two EFDs. For a sufficiently large doping density, solutions including CDLs can be similarly constructed [37]. For undoped weakly coupled SLs, Bonilla *et al.* [18] constructed static domain solutions using similar theoretical ideas, though with the density of photoexcited carriers playing the role of the doping density in doped SLs.

For lower voltages on the first plateau, the term containing n_{i+1} in Eq. (8.33) cannot be ignored, and the stationary solutions solve a second order difference equation. In this case, it is more convenient to numerically solve the time-dependent equations and find the stable stationary profiles. The results are similar for different discrete models, and the corresponding field profiles and current–voltage characteristics are similar to the ones obtained with the mapping in Eq. (8.52) for plateaus with higher voltages. The only qualitative difference is that the maximum current for field profiles in the first plateau is smaller than the current J_M corresponding to a uniform profile with $n_i = N_D$ and $F_i = F$. Note that, except for small regions close to the boundaries, the constructed static solutions coincide with wave fronts pinned at the SL period $i = i_0$ corresponding to the CAL. Prengel *et al.* [19] numerically solved a discrete model for voltages on the first plateau, and Kastrup *et al.* [9] compared these solutions to experimental data. Similar results can be obtained with the tunneling current in Eq. (8.6), as shown in Figure 24 of [14], using an expression for the forward drift velocity slightly different from the one in Eq. (8.7).

For all plateaus, plotting the current–voltage diagram of the stable solution branches yields a succession of disconnected curves all having the same maximum and minimum values of the current. For a given voltage, two or more stable stationary solutions having different current values are possible. They are found by solving the time-dependent problem with different step-like initial conditions. One simple way to find these branches is to follow the same procedure in the numerical simulations as was used in the experiments: systematically sweep up or down the current–voltage characteristic [9]. If we want to determine the influence of external parameters (doping density, temperature, etc.) on the stable stationary or time-dependent solutions and their current–voltage diagram, it is better to use numerical continuation algorithms [20, 38].

8.6 Relocation of EFDs

In 1998, Luo *et al.* [39] published experimental data on how a domain wall (belonging to a stable stationary electric field profile with two domains) relocates if the voltage across the SL is suddenly changed. These experiments have been recently explained by numerically simulating a discrete model with the tunneling current given by Eq. (8.6) [36]. New experiments have further confirmed the theory [40]. Let us now report in some detail the numerical simulations of [36], providing a somewhat more elaborated theoretical interpretation. A more complete theory that takes into account the switching time needed to connect the final voltage can be found in [41]. For the sake of simplicity, the contact functions in Eqs. (8.31) and (8.32) will be selected as $j_e = \sigma F_0$ and $j_c = \sigma F_N$. Then,

$$J_{0\to 1} = \sigma F_0 \,, \tag{8.54}$$

$$J_{N\to N+1} = \sigma F_N n_N \,. \tag{8.55}$$

The contact conductivity σ is selected so that charge dipole waves may propagate, which also occurs with general contact current functions depending on the contact doping [21]. The factor n_N in the expression for $J_{N\to N+1}$ avoids negative electron densities at the collector. We numerically simulate a 40-well SL with a barrier width of $d = 4.0\,\text{nm}$, a well width $w = 9.0\,\text{nm}$, doping of $N_D = 1.5 \times 10^{11}\,\text{cm}^{-2}$, $\gamma_{Ci} = 8\,\text{meV}$ (independent of the miniband index μ) and a cross section $A = 15\,000\,\mu\text{m}^2$, at a temperature $T = 5\,\text{K}$. The parameters are those of the SL in [39]. Reasonable agreement with the overall shape of the experimental current–voltage characteristics in [39] is found if we adopt $\sigma = 0.01(\Omega\,\text{m})^{-1}$.

For the homogeneous case, that is, $n_i = n_{i+1}$, the current density Eq. (8.6) depends on the electric field as in the typical N-shaped curve. The corresponding stationary current–voltage characteristics for inhomogeneous solutions is the typical sawtooth pattern of Figure 8.14 with upper and lower branches corresponding to up- and down-sweep of the external voltage, respectively.

Here, we show the SL response to voltage switching starting from a point on the upper branch of the first plateau of the I–V characteristic at $V_{\text{dc}} = 0.75\,\text{V}$ (Figure 8.14). After switching to a final voltage $V_f = V_{\text{dc}} + V_{\text{step}}$, the current will evolve towards a value on the stationary I–V characteristic corresponding to one of the branches at voltage V_f (in general, there are several such branches due to multistability). We find that the final stationary current is on the upper branch if $V_{\text{step}} = 0.1\,\text{V}$, while it is on the lower branch if $V_{\text{step}} = 0.18\,\text{V}$; see the arrows in Figure 8.14a. Thus, fast switching allows us to reach the lower branch by sufficiently increasing the voltage. This is a striking result. In conventional up- and down-sweep, the point on the lower branch at 0.93 V can only be reached by increasing the voltage to more than 1.1 V and then decreasing to 0.93 V.

In Figure 8.15a, we depict the current response to different positive values of V_{step} versus time. For

$$V_{\text{step}} < V_{\text{crit}} \,, \tag{8.56}$$

with $V_{\text{crit}} \approx 0.175\,\text{V}$, the current monotonically relaxes to its final value. There is a fundamentally different current response if Eq. (8.56) does not hold. Instead of monotonically relaxing, the current first drops to a level well below the lower stationary branch. Then, the current response exhibits a fast repetitive double-peak

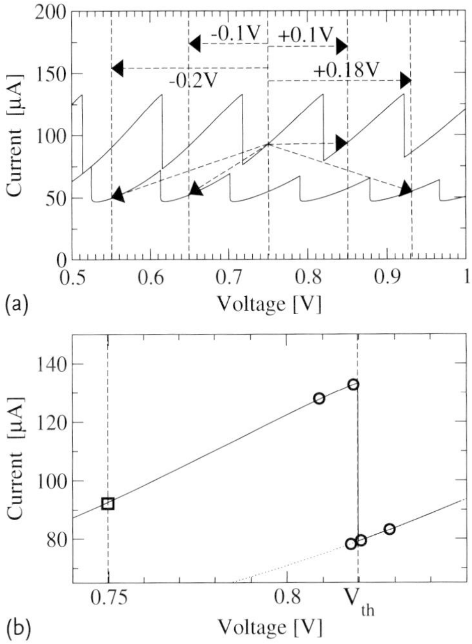

Fig. 8.14 Simulated sawtooth current–voltage characteristic of a 40-well SL (4.0 nm AlAs barriers, 9.0 nm GaAs wells). Upper branches correspond to voltage up-sweep and lower branches to down-sweep. The arrows in (a) indicate the starting and end points of the voltage steps discussed in the text. (b) gives an enlarged view of the initial operating point (box) as well as of the different final points (circles) considered. Reprinted from [36].

pattern until about 3 μs. Subsequently, following one larger spike, only single peaks occur. The spiky structure ends about 7 μs after the voltage switch, and the current evolves to a stationary value on the lower branch. The total number of peaks is roughly equal to the number of wells in the SL. The frequency of the peak burst is about 15 MHz. This behavior does not significantly change as long as Eq. (8.56) is violated, even for very different values of V_{step}. This effect is very similar to the experimental observations in [39]. The quantitative difference is that the experimental total relaxation time is only about 2 μs. Such values could be achieved numerically by choosing a larger scattering width $\gamma_{Ci} \approx 20\,\text{meV}$.

How can one describe this behavior? Let us use dimensionless units as follows. Consider the tunneling current density of Eq. (8.6), $J_{i\to i+1} = \mathcal{J}(n_i, n_{i+1}, F_i; T)$ evaluated at $n_i = n_{i+1} = N_D^w$ for a fixed $F_i = F$ and T. This curve is N-shaped, similar to the drift velocity $v(F)$. It has a local maximum at $F = F_M$ and J_M, and we can define $v_M = J_M l/N_D$. Let us define dimensionless units as in Section 8.4 with these values so that $v(F) = \mathcal{J}(N_D, N_D, F_M F; T)l/(N_D v_M)$. Furthermore, let us define $F^{(1)}(J)$, $F^{(2)}(J)$ and $F^{(3)}(J)$ to be the solutions of $v(F) = J$, ordered from smaller to higher field values.

As in Section 8.4, there are accumulation wave fronts joining $F^{(1)}(J)$ to $F^{(3)}(J)$ that have monotonically increasing field profiles with $F_i < F_{i+1}$; they move with

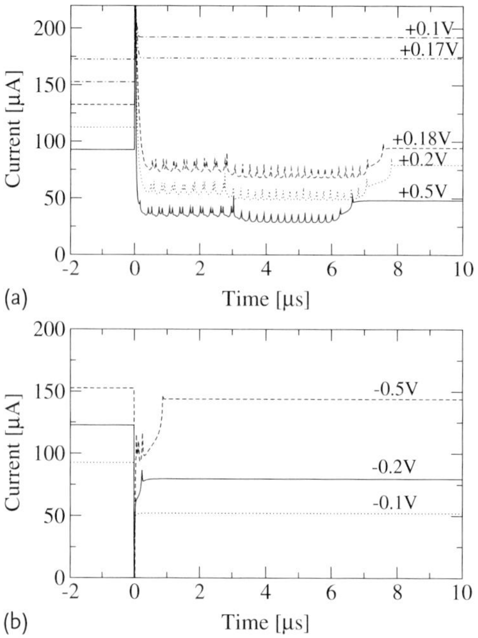

Fig. 8.15 Current response vs. time for various (a) positive and (b) negative voltage steps at $t = 0$. For $t < 0$, the voltage is $V_i = 0.75$ V. The curves are shifted vertically in units of 20 µA in (a) and 30 µA in (b) for clarity. Reprinted from [36].

velocities $c_+(J)$, which may be positive, zero or negative (compare with Figure 8.6). Furthermore, there are depletion wave fronts joining $F^{(3)}(J)$ to $F^{(1)}(J)$ that have monotonically decreasing wave profiles $F_i > F_{i+1}$. These wave fronts with a decreasing field profile have not been considered until now. They always move with positive velocities (no matter the positive value of J we take), which we shall denote by $c_-(J)$. The dimensionless velocities $c_+(J)$ and $c_-(J)$ are calculated with the full sequential tunneling current (8.6) instead of using the DDD model as in Section 8.4, and the result is depicted in Figure 8.16. Approximations to $c_+(J)$ can be obtained from the active well theory for J close to J_1 or J_2, or even from the continuum limit $\nu \to 0$ (provided the doping ν is not large). In the latter case, $c_-(J) \approx J$ (corresponding to $n \approx 0$ at the leading edge of a pulse) and the possibility $c_+ \leq 0$ is lost. The latter fact shows that the continuum limit may be a poor approximation of wave front dynamics if the current is large enough, unless the doping is sufficiently low.

As explained before, electron accumulation monopoles may have positive, zero, or negative velocities, as the current is increased. The interval (I_l, I_u) of the dimensional current density, corresponding to stationary monopoles, depends on the location and size of the peak and bottom values of the sequential tunneling current Eq. (8.6). For the DDD model, $I_l = eN_D v_M J_1/l$, $I_u = eN_D v_M J_2/l$. For depletion waves, $n_i \ll N_D$ and $F_i - F_{i-1} \approx -eN_D/\varepsilon$, we can ignore the tun-

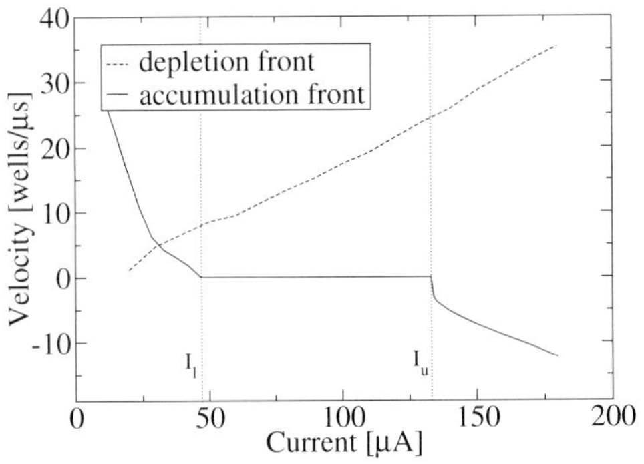

Fig. 8.16 Front velocity vs. current for electron depletion and accumulation fronts. Reprinted from [36].

neling current $J_{i \to i+1}$ in Eq. (8.3). Therefore, the front velocity is approximately $J/(eN_D)$, that is, a linear function of the current [32, 42]. The point in Figure 8.16 where the velocities of accumulation and depletion fronts intersect is of special interest. This point determines the velocity and current at which a dipole wave consisting of a leading depletion front and a trailing accumulation front can move rigidly [42].

Let us now consider a switching process where condition (8.56) is fulfilled and $V_{\text{step}} > 0$. The dynamical evolution of the electron densities n_i is depicted in Figure 8.17a for $V_{\text{step}} = 0.1$ V. We observe that the charge monopole region of high electron density (light region) is shifted upstream (against the field eF) towards the emitter. This shifting occurs on a time scale of 0.1 µs, explaining the fast monotonic relaxation if Eq. (8.56) is fulfilled. The switching to a higher external voltage has the effect that all fields in the superlattice are increased and the current instantaneously rises above I_u. According to Figure 8.16, this gives rise to a negative velocity of the accumulation front.

If condition (8.56) is violated, the switching scenario is more complicated. For $V_{\text{step}} = 0.18$ V and $V_{\text{step}} = 0.50$ V, the evolution of the electron density profile is depicted in Figures 8.17b and c, respectively. Before switching V_{step}, there is a charge accumulation front inside the sample corresponding to the domain wall separating two coexisting stationary high field domains. After switching the voltage, the charge dynamics in the superlattice exhibits four different phases: (i) upstream shift of the accumulation front and generation of new fronts at the emitter, (ii) coexistence of three fronts in downstream motion, (iii) downstream motion of two fronts, and (iv) upstream motion of the accumulation front while the depletion front is pinned at the collector. These four phases will now be considered in detail.

Phase (i): Shortly after switching the voltage step, the original preexisting electron accumulation layer moves upstream towards the emitter. Simultaneously, a charge dipole wave appears at the emitter. Its leading depletion front moves towards the collector while its amplitude increases. The trailing electron accumulation front of the dipole is pinned at the first SL well.

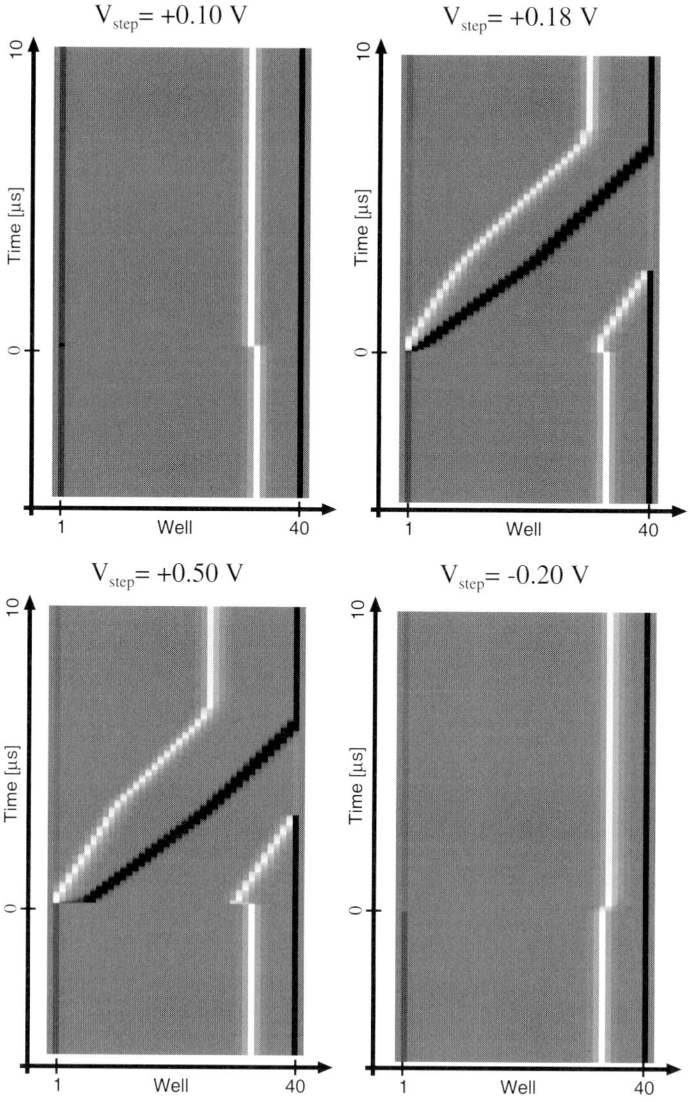

Fig. 8.17 Evolution of electron densities in the quantum wells during the switching process for various voltage steps. White indicates high electron density (accumulation front) and black indicates low electron density (depletion front). In the grey area, the electron density is $\approx N_D$. Well #1 is located at the emitter and well $\#N = 40$ at the collector. At $t = 0$, a voltage step V_{step} is applied, starting from $V_i = 0.75$. Reprinted from [36].

The mechanism for the generation of a dipole at the emitter is as follows: In a stationary situation, the current through the emitter barrier is equal to the current through the first SL barrier. The field at the emitter can be calculated from

Eq. (8.55). We assume a dimensionless contact conductivity σ such that the dimensionless boundary current at the emitter, σF, and the homogeneous, dimensionless current-field characteristic, $v(F)$, intersect on the second branch of the latter, at a critical value J_c (We then have to choose the slope of the emitter current to be lower than the slope of the homogeneous low-field current-field characteristic). If $0 < J < J_c$, the field at the emitter barrier is larger than that at the first barrier, $F_0 > F_1$. Then, the Poisson equation predicts electron depletion, $n_1 < 1$. If we could suddenly change the current to a value larger than J_c, the field F_1 would increase according to Eq. (8.3), in order to attain a value on the third branch of the characteristic curve. This would produce an electron accumulation layer at this well followed by the depletion layer which was there before changing J. The net outcome of this mechanism would be the creation of a dipole. As seen in Figure 8.16, a depletion layer separated from the contacts has to move towards the collector at a speed $c_-(J)$, roughly proportional to J. As the depletion wave front moves, it leaves a high field region behind that extends all the way to the injecting region. The width of this region increases as its leading edge, the depletion wave front, advances. The extra area gained by this region has to be compensated by lowering the current in order for both the low field and the high field domains to occur in regions of positive differential conductivity. This must be compensated in order to keep the total voltage constant. Once the current has become smaller than J_c, the field in the immediate neighborhood of the injecting region should correspond to another depletion layer. This means that in the mean time an accumulation wave front forming the back of the pulse has been created. Now, the dipole is fully detached from the injecting region and a new phase begins.

Phase (ii): After about 0.2 µs, the current has dropped below I_l, which by Figure 8.16 means that all fronts have positive velocities. For an instantaneous value of the current density J in dimensionless units, accumulation fronts move with velocity $c_+(J)$, whereas depletion fronts move with velocity $c_-(J)$. Let us ignore the fast time scale responsible for the current spikes in Figure 8.15a and try to find an equation for the envelope of the current time trace. Then, we can consider that the field profile adjust adiabatically to the instantaneous value of the envelope of the current time trace, $J(t)$, which evolves slowly. The electric field profile consists of an advancing dipole wave with field $F^{(3)}(J)$ enclosed by an accumulation wave front centered at $i = m_+^{(1)}(t)$, and a depletion wave front centered at $i = m_-^{(1)}(t)$. It also consists of an accumulation wave front at $i = m_+^{(2)}(t)$ that encroaches on a high field region with $F = F^{(3)}(J)$ near the collector contact. We can mark the center of a wave front at a time t as the well at which there is a local maximum, or minimum in the case of a depletion wave front, of the charge. Then the functions $m_\pm^{(j)}(t)$ are integer valued, because the charge inside the wave front peaks at different wells as time changes. Consider the lifetime of a given wave front and denote by $t_1, t_2, \ldots$ the times at which $m_\pm^{(j)}(t)$ changes. We may define a local velocity by $c_\pm^i = [m_\pm^{(j)}(t_{i+1}) - m_\pm^{(j)}(t_i)]/(t_{i+1} - t_i)$. Let us assume that we average these $c_\pm^i$ over a time interval that is short compared to the time scale typical of wave front motion, but sufficiently long compared to the mean value of $(t_{i+1} - t_i)$ for the wave front to have advanced over many wells. Furthermore, as the envelope of the current time trace varies slowly, the average of the $c^i_\pm$ can be approximated by $c_\pm(J)$, where J is the instantaneous value of the envelope of the current time trace. The average wave front velocities are

$$\left\langle \frac{dm^{(j)}_+}{dt} \right\rangle = c_+(J)\,, \quad \left\langle \frac{dm^{(j)}_-}{dt} \right\rangle = c_-(J)\,, \quad j = 1, 2\,. \tag{8.57}$$

We can now easily calculate the dimensionless voltage ϕ corresponding to the field profile considered above with one advancing pulse and one accumulation wave front at the end of the SL. Ignoring transition regions, we obtain

$$\phi = \frac{1}{N}\sum_{j=1}^{N} E_i = F^{(1)}(J) + [F^{(3)}(J) - F^{(1)}(J)]\left(\frac{N - m^{(2)}_+}{N} + \frac{m^{(1)}_- - m^{(1)}_+}{N}\right)\,. \tag{8.58}$$

We now differentiate this equation with respect to t, use $d\phi/dt = 0$, average over short time intervals as indicated above, and use Eq. (8.57). Then we obtain the following equation for J, the envelope of the current time trace,

$$\frac{dJ}{dt} = \frac{2c_+(J) - c_-(J)}{N} \frac{[F^{(3)}(J) - F^{(1)}(J)]^2}{\frac{F^{(3)}(J)-\phi}{v'(F^{(1)}(J))} + \frac{\phi - F^{(1)}(J)}{v'(F^{(3)}(J))}}\,. \tag{8.59}$$

Notice that if we had started from a field profile comprising n_+ moving accumulation wave fronts and n_- depletion wave fronts, the same arguments would lead to the more general equation

$$\frac{dJ}{dt} = \frac{n_+c_+(J) - n_-c_-(J)}{N} \frac{[E^{(3)}(J) - E^{(1)}(J)]^2}{\frac{E^{(3)}(J)-\phi}{v'(E^{(1)}(J))} + \frac{\phi - E^{(1)}(J)}{v'(E^{(3)}(J))}}\,. \tag{8.60}$$

J evolves on the time scale t/N, which is slow provided that N is large. Clearly, J, as given by Eq. (8.59), changes until it reaches a value $J^\dagger$ such that $2c_+(J) = c_-(J)$. This means that the original accumulation layer and the trailing accumulation front start advancing towards the collector with the same velocity $c_+(J^\dagger)$, while the positively charged leading front of the dipole moves towards the collector at a higher velocity $c_-(J^\dagger)$. Numerical simulations of different discrete models [13, 36, 43] show that each time that an accumulation layer advances by a SL period, a spike of the current appears. An asymptotic theory explaining this fact is still missing. If we accept it at face value, the double-peak structure observed in the numerical simulation of Figure 8.15a means that two accumulation wave fronts exist during that part of the period. Notice that the transient region where the current exhibits double spikes has a flat appearance, indicating the constant *mean value* of the current, $J^\dagger$. This is further corroborated by Figure 8.17b and c. In these figures, time traces of the positions of all wave fronts are recorded. Velocities are the reciprocal of the slopes. Notice that the velocity of the depletion wave front constituting the leading edge of the dipole wave $c_{-}(J)$ is larger than the velocity of the two accumulation wave fronts that coexist during a short time interval after $t = 0$. When the accumulation wave front closer to the receiving region exits, Phase (ii) ends and a new Phase (iii) starts. As observed in Figure 8.17b and c and explained below, the slopes of accumulation and depletion wave fronts enclosing the high field region become identical and their corresponding velocity is smaller than $c_{-}(J^{\dagger})$ during Phase (ii).

Please notice that the above explanation is heuristic and that a consistent asymptotic argument should replace it in the future. Such an argument should explain the spikes in the current time trace and still retain Eq. (8.60), conveniently reinterpreted as necessary.

Phase (iii): After the original accumulation layer has reached the collector at $t_o \approx 3\ \mu s$, there is only one accumulation layer and one depletion layer present in the sample, giving rise to a single-spike structure of the current response as depicted in Figure 8.15a. The same reasoning as in (ii) leads us to Eq. (8.60) with $n_{+} = n_{-} = 1$. After a short transient, the velocities of the positively and negatively charged fronts should become equal, $c_{+}(J) = c_{-}(J)$. This occurs at a current $J = J^*$ corresponding to the crossing point of the two front velocities, as depicted in Figure 8.16. In comparison to phase (ii), the velocity of the accumulation front has almost doubled, while the velocity of the depletion front has decreased slightly (see Figure 8.17b and c).

Phase (iv): The CDL is pinned at the collector while the CAL advances to its final position, which it reaches in a short time. During this stage, the current increases, as described by Eq. (8.60), with $n_{+} = 1$, $n_{-} = 0$.

After these four stages of its evolution, the accumulation front finally reaches a stable stationary state. It becomes the domain wall separating two stationary high field domains. Of all such possible stationary solutions at voltage $V_{dc} + V_{step}$, the one having an accumulation layer closer to the emitter is reached. This final situation on a low current branch of the current–voltage characteristics at voltage $V_f = V_{dc} + V_{step}$ could also be reached by a conventional down-sweeping of the current–voltage characteristics. In the latter case, the electron accumulation layer also moves towards the collector.

For $V_{step} < 0$, the electron accumulation layer always travels towards the collector and stops at a position corresponding to the domain wall separating low and high field domains of a stationary solution. This solution is the same as that which could be reached by down-sweeping to the final voltage, as shown in Figure 8.17d. Contrary to the case of positive V_{step}, the resulting current response shows no threshold-like behavior (see Figure 8.15b). Since all fields decrease during the switching process, no dipole wave can be generated at the emitter.

8.7
Self-Sustained Oscillations of the Current

Several years ago, self-sustained oscillations of the current in weakly coupled SLs were experimentally observed and interpreted on the basis of the discrete drift model [44]. Self-oscillations are caused by recycling and motion of waves in a dc voltage biased SL. Depending on contact conditions, these waves may be charge accumulation layers (monopoles or wave fronts) or electric field pulses (charge dipoles) [13]. Both types of waves are found in the DDD model assuming that the injecting contact is described by Ohm's law with appropriate resistivity [36], a result common to the Gunn effect in bulk GaAs [32]. The asymptotic theory for the Gunn effect can be used to analytically describe self-oscillations in the continuum limit [32, 35, 42]. This description leaves out effects due to the discrete nature of the equations, such as current spikes [13]. In this section, we shall describe the asymptotic theory of self-oscillations, and the effect of parameters such as doping, temperature, photoionization, and so on. on oscillation frequency and amplitude.

8.7.1
Asymptotic Theory

In the continuum limit where $\nu \to 0$ and $L \equiv N\nu \gg 1$, the equations for electric field and current density are given by Eq. (8.46), and the voltage bias condition by Eq. (8.30). This becomes

$$\frac{1}{L}\int_0^L F(x,t)dx = \phi \,. \tag{8.61}$$

Appropriate boundary conditions may be Eqs. (8.54), together with Ampère's law (8.3) at $i = 0$ and at $i = N$. In dimensionless units, these equations are

$$\begin{aligned} \frac{\partial F(0,t)}{\partial t} + \sigma\, F(0,t) &= J\,, \\ \frac{\partial F(L,t)}{\partial t} + \sigma\, n(L,t)\, F(L,t) &= J\,, \end{aligned} \tag{8.62}$$

where σ is the dimensionless contact conductivity. Equation (8.46) is the hyperbolic limit of the Kroemer model for the Gunn effect in bulk n-GaAs with zero diffusivity. Knight and Peterson had already shown in 1966 that at constant J, this equation may develop shock waves (our approximations to the wave fronts) [33]. The difference with the model for the Gunn effect is that the shock velocities are given by Eqs. (8.47), (8.50) and (8.51) in Section 8.4 instead of the usual equal-area rule for the Gunn effect [33]. Except for this difference and the peculiarities introduced by our boundary conditions, we can use the asymptotic theory of the Gunn effect to understand current self-oscillations in a SL. In an incomplete form, this theory was introduced in [32] for Ohmic boundary conditions, and later elaborated for different boundary conditions in [42].

The key to understand self-oscillations is to realize that the monopole or dipole waves are small compared to the dimensionless SL length ($L = N\nu = NeN_D/(\varepsilon F_M) \gg 1$) in the limit we consider. Outside the wave fronts, we can therefore rescale time and length as $s = t/L$ and $y = x/L$, so that Eqs. (8.46) and

(8.61) become

$$J - v(F) = \frac{1}{L}\left[\frac{\partial F}{\partial s} + v(F)\,\frac{\partial F}{\partial y}\right], \tag{8.63}$$

$$\int_0^1 F(y, s)\,dy = \phi\,, \tag{8.64}$$

where $1/L \ll 1$ is a small parameter. Notice that according to the table in Section 8.4, the unit of time is $l/(v_M \nu)$. Then the dimensionless time t is $\nu v_M t_d/l$ where t_d is time measured, for instance, in seconds. The "slow" dimensionless time s is obtained dividing t by $L = N\nu$, so that it is $s = v_M t_d/(Nl)$. Nl/v_M is the order of magnitude of the time an electron would need to traverse the SL, while $l/(v_M \nu)$ gives the order of magnitude of the time an electron would take to traverse one SL period. The ratio of these two times is $L = N\nu \gg 1$. Similarly, the "slow" spatial variable $y = x/L = i/N$, where i is the index of the SL period we are considering (i. e., the discrete variable in Section 8.4).

As $L \gg 1$, Eq. (8.64) yields $v(F) = J$ outside the wave fronts, that is, the field there is either $F^{(1)}(J)$ or $F^{(3)}(J)$. The current density J evolves on the slow time scale $J = J(s)$, so that we may consider J as a constant when constructing the wave fronts. These will be described in Section 8.4. We shall now asymptotically describe one time period of the current self-oscillation, starting from a given field configuration inside the SL. The initial profile will evolve with time adiabatically following the current, and our first goal will be to find an evolution equation for J. The easiest imaginable dynamic situation as an initial field profile is that we have only one monopole or one dipole inside the SL.

Monopole. If a monopole joining $F^{(1)}(J)$ to $F^{(3)}(J)$ is at $x = X^+(t)$, or equivalently at $y = X^+/L = Y^+$, the bias Eq. (8.61) is given by

$$\phi \sim F^{(1)}(J)\,Y^+ + F^{(3)}(J)\,(1 - Y^+)\,. \tag{8.65}$$

If we time-differentiate this equation and use $dX^+/dt = dY^+/ds = C(J)$ together with Eq. (8.65), we obtain

$$\frac{dJ}{ds} = \frac{[F^{(3)}(J) - F^{(1)}(J)]^2}{\frac{F^{(3)}(J)-\phi}{v'(F^{(1)}(J))} + \frac{\phi - F^{(1)}(J)}{v'(F^{(3)}(J))}}\,C(J)\,. \tag{8.66}$$

Since $C(J) > 0$, the current increases as the monopole moves. Notice that Eq. (8.66) is a particular case of Eq. (8.60) with $n_+ = 1$, $n_- = 0$, $c_+ = C(J)/\nu$ and $s = t/L$.

Dipole. A dipole consists of a region where the field is $F^{(3)}(J)$, moving towards the anode. Its trailing edge is a monopole at $x = X^+(t)$, or equivalently at $y = X^+/L = Y^+$, joining $F^{(1)}(J)$ to $F^{(3)}(J)$. The dipole leading front is a region depleted of electrons, $F = -x + \int J\,dt$, and is located at $y = Y^-$. Its velocity is clearly $dY^-/ds = J$. The bias Eq. (8.61) is now

$$\phi \sim F^{(1)}(J) + [F^{(3)}(J) - F^{(1)}(J)]\,(Y^- - Y^+)\,. \tag{8.67}$$

If we time-differentiate this equation and use $dY^+/ds = C(J)$ and $dY^-/ds = J$ together with Eq. (8.65), we obtain

$$\frac{dJ}{ds} = \frac{[F^{(3)}(J) - F^{(1)}(J)]^2}{\frac{F^{(3)}(J)-\phi}{v'(F^{(1)}(J))} + \frac{\phi - F^{(1)}(J)}{v'(F^{(3)}(J))}} [C(J) - J], \tag{8.68}$$

which is a particular case of Eq. (8.60) with $n_+ = n_- = 1$, $c_+ = C(J)/v$, $c_- = J/v$ and $s = t/L$. Now the current evolves towards $J = J^*$ such that $C(J^*) = J^*$. We observe that the dipole moves at constant current with velocity given by the equal area rule Eq. (8.48), $V(F^{(1)}, F^{(3)}) = J$. After $Y^- = 1$, there is only one monopole in the SL and we return to Eq. (8.66). Again, the current increases as the monopole moves.

Instability at the injecting contact. Near the injecting contact at $y = 0$, we have a boundary layer where the field profile adiabatically follows $J(s)$ according to

$$\frac{\partial F}{\partial x} \sim \frac{J}{v(F)} - 1, \quad F(0, s) = J/\sigma. \tag{8.69}$$

Notice that we have ignored $\partial F/\partial t = L^{-1} \partial F/\partial s \ll 1$. Typically, the contact conductivity is chosen so that either, (i) $\sigma > 1$ and the straight line $j = \sigma F$ does not intersect the curve $j = v(F)$, or (ii) $0 < \sigma < 1$ and $j = \sigma F$ intersects $j = v(F)$ at (E_c, J_c) with $1 < F_c < F_m$ on the second branch of $v(F)$.

As x departs from the cathode, $F(x, s)$ has to reach the constant solution of $J - v(F) = 0$ which is found to the left of the accumulation layer moving towards the anode, and is either a monopole or the back of a dipole, depending on the value of the contact resistivity. If $v_m < J < J_c$, the appropriate field value to the left of the accumulation layer is $F^{(1)}(J)$. Thus $F(x, s)$ either increases monotonically from J/σ to $F^{(1)}(J)$ in case (i), or it decreases monotonically from J/σ to $F^{(1)}(J)$ in case (ii). We assume that we just have one monopole left in the SL. In case (ii) this means that the leading front of the dipole has already reached the SL end. Then, $J(s)$ increases according to Eq. (8.67) until it surpasses $J = 1$ in case (i), or $J = J_c$ in case (ii).

In the first case, $E(x, s)$ keeps the same shape, only now the field at the left of the monopole and outside the cathode layer would try to leave $F = 1$ and tend towards $F^{(3)}(J)$, with $J \approx 1$. This situation is unstable, a new monopole is created at the cathode, and J decreases below 1 [32, 35]. For a very short time, the newly created monopole joins $F^{(1)}(J)$ to $F^{(2)}(J)$, while the old one joins $F^{(2)}(J)$ to $F^{(3)}(J)$. Then the old monopole disappears rapidly. It is important to note that the charge inside the new monopole is not appreciable until it has departed sufficiently far from the cathode region, as shown in Figure 8.18. Details of asymptotic calculations can be found in [35]. Notice that in the last reference, the boundary condition at the cathode $x = 0$ is $\partial F/\partial x = c$, and that part of the calculation has to be slightly modified if the boundary condition (8.62) is used instead.

In case (ii), $F^{(1)}(J)$ still exists at $J = J_c < 1$. As J surpasses J_c, the solution of the contact equation (8.69) would try to increase towards $F^{(3)}(J)$ as x increases, but far from the cathode the field is still $F^{(1)}(J)$. As explained in the previous section,

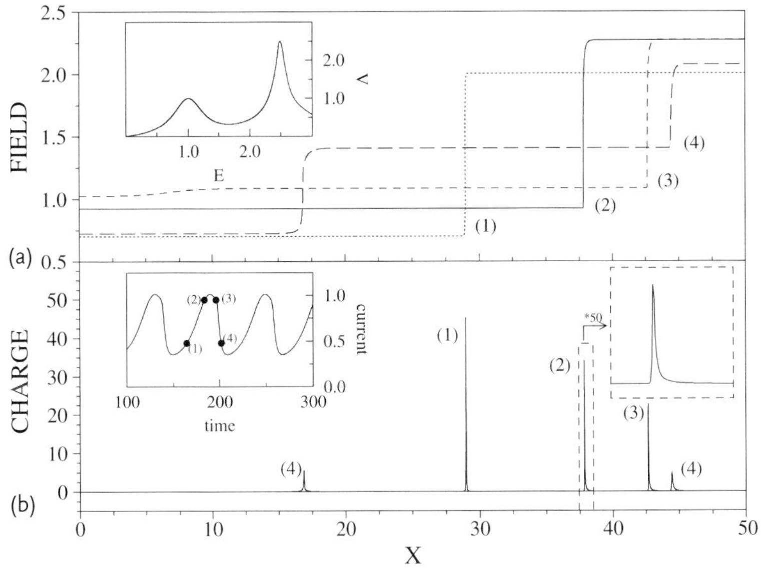

Fig. 8.18 (a) Time evolution of the electric field profile on the SL using the velocity curve shown in the inset. (b) Charge density profiles $n - 1 = \partial F(x,t)/\partial x$ showing the location of the wave front for different times. The total current density versus time is shown in the leftmost inset, in which we have marked the times corresponding to the profiles depicted in part (a). The rightmost inset clearly shows a monopole with a right tail. Reprinted from [35].

this is an unstable situation that gives rise to dipole creation at the cathode [32, 42]. The details of the fast wave nucleation at the cathode can be found in [42]. A slight modification in the calculations is needed again if we use the boundary condition of Eq. (8.62) instead of the one used in the cited reference. After a new wave is created, J decreases. Then, inside the SL there are either two monopoles in case (i), or one monopole and a dipole in case (ii). In the latter case, provided ϕ is sufficiently large, we obtain Eq. (8.59) with $s = t/L$, $c_+ = C(J)/\nu$ and $c_- = J/\nu$. Then J tends to $J = J^\dagger$ such that $2C(J) = J$. Provided $J^\dagger < J_c$, J stays at the value $J^\dagger$ until the old monopole exits the sample, and we are back at the initial situation. The description of one period of the self-oscillation in case (i) is somewhat more complicated during the stage where two monopoles coexist [35]. In any case, one of the monopoles is eventually destroyed and we have completed one period of the oscillation. If $J > J^\dagger$ in case (ii), it is theoretically possible to create several dipoles and to produce a chaotic current signal [45]. This seems to have been numerically observed by Amann *et al.* [46], who used the discrete model with a general constitutive relation for the tunneling current as in Eq. (8.6). Whether this possibility can be realized in an experimental situation remains to be seen.

It is important to highlight the far-reaching difference between monopole and dipole mediated self-oscillations we have mentioned before. Dipoles are created quickly, and therefore contain noticeable charge accumulation and depletion layers that travel through almost the entire SL. On the other hand, monopoles are also created at the injecting contact, but the charge accumulation in them becomes noticeable only after a certain buildup time. This means that monopoles apparently traverse part of the SL during an oscillation period, whereas dipoles traverse the whole SL. Obviously this difference has important consequences that may be experimentally testable, that is, monopole mediated oscillations have higher frequencies. Furthermore, numerical simulations of discrete models show that each time a charge accumulation layer jumps from well to well, a current spike is produced. Then the number of current spikes per oscillation period seen in simulations of discrete models, but not in the continuum limit, is smaller for monopole self-oscillations than in the case of dipole mediated oscillations. For further details see the numerical results in [13] and Figure 8.4.

8.7.2 Dependence of the Oscillations on Control Parameters

It is important to learn how the frequency and amplitude of current self-oscillations depend on the configuration of the SL, such as doping, barriers and well widths and the number of SL periods. It also depends on other parameters such as voltage, temperature, magnetic field, photoexcitation, and so on, all of which can be controlled. We have previously explained how the boundary condition at the cathode selects whether monopole or dipole recycling and motion characterize the self-oscillation. In essence, a cathode condition that imposes a charge accumulation layer near the cathode for all values of the current selects monopole recycling. In addition to Ohm's law with low resistivity, an excess of electrons in the first SL period, namely,

$$n_1 - 1 = c > 0 \Longrightarrow F_0 = F_1 - c\,\nu \tag{8.70}$$

has also been widely used to select monopoles [7, 23, 35, 47].

8.7.2.1 Doping Density

Figure 8.19 shows a doping–voltage phase diagram for the second plateau, thereby zero diffusivity, of a 20-well SL, assuming that the cathode condition is Eq. (8.70) with $c = 10^{-4}$ [38]. For dimensionless doping lower than the minimum of the solid line, the SL evolves towards an almost uniform stationary state. For larger doping with $\nu < \nu_{TB}$, we have a nearly uniform stable stationary state outside a certain bias interval $(\phi_\alpha, \phi_\omega)$. At these points, a branch of self-oscillations stably bifurcates supercritically, starting with the zero amplitude and nonzero frequency of a Hopf bifurcation. For $\nu > \nu_{TB}$, a horizontal line intersects a number of different curves: (i) Hopf bifurcations, (ii) saddle-node bifurcations, (iii) homoclines.

Intersecting several Hopf bifurcation curves may indicate that there are intervals of self-oscillations alternating with intervals where the stable solution is a stationary state. The stationary state is a pinned wave front, or monopole, separating two electric field domains at which F_i is either $F^{(1)}(J)$ or $F^{(3)}(J)$. The wave front is pinned at a well M such that $\phi \approx F^{(1)}(J)\,(M/N) + F^{(3)}(J)\,(1 - M/N)$. There may

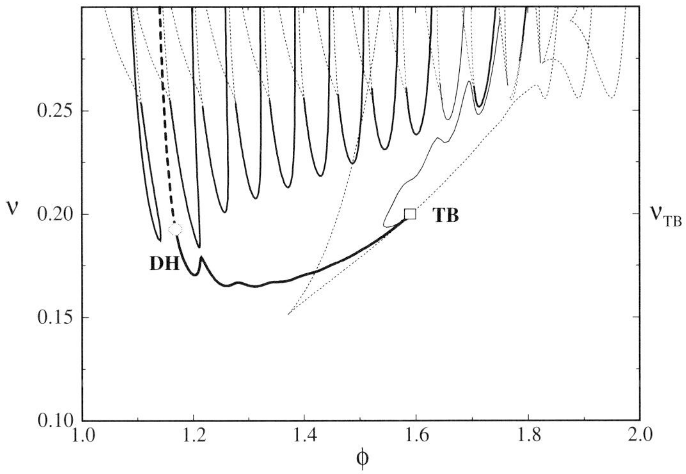

Fig. 8.19 Total phase diagram of the model for $N = 20$ and $c = 10^{-4}$. The dotted lines are curves of stationary saddle-nodes. For the sake of clarity, we have plotted only the main line of homoclinic orbits which sprout from the Takens–Bogdanov point TB (thin solid line). At a Takens–Bogdanov point, lines of Hopf bifurcations, saddle-node bifurcations, and homoclinic orbits intersect tangentially). We have not shown other homoclinic orbits. There is one curve of homoclinic orbits for each Hopf curve. Reprinted from [38].

be at most N intervals of stationary wave fronts. Between two such intervals, there may be a bias interval of stable self-oscillations in which monopole recycling and motion occurs about the Mth well. In some of the interior regions observed for such doping values, there may be coexisting multistable states with hysteresis cycles in their transitions from one to the next. The last interval of self-oscillations ends at a homocline, that is, the oscillation frequency tends to zero while the amplitude remains finite.

Another interesting feature of the phase diagram is the dashed line of Hopf bifurcations above the point marked DH. On this line, the Hopf bifurcation is subcritical, that is, an unstable branch of self-oscillations bifurcates for $\phi < \phi_\alpha$. Typically this branch coalesces with a branch of stable oscillations at a smaller bias ϕ_{LP}. Then there is an interval where both self-oscillations of finite amplitude and frequency and a stationary state are stable, that is, an interval of bistability. Driving the bias adiabatically, we can obtain a hysteresis cycle. Finally, for large enough doping the self-oscillations disappear, and we have multistability of stationary solutions corresponding to coexistence of domains separated by a pinned wave front.

The previous phase diagram may change substantially if we change the number of SL periods or c. For example, the branch of self-oscillations may disappear at the high bias region either at a Hopf bifurcation (finite frequency) or at a homoclinic orbit (zero frequency), depending on the values of these parameters. The frequency may decrease or increase with increasing bias. The first situation was observed in [7], and the second in [48].

8.7.2.2 **Temperature**

Temperature changes both the Fermi functions and the scattering amplitudes in the expressions for the tunneling currents. As a consequence, drift velocity and diffusion coefficients in the DDD model can be substantially changed, which in turn can drastically affect self-oscillations.

Figure 8.20 depicts the field-dependent drift velocity at different temperatures for SL parameter values of [50]: 40 periods of 14 nm GaAs and 4 nm AlAs and well doping $N_D = 2\times10^{11}\ \text{cm}^{-2}$. It has been calculated from the microscopic tunneling current density by the procedure explained in [21]. The only adjustable parameter in the sequential tunneling formulas is the Lorentzian half-width of the scattering amplitudes, γ. To estimate them, we have considered that the voltage difference, ΔV, between the peaks of two consecutive branches on the second plateau of the static I–V characteristic is

$$\Delta V \approx \mathcal{E}_{C3} - \mathcal{E}_{C2} - 2\eta\gamma\ . \tag{8.71}$$

Here $\mathcal{E}_{Ci}$ is the i-th energy level of a given well. For $\gamma = 0$, the field profile on the second plateau corresponds to two coexisting electric field domains with fields $(\mathcal{E}_{C2} - \mathcal{E}_{C1})/(e\,l)$ and $(\mathcal{E}_{C3} - \mathcal{E}_{C1})/(e\,l)$. The domain walls corresponding to two adjacent branches in the I–V diagram are located in adjacent wells. Then the voltage difference should be $\Delta V \approx (\mathcal{E}_{C3} - \mathcal{E}_{C2})/e$. In the presence of scattering, resonant peaks have finite widths which we take as $2\eta\gamma$, thereby obtaining Eq. (8.71). 2η is an adjustable parameter of the order of unity [51]. By using this formula and the measured current in [50] (Figures 1–3), we find $\gamma = 18\ \text{meV}$ at 1.6 K and $\gamma = 23\ \text{meV}$ at 140 K for $\eta \approx 0.6$. Linear interpolation yields the temperature dependence of γ in the range we are interested in.

Note that the first peak of the velocity in Figure 8.20 rapidly disappears as the temperature increases for this particular sample. In addition, note the shoulder

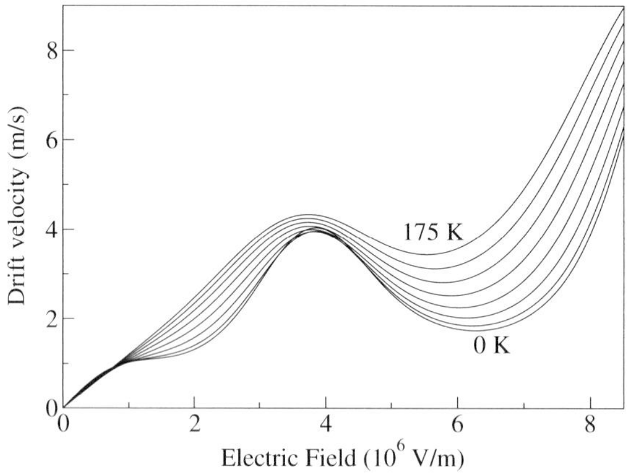

Fig. 8.20 Drift velocity versus electric field for different temperatures (starting at 0 K up to 175 K in 25 K steps) for a 40-well 14 nm GaAs/4 nm AlAs SL. Well doping is $N_D^w = 2\times10^{11}\ \text{cm}^{-2}$. Reprinted from [49].

feature at about 106 V/m for temperatures below 50 K. The former result might change if we assume different scattering amplitudes for each of the two first subbands of the wells. Moreover, the different extrema of the velocity curves shifts to lower field values as the temperature increases. Thus, formation of electric field domains and current self-oscillations are expected for voltages on the second and higher plateaus. Multistable solution branches of the current–voltage characteristic curve should also shift to lower voltages and higher currents as the temperature increases, as observed in experiments [52]. These effects could not be obtained from the fitted drift velocity in [52]. As the diffusion coefficient decreases very rapidly with the field, we can safely set $D \equiv 0$ in our DDD model for the experimentally observed voltage range. The relevant model is thus the well-known discrete drift model of [23], with the drift velocity of Figure 8.20 and the boundary condition (8.70).

Results of numerical calculations are shown in Figure 8.21. We observe that the I–V curve presents intervals in which the average current increases with voltage, followed by intervals in which the average current decreases. At lower temperatures the intervals of increasing current are wider, whereas the opposite occurs at higher temperatures. Correspondingly, the frequency of the self-oscillations in such an interval starts increasing but it drops to a smaller value than the initial one at the upper limit of the interval. The amplitude of the self-oscillations (not shown here) vanishes at the upper and lower limits of each voltage interval. This suggests that the branches of self-oscillations begin and end at supercritical Hopf bifurcations.

As the temperature increases, the region of negative differential conductivity in Figure 8.20 becomes smoother and the frequency of the self-oscillations increases (Figure 8.21b). At low temperatures, the electric field profiles consist of basically two stationary domains joined by a domain wall. The I–V characteristic curve has multiple branches corresponding to stationary domains with the domain wall locat-

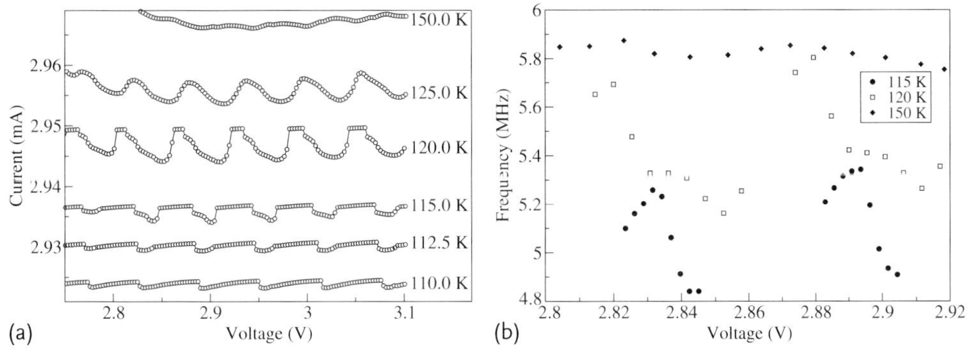

Fig. 8.21 (a) I–V characteristics for different temperatures, showing stationary (dynamic) states with full (empty) circles. The boundary condition parameter $c = 10^{-3}$ has been used in the numerical simulations. The curve corresponding to 150 K has been shifted −0.04 mA for clarity. Lines are plotted only for eye-guiding purposes. (b) Current oscillation frequency versus voltage for some dynamic dc bands of the curves shown in (a). Reprinted from [49].

ed at different wells. This situation resembles that obtained as voltage and doping are varied, provided doping and inverse temperature are assimilated. Note in Figure 8.21 that there are voltage intervals where the oscillation frequency increases with voltage, while the average current decreases with voltage. This behavior was called *anomalous* by Wang *et al.* [48], but it can be directly explained by the discrete drift model equations. See [49] for details.

8.7.2.3 Effect of Other Parameters on Self-Oscillations

The effect of other parameters such as photoexcitation or an external magnetic field on electric field domains [53] or self-oscillations [54–56] has been explored both experimentally and theoretically, although not to the same extent as for the cases of doping and temperature.

The influence on self-oscillations of a magnetic field B transverse to the SL growth direction has been studied experimentally and theoretically by Sun *et al.* [55, 56]. They found that increasing the magnetic field was qualitatively similar to increasing the temperature in a doped SL that presented a multistable field domain configuration. Their SL presented such a configuration at $B = 0\,T$. Increasing B both shifted the plateaus to larger voltages and diminished the length of the branches in the I–V characteristic curve and the peak current. Above a critical field, the I–V curve became flat and self-oscillations started. If a higher critical B was surpassed, the self-oscillations disappeared and so did the corresponding plateau. These observations were explained by simulations of the discrete drift model with parameters corresponding to the second plateau of the SL [56]. For a drift velocity, Sun *et al.* used the Kazarinov–Suris expression [24] in which the Lorentzian width increased with B and its center, eFl, changed to $eFl - e^2l^2B^2/(2m^*)$. It would be interesting to compare experimental results with similar simulations using the drift velocity Eq. (8.14).

Photoexcitation acts qualitatively on self-oscillations, similar to doping [18, 44, 57, 58]. Discrete models are somewhat more complicated. We have to consider the two-dimensional hole density in the Poisson equation and add a rate equation for it [11, 18]. Phase diagrams of photoexcitation vs. voltage and frequency dependence of the self-oscillations have been studied experimentally [58], but more theoretical work is needed in order to better explain them [59].

8.8
Spin Transport in Dilute Magnetic Semiconductor Superlattices

An impressive success of spintronic applications has been typically realized in metal-based structures which utilize magnetoresistive effects for substantial improvements in the performance of computer hard drives and magnetic random access memories [60]. Correspondingly, the theoretical understanding of spin-polarized transport is usually restricted to the metallic regime in linear response, which, while providing a good description for data storage and magnetic memory devices, is not sufficient for signal processing and digital logic. In contrast, much less is known about possible applications of semiconductor-based spintronics and spin-polarized transport in related structures which could utilize strong intrinsic nonlinearities in current–voltage characteristics to implement spin-based logic. Semiconducting materials offer the possibility of new device functionalities not realizable in metallic systems. Equilibrium carrier densities can be varied through a wide range of doping. In heterostructures and quantum dots, nanosecond spin dynamics persist at room temperature. Furthermore, because the typical carrier densities in semiconductors are low compared to metals, electronic properties are easily tunable by gate potentials [61].

While optical excitation is used to create spin polarization in most spin dynamic experiments, electrical injection and detection of spin currents are more appropriate for practical applications. The large mismatch in conductivity and spin relaxation time between metals and semiconductors produces very small efficiencies in ferromagnetic/semiconductor junctions when used as spin injectors [62]. This has spurred research on diluted magnetic semiconductors (DMS) that can be associated more easily with nonmagnetic semiconductors for spin injection [63, 64]. Using contacts based in Mn, it has been shown that DMS are efficient spin injectors [65]. Electrical injection of spin polarized current makes DMS nanostructures interesting potential spintronic devices. In DMS, spin plays an important role in electron dynamics, particularly in II–VI based semiconductor superlattices (SL) doped with Mn^{2+} ions [66]. Recently [67–69], nonlinear transport through DMS SLs has been investigated. The interplay between the nonlinearity of the current–voltage characteristics and the exchange interaction produces interesting spin dependent features: multistability of steady states with different polarization in the magnetic wells [67], time-periodic oscillations of the spin-polarized current and induced spin polarization in nonmagnetic wells by their magnetic neighbors [68, 69], among others. The high sensitivity of these systems to external fields points out to their potential application as magnetic sensors [67].

A typical sample consists of an n-doped ZnSe/(Zn,Cd,Mn)Se weakly coupled MQWS. The spin for the magnetic ion Mn^{2+} is $S = 5/2$ and the exchange interaction between the Mn local moments and the conduction band electrons is ferromagnetic in II–VI QWs. Using the virtual crystal and mean field approximations, the effect of the exchange interaction makes the subband energies spin dependent in those QWs that contain Mn ions: $E_j^{\pm} = E_j \mp \Delta/2$ where $\Delta = 2J_{sd} N_{Mn} S\, B_S(g\mu_B B S/(k_B T_{\text{eff}}))$ for spin $s = \pm 1/2$, and B, J_{sd}, N_{Mn}, k_B, and T_{eff} are the external magnetic field, the exchange integral, the density of magnetic impurities, the Boltzmann constant and an effective temperature which accounts for Mn interactions, respectively [67, 70]. We model spin-flip scattering coming from spin–orbit or hyperfine interaction by means of a phenomenological scattering time τ_{sf}, which is larger than impurity and phonon scattering times: $\tau_{\text{scat}} = \hbar/\gamma < \tau_{\text{sf}}$. Vertical transport in the nanostructure is spin-independent sequential tunneling between adjacent QWs so that when electrons tunnel to an excited state, they instantaneously relax by phonon scattering to the ground state with the same spin polarization. Lastly, electron–electron interaction is considered within the Hartree mean field approximation.

The equations describing our model are: [69, 70]

$$F_i - F_{i-1} = \frac{e}{\varepsilon}\left(n_i^+ + n_i^- - N_D\right), \tag{8.72}$$

$$e\,\frac{dn_i^\pm}{dt} = J_{i-1\to i}^\pm - J_{i\to i+1}^\pm \pm \frac{1}{\tau_{\rm sf}}\left(n_i^- - \frac{n_i^+}{1+e^{\frac{E_{1,i}^- - \mu_i^+}{\gamma_\mu}}}\right), \tag{8.73}$$

where $i = 1,\ldots,N$. n_i^+, n_i^- and $-F_i$ are the two-dimensional (2D) spin-up and spin-down electron densities, and the average electric field at the i-th SL period (which starts at the right end of the $(i-1)$-th barrier and finishes at the right end of the i-th barrier), respectively.

The voltage bias condition for the applied voltage V is

$$\sum_{i=0}^{N} F_i l = V. \tag{8.74}$$

We have denoted the spin-dependent subband energies ($\mathcal{E}$) (measured from the bottom of the i-th well) by $E_{j,i}^\pm = E_j \mp \Delta_i/2$, with $\Delta_i = \Delta$ or 0, depending on whether the i-th well contains magnetic impurities. N_D, ε, $-e$, $l = d + w$, and $-J_{i\to i+1}^\pm$ are the 2D doping density at the QWs, the average permittivity, the electron charge, the width of a SL period (d and w are barrier and well widths), and the tunneling current density across the i-th barrier, respectively.

For electrons with spin $\pm 1/2$, the chemical potentials at the i-th SL period, $\mu_i^\pm$, are related to the electron densities by

$$n_i^\pm = \rho \ln\left[1 + \exp\left(\frac{\mu_i^\pm - E_{1,i}^\pm}{k_B T}\right)\right], \tag{8.75}$$

where $\rho = m^* k_B T/(2\pi\hbar^2)$ and m^* is the effective electron mass.

For numerical convenience, the right-hand side of Eq. (8.73) contains a smoothed form of the scattering term used in [67]. As $\gamma_\mu \to 0$, our scattering term becomes $\pm(n_i^- - n_i^+)/\tau_{\rm sf}$ for $\mu_i^+ > E_{1,i}^-$ (equivalently, $\mu_i^+ - E_{1,i}^+ > \Delta$), and $\pm n_i^-/\tau_{\rm sf}$ otherwise [67].

By time-differencing Eq. (8.72) and inserting (8.73) in the result, we obtain the following form of Ampère's law,

$$\varepsilon\,\frac{dF_i}{dt} + J_{i\to i+1} = J(t), \tag{8.76}$$

where $J_{i\to i+1} = J_{i\to i+1}^+ + J_{i\to i+1}^-$ and $J(t)$ is the total current density. The total current density $J(t)$ is independent of i, as it can be written as

$$J(t) = \frac{1}{N+1}\sum_{i=0}^{N} J_{i\to i+1} \tag{8.77}$$

by adding Eq. (8.76) for all i and using that time-differencing Eq. (8.74) yields $dV/dt = l\sum_{i=0}^{N} dF_i/dt = 0$.

Tunneling currents $J^{\pm}_{i\to i+1}$ are calculated by the Transfer Hamiltonian method, taking into account that spin up and down electrons have different energies:

$$J^{\pm}_{i\to i+1} = \frac{e\,\nu^{(f)\pm}(F_i)}{l}\left\{n_i^{\pm} - \rho\ln\left[1 + e^{-\frac{eF_i l}{k_B T}}\left(e^{\frac{n^{\pm}_{i+1}}{\rho}} - 1\right)\right]\right\}, \qquad (8.78)$$

for $i = 1, \ldots, N-1$, provided that scattering-induced broadening of energy levels is much smaller than subband energies and chemical potentials; see Eq. (8.6).

According to (8.7), the spin-dependent "forward tunneling velocity" $\nu^{(f)\pm}$ is a sum of Lorentzians of width 2γ (the same value for all subbands, for simplicity) centered at the resonant field values $F^{\pm}_{j,i} = (E^{\pm}_{j,i+1} - E^{\pm}_{1,i})/(el)$:

$$\nu^{(f)\pm}(F_i) = \frac{\hbar^3 l\gamma}{2\pi^2 m^{*2}}\sum_{j=1}^{2}\frac{\mathcal{T}_i(E^{\pm}_{1,i})}{\left(E^{\pm}_{1,i} - E^{\pm}_{j,i+1} + eF_i l\right)^2 + (2\gamma)^2}. \qquad (8.79)$$

Here, $\mathcal{T}_i$ is proportional to the dimensionless transmission probability across the i-th barrier; see Eq. (8.8).

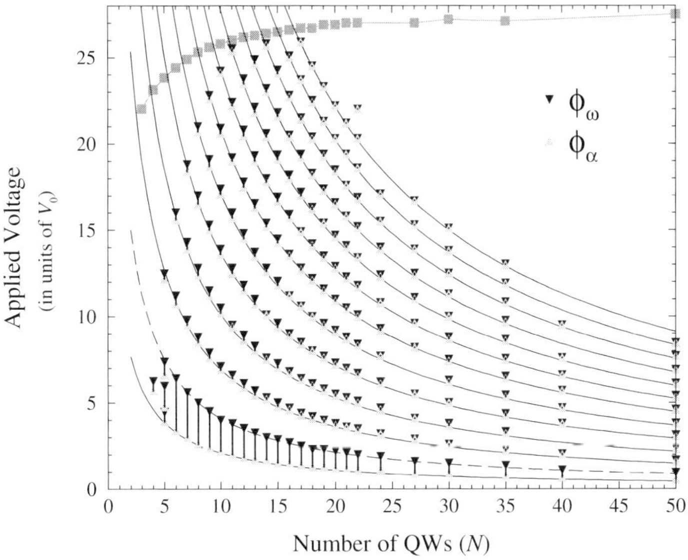

Fig. 8.22 Phase diagram of average electric field $\phi = V/[F_M(N+1)l]$ vs. N for a MQWS containing magnetic impurities only in its first QW and having $N_D = 10^{10}\,\text{cm}^{-2}$, $F_M = 0.64\,\text{kV/cm}$, $J_M = 0.409\,\text{A/cm}^2$, $\Delta = 12\,\text{meV}$. The SSCOs begin at $\phi = \phi^N_{\alpha,k}$ (lines marked with triangles) and end at $\phi = \phi^N_{\omega,k}$ (lines marked with inverted triangles). Solid lines are given by the formula $\phi^N_{\alpha,k} = C_k k/(N+1)$, where $C_1 = 23$ and $C_k = 38$ for $k \geq 2$. In the regions of the phase diagram where there are no SSCOs, the stable solutions are stationary states. The upper line joining solid squares marks the bias values at which the current in the current–voltage characteristics drops to a value corresponding to an almost uniform state with zero charge density. Reprinted from [70].

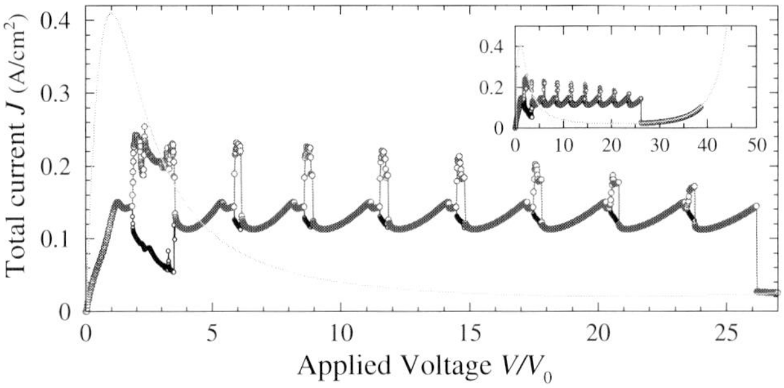

Fig. 8.23 Current–voltage characteristics for the MQWS with $N = 12$ and $\Delta = 12$ meV. The maximum and minimum of the SSCOs has been represented in each interval $(\phi^N_{\alpha,k}, \phi^N_{\omega,k})$ with $k = 1, \ldots, 8$. The solid line is $J_{i\to i+1}$ for $F_i = V/V_0$, $n^\pm = N_D/2$, with $V_0 = F_M(N+1)l = 12.5$ mV. Inset: Same for a wider voltage range. Reprinted from [70].

The tunneling current density $J_{i\to i+1} = J^+_{i\to i+1} + J^-_{i\to i+1}$ is a function of F_i, $n_i^\pm$ and $n_{i+1}^\pm$. For constant values $n_i^\pm = N_D/2$ and $F_i = F$, the tunneling current density at a nonmagnetic QW has a maximum J_M at a value F_M of the field. In terms of F_M, the voltage bias condition can be written as a condition for the average field ϕ:

$$\phi \equiv \frac{V}{(N+1)F_M l} = \frac{1}{(N+1)F_M} \sum_{i=0}^{N} F_i \,. \tag{8.80}$$

As boundary tunneling currents for $i = 0$ and N, we use Eq. (8.78) with $n_0^\pm = n_{N+1}^\pm = \kappa N_D/2$ (identical normal contacts with $\kappa \geq 1$) [67]. Initially, we set $F_i = V/[l(N+1)]$ (then $\phi = F_i/F_M$), and $n_i^\pm = N_D/2$ (normal QWs).

The spin transport model (8.72)–(8.80) was numerically solved and analyzed in [69, 70] following the same ideas as in the case of weakly coupled III–V semiconductor SLs. It is possible to find both static electric field domains which may be bistable and self-sustained spin polarized current oscillations. Figure 8.22 shows the phase diagram displaying regions of stationary and self-sustained current oscillations (SSCO) for SL with a different number of periods. Figure 8.23 shows the current–voltage characteristics for a 12-period SL. More information can be found in [70].

References

1 Allan, G., Bastard, G., Boccara, N., Lannoo, M. and Voos, M. (eds) (1986) *Heterojunctions and Semiconductor Superlattices*, Springer, Berlin.

2 Bastard, G. (1988) *Wave Mechanics Applied to Semiconductor Superlattices*, Halsted Press, New York.

3 Grahn, T.H. (ed.) (1995) *Semiconductor Superlattices: Growth and Electronic Properties*, World Scientific, Singapore.
4 Bonilla, L.L. and Grahn, T.H. (2005) Nonlinear dynamics of semiconductor superlattices. *Rep. Prog. Phys.*, **68**, 577–683.
5 Schneider, H., von Klitzing, K., and Ploog, H.K. (1989) Resonant and nonresonant tunneling in GaAs/AlAs multi quantum well structures. *Superlattices Microstruct.*, **5**, 383–396.
6 Grahn, T.H., Haug, J.R., Müller, W., and Ploog, K. (1991) Electric-field domains in semiconductor superlattices: A novel system for tunneling between 2D systems. *Phys. Rev. Lett.*, **67**, 1618–1621.
7 Kastrup, J., Grahn, H.T., Hey, R., Ploog, K., Bonilla, L.L., Kindelan, M., Moscoso, M., Wacker, A., and Galán, J. (1997) Electrically tunable GHz oscillations in doped GaAs-AlAs superlattices. *Phys. Rev. B*, **55**, 2476–2488. Copyright (1997) by the American Physical Society.
8 Grahn, T.H., Schneider, H., and von Klitzing, K. (1990) Optical studies of electric field domains in GaAs-$Al_x Ga_{1-x}$As superlattices. *Phys. Rev. B*, **41**, 2890–2899.
9 Kastrup, J., Grahn, T.H., Ploog, K., Prengel, F., Wacker, A., and Schöll, E. (1994) Multistability of the current–voltage characteristics in doped GaAs-AlAs superlattices. *Appl. Phys. Lett.*, **65**, 1808–1810.
10 Kantelhardt, W.J., Grahn, T.H., Ploog, H.K., Moscoso, M., Perales, A., and Bonilla, L.L. (1997) Spikes in the current self-oscillations of doped GaAs/AlAs superlattices. *Phys. Status Solidi (b)*, **204**, 500–503.
11 Grahn, T.H., Kastrup, J., Ploog, K., Bonilla, L.L., Galán, J., Kindelan, M., and Moscoso, M. (1995) Self-oscillations of the current in doped semiconductor superlattices. *Japan. J. Appl. Phys.*, **34**, 4526–4530.
12 Miller, D.A.B., Chemla, D.S., Damen, T.C., Gossard, A.C., Wiegmann, W., Wood, T.H., and Burrus, C.A. (1984) Band-edge electroabsorption in quantum well structures: the quantum-confined Stark effect. *Phys. Rev. Lett.*, **53**, 2173–2176.
13 Sánchez, D., Moscoso, M., Bonilla, L.L., Platero, G., and Aguado, R. (1999) Current self-oscillations, spikes and crossover between charge monopole and dipole waves in semiconductor superlattices. *Phys. Rev. B*, **60**, 4489–4492.
14 Wacker, A. (2002) Semiconductor superlattices: A model system for nonlinear transport. *Phys. Rep.*, **357**, 1–111.
15 Likharev, K.K., Bakhvalov, S.N., Kazacha, S.G., and Serdyukova, I.S. (1989) Single-electron tunnel junction array: an electrostatic analog of the Josephson transmission line. *IEEE Trans. Magn.*, **MAG-25**, 1436–1439.
16 Laikhtman, B. (1991) Current-voltage instabilities in superlattices. *Phys. Rev. B*, **44**, 11260–11265.
17 Laikhtman, B. and Miller, D. (1993) Theory of current–voltage instabilities in superlattices. *Phys. Rev. B*, **48**, 5395–5412.
18 Bonilla, L.L., Galán, J., Cuesta, J.A., Martínez, F.C., and Molera, M.J. (1994) Dynamics of electric field domains and oscillations of the photocurrent in a simple superlattice model. *Phys. Rev. B*, **50**, 8644–8657.
19 Prengel, F., Wacker, A., and Schöll, E. (1994) A simple model for multistability and domain formation in semiconductor superlattices. *Phys. Rev. B*, **50**, 1705–1712.
20 Aguado, R., Platero, G., Moscoso, M., and Bonilla, L.L. (1997) Microscopic model for sequential tunneling in semiconductor multiple quantum wells. *Phys. Rev. B*, **55**, R16053–R16056.
21 Bonilla, L.L., Platero, G., and Sánchez, D. (2000) Microscopic derivation of transport coefficients and boundary conditions in discrete drift-diffusion models of weakly coupled superlattices. *Phys. Rev. B*, **62**, 2786–2796.
22 Bonilla, L.L. (2002) Theory of nonlinear charge transport, wave propagation and self-oscillations in semiconductor superlattices. *J. Phys. Condens. Matter*, **14**, R341–R381. Copyright (2002), Institute of Physics.
23 Bonilla, L.L. (1995) Dynamics of electric field domains in superlattices. In: *Nonlinear Dynamics and Pattern Formation in Semiconductors and Devices* (ed. F.-J.

Niedernostheide), Springer, Berlin, pp. 1–20.

24 Kazarinov, R.F. and Suris, R.A. (1972) Electric and electromagnetic properties of semiconductors with a superlattice. *Sov. Phys. Semicond.*, **6**, 120–131, [Fiz. Tekh. Poluprovodn. **6**, 148–162 (1972)].

25 Arana, J.I., Bonilla, L.L., and Grahn, H.T. (2008) Homogeneous nucleation of opposite moving dipole domains and current self-oscillations in undoped photoexcited superlattices. *Physica E*, **40**(5), 1209–1211.

26 Carpio, A., Bonilla, L.L., Wacker, A., and Schöll, E. (2000) Wavefronts may move upstream in doped semiconductor superlattices. *Phys. Rev. E*, **61**, 4866–4876.

27 Carpio, A., Chapman, S.J., Hastings, S., McLeod, J.B. (2000) Wave solutions for a discrete reaction-diffusion equation. *Eur. J. Appl. Math.*, **11**, 399–412.

28 Carpio, A., Bonilla, L.L., and Dell'Acqua, G. (2001) Motion of wave fronts in semiconductor superlattices. *Phys. Rev. E*, **64**, 036204.

29 Carpio, A. and Bonilla, L.L. (2001) Wave front depinning transition in discrete one-dimensional reaction-diffusion systems. *Phys. Rev. Lett.*, **86**, 6034–6037.

30 Bender, C.M. and Orszag, S.A. (1978) *Advanced Mathematical Methods for Scientists and Engineers*, MacGraw Hill, New York.

31 Bonilla, L.L (1987) Stable probability densities and phase transitions for mean-field models in the thermodynamic limit. *J. Statist. Phys.*, **46**, 659–678.

32 Higuera, J.F. and Bonilla, L.L. (1992) Gunn instability in finite samples of GaAs. II. Oscillatory states in long samples. *Physica D*, **57**, 161–184.

33 Knight, W.B. and Peterson, A.G. (1966) Nonlinear analysis of the Gunn effect. *Phys. Rev.*, **147**, 617–621.

34 Murray, D.J (1970) On the Gunn effect and other physical examples of perturbed conservation equations. *J. Fluid Mech.*, **44**, 315–46.

35 Bonilla, L.L., Kindelan, M., Moscoso, M., and Venakides, S. (1997) Periodic generation and propagation of traveling fronts in dc voltage biased semiconductor superlattices. *SIAM J. Appl. Math.*, **57**, 1588–1614.

36 Amann, A., Wacker, A., Bonilla, L.L., and E. Schöll (2001) *Phys. Rev. E*, **63**, 066207.

37 Wacker, A., Moscoso, M., Kindelan, M., and Bonilla, L.L. (1997) Current–voltage characteristic and stability in resonant-tunneling n-doped semiconductor superlattices. *Phys. Rev. B*, **55**, 2466–2475.

38 Moscoso, M., Galán, J., and Bonilla, L.L. (2000) Bifurcation behavior of a superlattice model. *SIAM J. Appl. Math.*, **60**, 2029–2057.

39 Luo, J.K., Grahn, T.H., and Ploog, H.K. (1998) Relocation time of the domain boundary in weakly-coupled GaAs/AlAs superlattices. *Phys. Rev. B*, **57**, R6838–R6841.

40 Rogozia, M., Teitsworth, W.S., Grahn, T.H., and Ploog, H.K. (2002) Relocation dynamics of domain boundaries in semiconductor superlattices. *Phys. Rev. B*, **65**, 205303.

41 Bonilla, L.L., Escobedo, R., and Dell'Acqua, G. (2006) Voltage switching and domain relocation in semiconductor superlattices. *Phys. Rev. B*, **73**, 115341.

42 Bonilla, L.L., Cantalapiedra, R.I., Gomila, G., and Rubí, M.J. (1997) Asymptotic analysis of the Gunn effect with realistic boundary conditions. *Phys. Rev. E*, **56**, 1500–1510.

43 Sánchez, D., Platero, G., and Bonilla, L.L. (2001) *Phys. Rev. B*, **63**, 201306(R).

44 Merlin, R., Kwok, H.S., Norris, B.T., Grahn, T.H., Ploog, K., Bonilla, L.L., Galán, J., Cuesta, A.J., Martínez, F.C., and Molera, M.J. (1995) Dynamics of resonant tunneling domains in superlattices: theory and experiment. *Proc. 22nd ICPS* (ed. D.J. Lockwood). World Scientific, pp. 1039–1042.

45 Cantalapiedra, R.I., Bergmann, J.M., Bonilla, L.L., and Teitsworth, W.S. (2001) Chaotic motion of space charge waves in semiconductors under time-independent voltage bias. *Phys. Rev. E*, **63**, 056216.

46 Amann, A., Schlesner, J., Wacker, A., and Schöll, E. (2002) Dynamic scenarios of multistable switching in semiconductor superlattices. *Phys. Rev. B*, **65**, 193313.

47 Bulashenko, M.O. and Bonilla, L.L. (1995) Chaos in resonant-tunneling superlattices. *Phys. Rev. B*, **52**, 7849–7852.

48 Wang, R.X., Wang, J.N., Sun, B.Q., and Jiang, D.S. (2000) Anomaly of the current self-oscillation frequency in the sequential tunneling of a doped GaAs/AlAs superlattice. *Phys. Rev. B*, **61**, 7261–7264.

49 Sánchez, Bonilla, L.L., and Platero, G. (2001) Temperature dependence of current self-oscillations and electric field domains in sequential tunneling doped superlattices. *Phys. Rev. B*, **64**, 115311.

50 Wang, J.N., Sun, B.Q., Wang, R.X., Wang, Y., Ge, W., and Wang, H. (1999) Dynamic dc voltage band observed within each current branch in the transition from static to dynamic electric-field domain formation in a doped GaAs/AlAs superlattice. *Appl. Phys. Lett.*, **75**, 2620–2622.

51 Choi, K.K., Levine, F.B., Malik, J.R., Walker, J., and Bethea, G.C. (1987) Periodic negative conductance by sequential resonant tunneling through an expanding high-field superlattice domain. *Phys. Rev. B*, **35**, 4172–4175.

52 Li, C.-Y., Sun, B.-Q., Jiang, D.-S., and Wang, J.-N. (2001) Analysis of the temperature-induced transition to current self-oscillations in doped GaAs/AlAs superlattices. *Semicond. Sci. Technol.*, **16**, 239–242.

53 Luo, J.K., Friedland, K.-J., Grahn, T.H., and Ploog, H.K. (2000) Magnetotransport investigations of structural irregularities in a weakly-coupled GaAs/AlAs superlattice under domain formation. *Phys. Rev. B*, **61**, 4477–4480.

54 Luo, J.K., Friedland, K.-J., Grahn, T.H., and Ploog, H.K. (2000) Magnetic-field effects on undriven chaos in a weakly-coupled GaAs/AlAs superlattice. *Appl. Phys. Lett.*, **76**, 2913–2915.

55 Sun, B., Wang, J., Ge, W., Wang, Y., Jiang, D., Zu, H., Wang, Z., Deng, Y., and Feng, S. (1999) *Phys. Rev. B*, **60**, 8868.

56 Sun, B., Wang, J.N., Jiang, D.S., Wu, J.Q., Wang, Y.Q., and Ge, W.K. (2000) *Physica B*, **279**, 220.

57 Ohtani, N., Egami, N., Grahn, T.H., Ploog, H.K., and Bonilla, L.L. (1998) The transition between static and dynamic electric-field domain formation in weakly-coupled GaAs/AlAs superlattices. *Phys. Rev. B*, **58**, R7528–R7531.

58 Ohtani, N., Egami, N., Grahn, T.H., and Ploog, H.K. (2000) Phase diagram of static and dynamic domain formation in weakly-coupled GaAs/AlAs superlattices. *Phys. Rev. B*, **61**, R5097–R5100.

59 Perales, A., Bonilla, L.L., Moscoso, M., and Galán, J. (2001) Spatio temporal structures in undoped photoexcited semiconductor superlattices. *Int. J. Bifurc. Chaos*, **11**(11), 2817–2822.

60 Bandyopadhyay, S. and Cahay, M. (2008) *Introduction to Spintronics*, 2nd edn., CRC Press, Boca Raton.

61 Wolf, A.S., Awschalom, D.D., Buhrman, A.R., Daughton, M.J., von Molnár, S., Roukes, M.L., Chtchelkanova, Y.A., and Treger, M.D. (2001) Spintronics: A spin-based electronics vision for the future. *Science*, **294**, 1488–1495.

62 Schmidt, G., Ferrand, D., Molenkamp, L.W., Filip, A.T., van Wees, B.J. (2000) Fundamental obstacle for electrical spin injection from a ferromagnetic metal into a diffusive semiconductor. *Phys. Rev. B*, **62**, R4790–4793.

63 Zutic, I., Fabian, J., and Erwin, S.C. (2007) Bipolar spintronics: From spin injection to spin-controlled logic. *J. Phys.: Condens. Matter*, **19**, 165219.

64 Zutic, I., Fabian, J., Das Sarma, S. (2004) Spintronics: Fundamentals and applications. *Rev. Mod. Phys.*, **76**, 323–410.

65 Slobodskyy, A., Gould, C., Slobodskyy, T., Schmidt, G., Molenkamp, W.L., Sánchez, D. (2007) Resonant tunneling diode with spin polarized injector. *Appl. Phys. Lett.*, **90**, 122109.

66 Crooker, A.S., Tulchinsky, A.D., Levy, J., Awschalom, D.D., Garcia, R., Samarth, N. (1995) Enhanced spin interactions in digital magnetic heterostructures, *Phys. Rev. Lett.*, **75**, 505–508.

67 Sánchez, D., MacDonald, A.H., and Platero, G. (2002) Field-domain spintronics in magnetic semiconductor multiple quantum wells. *Phys. Rev. B*, **65**, 035301.

68 Béjar, M., Sánchez, D., Platero, G., and MacDonald, A.H. (2003) Spin-polarized current oscillations in diluted magnetic semiconductor multiple quantum wells. *Phys. Rev. B*, **67**, 045324.

69 Bonilla, L.L., Escobedo, R., Carretero, M., and Platero, G. (2007) Multiquantum well spin oscillator. *Appl. Phys. Lett.*, **91**, 092102.

70 Escobedo, R., Carretero, M., Bonilla, L.L., and Platero, G. (2009) Self-sustained spin-polarized current oscillations in multi-quantum well structures. *New J. Phys.*, **11**, 013033.

9
Nonlinear Wave Methods for Related Systems in the Physical World

9.1
Introduction

In this book, we have focused on electronic transport at high fields in bulk semiconductors and semiconductor superlattices. We have also focused on one-dimensional models in which it is assumed that the principal dynamics to be described is in the direction of the applied voltage bias. Furthermore, it is generally assumed that directions normal to the applied bias (we generally call this the *lateral* direction) can be eliminated. This assumption is made not only to simplify the analysis process. Many observed phenomena in experimental systems are only slightly or not at all affected by the lateral charge dynamics. However, there are cases where lateral pattern formation effects can occur in semiconductor systems [1].

For the most part, the choice of specific topics presented in the previous chapters reflects the interests and past research of the authors. The systems that we have focused on also possess the virtue that there is high-quality experimental data which enables detailed comparison with the theoretical predictions. There are a number of other electronic transport and other physical systems for which related phenomena are observed and for which nonlinear wave methods can provide theoretical insight. Additionally, it seems that nonlinear wave methods in the context of effective drift-diffusion models will become increasingly important with the advent of nanoscale electronic transport systems, including both self-assembled arrays as well as those fabricated by a top-down approach such as electron-beam lithography. Finally, the wave methods described here can be applied to other types of physical systems altogether. The purpose of this chapter is to present a few of these other systems. However, it should be noted that we do *not* give a complete survey.

9.1.1
NNDC, SNDC, and ZNDC

One widely used method by which to classify semiconductor systems that undergo space-charge instabilities is to examine the shape of the static I–V curves. Here, we briefly summarize some key points. The reader may consult, for example, [2], for

Nonlinear Wave Methods for Charge Transport. Luis L. Bonilla and Stephen W. Teitsworth

ISBN: 978-3-527-40695-1

a more complete discussion of this nomenclature. The Gunn effect discussed in Chapter 6 and trap-controlled behavior discussed in Chapter 7 both provide examples of N-type negative differential conductivity, frequently abbreviated as NNDC. This is also known as *voltage-controlled* NDC because it is through the application of voltage bias that one controls the transport state of the system. In other words, the I–V curve is a single-valued function of applied voltage with the approximate shape of the letter "N", as shown in Figure 9.1a. On the other hand, if one applies a current bias for a NNDC system, it is clear that there is a regime of multiple states corresponding to the same current value, that is, for values of current between I_m and I_M in Figure 9.1a. Through our detailed studies in Chapters 6 and 7, we have seen how NNDC is frequently associated with solitary waves in the electric field vs. position along the bias direction of the samples.

Another possibility is the case of S-type NDC, or simply SNDC, in which the I–V curve has the approximate shape of the letter "S". SNDC is also referred to as *current-controlled* NDC because the I–V curve is a single-valued function of the current, as in Figure 9.1b. In this case, there is range of applied voltages $V_m < V < V_M$ for which there are multiple possible states of the system. It is often found that SNDC systems are unstable to the formation of static or time-varying *current filaments*. These current filaments are manifested by a region of high current density that connects the emitting and collecting contacts. In some cases, they may form pathways of irregular appearance not unlike a lightning bolt. Unlike field domains, the proper description of current filaments requires models that possess at least two spatial dimensions. Prominent examples of semiconductor systems that exhibit SNDC are impact ionization breakdown in n-doped GaAs epitaxial layers [3], heterostructure hot electron diodes (HHED) [4], and multi-layered diode structures such as the pnpn-diode [5].

Finally, we mention the possibility of Z-type NDC, or simply ZNDC, shown in Figure 9.1c. This feature may occur in systems where one would normally expect NNDC, but where space-charge buildup, in a quantum well for instance, can lead to positive feedback and bistability. We have already seen this behavior in Chapter 8 for the weakly-coupled superlattice that has quantum wells with sufficiently high doping so that there is bistability. Another system in which this behavior occurs is the double barrier resonant tunneling structure [6] and optically bistable quantum well systems [1]. ZNDC may lead to nonuniform and dynamic distribution of charge in the lateral direction. It should be noted that the dashed curve in

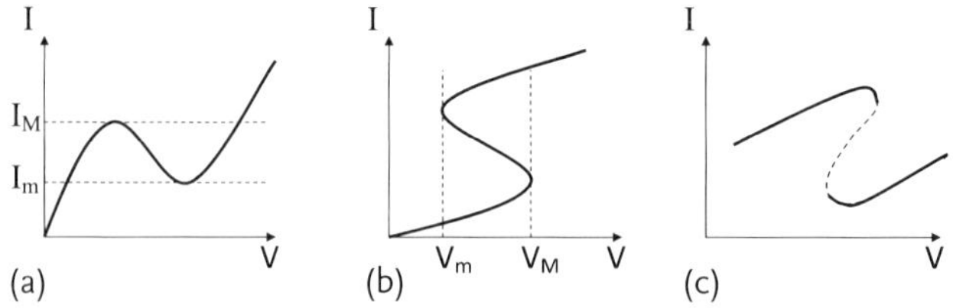

Fig. 9.1 (a) Illustration of I–V curve exhibiting NNDC, (b) I–V curve with SNDC, and (c) I–V curve with ZNDC.

Figure 9.1c implies the existence of an unstable steady state that connects the co-existing stable branches by a pair of saddle-node bifurcations. However, there are circumstances where such a steady unstable state may not exist, for instance, when the upper branch ends with a Hopf bifurcation as found in [7].

9.2 Superlattice Transport Model with Both Vertical and Lateral Dynamics

Recall that in Chapter 8, we presented a drift-diffusion model for transport in weakly-coupled superlattices (SLs) based on the microscopic transport process of sequential resonant tunneling between quasi-bound states of adjacent quantum wells. Central to this approach is to consider only the dynamics along the applied bias direction of the SL and to ignore the dynamics in the in-plane (lateral) direction. This is equivalent to treating each period as possessing a spatially-uniform charge density. For most experimental measurements on this system, this approximation is entirely reasonable as the lateral dynamics appears to play no significant role. However, there are a few exceptions, for example, when one is considering transient response on the scale of nanoseconds and faster. For example, recent theoretical work suggests that in stochastically-driven transitions from one pinned domain state to the next, there is a possibility that lateral dynamics can play a decisive role [6]. In 2005, Amann *et al.* [8] developed a theoretical framework which describes an extension of the one-dimensional SL model to include both lateral and vertical electronic dynamics. Subsequently, Xu *et al.* extended that framework to include the effects of a shunting side layer which is found to have the capability to stabilize a uniform field state in the SL against the formation of pinned domains [9]. In this section, we follow the treatment given in [9].

The coordinate system of the SL structure is shown in Figure 9.2. Each quantum well constitutes a slab that is parallel to the x–y plane, with cross sectional size L_x by L_y. There are N such quantum wells stacked on top of each other in the z direction, and these are sandwiched between an emitter layer and a collector layer. The SL period is l, and includes the width of one quantum well and one barrier layer. As in Chapter 8, the external voltage is applied in the z direction, across the emitter and the collector.

Because this is a weakly-coupled SL, the electrons are well-localized within each quantum well. It is also reasonable to assume that the electrons are at local equilibrium with local *two-dimensional* charge density at time t denoted by $n_i(x, y, t)$, where i is the well index. The charge continuity equation in the SL can be written as

$$e\,\dot{n}_i(x, y, t) = j_{\| i-1\to i} - j_{\| i\to i+1} - \nabla_\perp \cdot \boldsymbol{j}_{\perp i}\,, \tag{9.1}$$

where

$$\nabla_\perp = \hat{x}\frac{\partial}{\partial x} + \hat{y}\frac{\partial}{\partial y}\,, \tag{9.2}$$

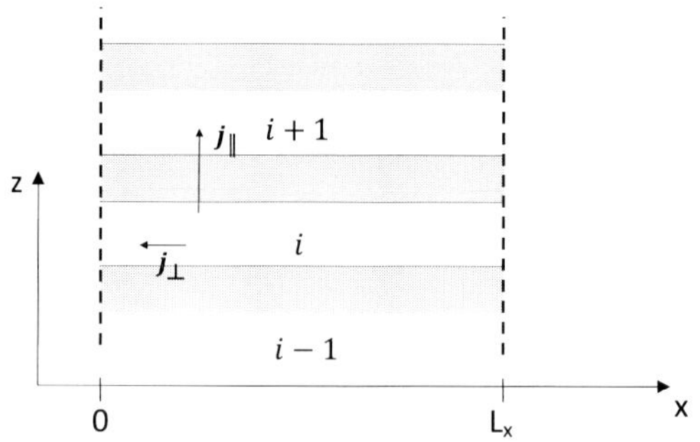

Fig. 9.2 Schematic illustration of the quantum wells and barriers in a superlattice, and also indicating the lateral extent in the *x*-direction.

and $j_{\parallel i-1\to i}$ denotes the three-dimensional vertical current density (standard SI units are Amps per square-meter) in z direction tunneling through the barrier between wells $i-1$ and i, and $\boldsymbol{j}_{\perp i}$ is the lateral two-dimensional current density (standard SI units are Amps per meter). Here $e < 0$ is the electron charge. For the sake of implicity, the y-dependence will be ignored and then Eq. (9.1) can be rewritten as

$$e\,\dot{n}_i(x,t) = j_{\parallel i-1\to i} - j_{\parallel i\to i+1} - \frac{\partial j_{\perp i}(x)}{\partial x}\,. \tag{9.3}$$

Here, the local vertical tunneling current $j_{\parallel i\to i+1}$ through each barrier is described by the sequential resonant tunneling model as described in detail in the preceding chapter. However, now, the tunneling current depends on the *local* electric field $F_{\parallel i}(x)$ across the barrier through which the tunneling occurs and the *local* electron charge densities $n_{i-1}(x)$ and $n_i(x)$. More explicitly, the tunneling current has the functional form $j_{\parallel i-1\to i}(x) = j_{\parallel i-1\to i}[F_{\parallel i}(x), n_{i-1}(x), n_i(x)]$.

As with the one-dimensional model, the tunneling current densities through the emitter and collector layers can be modeled by Ohmic boundary conditions, that is, $j_{\parallel 0\to 1}(x) = \sigma F_{\parallel 0}(x)$, and $j_{\parallel N\to N+1}(x) = \sigma F_{\parallel N}(x) n_N/N_D$, where σ denotes contact conductivity and N_D is the two-dimensional doping density per well.

The lateral dynamics is intrinsically determined by the lateral current density $\boldsymbol{j}_{\perp i}$ which may be taken to have the drift-diffusion form:

$$j_{\perp i}(x) = -e\mu n_i F_{\perp i} - e D_0 \frac{\partial n_i}{\partial x}\,, \tag{9.4}$$

where $F_{\perp i}(x)$ is the lateral component of the electric field at x in well i, μ is the mobility and D_0 is the diffusion coefficient. The generalized Einstein relation provides a convenient connection between μ and D_0 for arbitrary two-dimensional electron densities including the degenerate regime:

$$D_0(n_i) = \frac{n_i}{-e\rho_0(1-\exp[-n_i/(\rho_0 k_B T)])}\mu\,, \tag{9.5}$$

where we have defined the two-dimensional density of states $\rho_0 = m^*/(\pi\hbar^2)$, and m^* denotes the electron effective mass.

Both the lateral and vertical currents depend on the electrical fields which in turn depend on the scalar (electrostatic) potential $\phi_i(x, y)$. This potential can be determined from the Poisson equation

$$\Delta\phi_i(x, y) = (\Delta_\perp + \Delta_\parallel)\phi_i(x, y) = -\frac{e}{l\epsilon_r\epsilon_0}(n_i - N_D)\,, \tag{9.6}$$

where

$$\Delta_\perp\phi_i(x) \equiv \frac{\partial^2}{\partial x^2}\phi_i(x)\,, \tag{9.7}$$

$$\Delta_\parallel\phi_i(x) = \frac{\phi_{i-1}(x) - 2\phi_i(x) + \phi_{i+1}(x)}{l^2}\,. \tag{9.8}$$

Here, ϵ_r and ϵ_0 are the relative and absolute permittivity, respectively, and the second of these two equations is a discrete approximation to the Laplacian. Then, the field can be calculated as

$$F_{\parallel i}(x, y) = -\frac{\phi_{i+1}(x) - \phi_i(x)}{l}\,,$$

$$F_{\perp i}(x) = -\frac{\partial\phi_i(x)}{\partial x}\,. \tag{9.9}$$

A limitation of this approach is that one solves the Poisson equation using an approximation method assuming that the typical variations in the lateral direction are on a length scale much longer than the mean free path of the degenerate electrons. This seems reasonable for many experimental SL structures [8].

Turning to the question of boundary conditions in the lateral direction, we note that the normal components of electric field and electron current density should vanish at the side boundaries, which implies

$$F_{\perp i}(x = 0, L_x) = 0\,, \tag{9.10}$$

$$\frac{\partial n_i}{\partial x}(x = 0, L_x) = 0\,. \tag{9.11}$$

It is also interesting to note that the parallel component of the total current density integrated over x

$$J = \int_0^{L_x}\left(\epsilon_r\epsilon_0\dot{F}_{\parallel i} + j_{\parallel i\to i+1}\right)dx\,, \tag{9.12}$$

is the same for each period, just as the 1D total current density for the one-dimensional model of Chapter 8. To show this, we note that the Poisson equation can be written as

$$\nabla\cdot(\boldsymbol{F}_\perp + \boldsymbol{F}_\parallel) = \frac{e}{l\epsilon_r\epsilon_0}(n_i - N_D)\,, \tag{9.13}$$

which can be approximated as

$$\frac{F_{\parallel i} - F_{\parallel i-1}}{l} + \frac{\partial F_\perp}{\partial x} = \frac{e}{l\epsilon_r\epsilon_0}(n_i - N_D)\,. \tag{9.14}$$

Now, we substitute this last equation into Eq. (9.1) to obtain

$$l\epsilon_r\epsilon_0\frac{d}{dt}\left(\frac{F_{\|i}-F_{\|i-1}}{l}+\frac{\partial F_\perp}{\partial x}\right)=j_{\|i-1\to i}-j_{\|i\to i+1}-\frac{\partial j_{\perp i}(x)}{\partial x}\,. \quad (9.15)$$

Then, one integrates both sides of the preceding equation with respect to x from 0 to L_x. If we assume no current flow through the lateral boundaries, then we write the boundary conditions $F_\perp(0) = F_\perp(L_x) = 0$ and $j_{\perp m}(0) = j_{\perp m}(L_x) = 0$, and the lateral terms in the above equation integrate to zero. This yields

$$\epsilon_r\epsilon_0\frac{d}{dt}\int_0^{L_x} F_{\|i}dx+\int_0^{L_x} j_{\|i\to i+1}dx=\epsilon_r\epsilon_0\frac{d}{dt}\int_0^{L_x} F_{\|i-1}dx+\int_0^{L_x} j_{\|i-1\to i}dx\,, \quad (9.16)$$

which shows that the x-integral of the parallel component of the total current density is independent of the well index i. Simulations of this model have given insight into the role of lateral dynamics in switching between static current branches [8]. In particular, it is found that fronts of higher density n propagate laterally across the sample in the quantum well that contains an accumulation layer (cf., Chapter 8).

9.3
Semi-Insulating GaAs

Besides p-Ge studied in Chapter 7, there are other semiconductor materials displaying low-frequency Gunn-type oscillations due to repeated formation and motion of high field domains. One example is semi-insulating n-GaAs in which oscillations of a few Hz are caused by capture and emission of electrons onto deep defect-associated levels known as EL2 traps. See [10] for a review of experimental results and theory. The Sacks and Milnes model [11] has been used to study trap dynamics:

$$\frac{\varepsilon}{q}\frac{\partial \mathcal{E}}{\partial x}=N_{\mathrm{EL2}}-N_A+N_D-n-n_{\mathrm{EL2}}\equiv-\delta n_{\mathrm{EL2}}-n\,, \quad (9.17)$$

$$\varepsilon\frac{\partial \mathcal{E}}{\partial t}+q\,n\,v(\mathcal{E})+q\,D\frac{\partial n}{\partial x}=J\,, \quad (9.18)$$

$$\frac{\partial n_{\mathrm{EL2}}}{\partial t}=C_n(\mathcal{E})\,(N_{\mathrm{EL2}}-n_{\mathrm{EL2}})\,n-X_n(\mathcal{E})\,n_{\mathrm{EL2}}\,. \quad (9.19)$$

Equation (9.17) is the one-dimensional Poisson equation for the electric field E, where N_{EL2}, N_A, N_D, n and n_{EL2} are the densities of EL2 states, shallow acceptor, shallow donor, free electron and trapped electrons, respectively. ε is the permittivity and $q > 0$ is minus the charge of the electron. Equation (9.18) is the current balance equation (Ampère's law) establishing that the sum of the displacement and drift-diffusion currents equals the total current density J. x-differentiation of this equation yields the charge continuity equation. Equation (9.19) is the rate equation for the trapped electrons including capture and emission terms. The drift velocity is given by $v(\mathcal{E}) = \mu_0\mathcal{E}/[1 + (\mathcal{E}/\mathcal{E}_R)^2]$, whereas the capture and emission coefficients were adjusted to available experimental data by Bonilla *et al.* [12]. In particular, $C_n(\mathcal{E})$ is an increasing function which takes on very low values for electric fields below a certain threshold. Then, it increases rapidly and later levels on a plateau. The emission coefficient $X_n(\mathcal{E})$ mostly increases with $\mathcal{E}$.

Given the experimental values listed by Piazza *et al.* [13], the semi-insulating semiconductor is almost electrically neutral during the self-oscillations and the number of free electrons is much lower than that of trapped electrons. More precisely, $n \ll \delta n_{\mathrm{EL2}}$. Then, the leading order approximation to this system of equations can be obtained as follows [12]: (i) find n from Eq. (9.19) in terms of δn_{EL2}; (ii) expand the result up to first order in δn_{EL2} and insert it in Eq. (9.18); (iii) eliminate δn_{EL2} from this equation by approximating $\delta n_{\mathrm{EL2}} \sim -\partial\mathcal{E}/\partial x$; (iv) ignore terms of the order of $D\,\delta n_{\mathrm{EL2}}$. We obtain

$$-\frac{\varepsilon}{q}\frac{\partial^2\mathcal{E}}{\partial x\partial t} + \frac{\varepsilon(N_A - N_D)C_n}{qv}\frac{\partial\mathcal{E}}{\partial t} + \left\{\frac{D(N_{\mathrm{EL2}} - N_A + N_D)C_n\frac{d(X_n/C_n)}{d\mathcal{E}}}{v} - \frac{\varepsilon N_{\mathrm{EL2}}X_n}{q(N_A - N_D)}\right\}\frac{\partial\mathcal{E}}{\partial x} + (N_{\mathrm{EL2}} - N_A + N_D)\,X_n - \frac{C_n(N_A - N_D)}{v}J = 0\,. \tag{9.20}$$

This reduced equation can be used to determine states directly comparable to experimental data: stationary states and high-field charge dipole waves moving steadily far from the contacts. The latter were measured by Piazza *et al.* while observing the Gunn effect [13]. Stationary states are found by solving Eq. (9.20) with an appropriate boundary condition for a Schottky or Ohmic injecting contact, plus the dc voltage bias condition. On the other hand, high-field charge dipoles moving with speed c are certain solutions of Eq. (9.20) of the form $\mathcal{E} = \mathcal{E}(\chi)$, with $\chi = x - ct$ such that $\mathcal{E}(\pm\infty) = \mathcal{E}_1(J)$. For a given value of J, $\mathcal{E}_1(J)$ is the smallest positive constant solution of Eq. (9.20): $J(\mathcal{E}_1) - J = 0$, where

$$J(\mathcal{E}) = \frac{(N_{\mathrm{EL2}} - N_A + N_D)\,X_n(\mathcal{E})v(\mathcal{E})}{(N_A - N_D)\,C_n(\mathcal{E})}\,, \tag{9.21}$$

as depicted in Figure 9.3.

The function $J(\mathcal{E})$ is N-shaped for the low field values depicted in Figure 9.3 due to the capture coefficient $C_n(\mathcal{E})$. There is an interval of values of J for which three positive constant solutions are possible. For J, there is only one possible value of c such that there exists a solution of Eq. (9.20) satisfying $\mathcal{E}(\pm\infty) = \mathcal{E}_1(J)$. This solution (the high field charge dipole) is a homoclinic trajectory leaving and entering the saddle point $(\mathcal{E}_1(J), 0)$ on the phase plane $(\mathcal{E}, d\mathcal{E}/d\chi)$. The determination of the solitary wave follows the ideas explained in Chapter 7 for p-Ge. Figure 9.4 depicts $\mathcal{E}(\chi)$ as obtained numerically in [12]. It corresponds to the func-

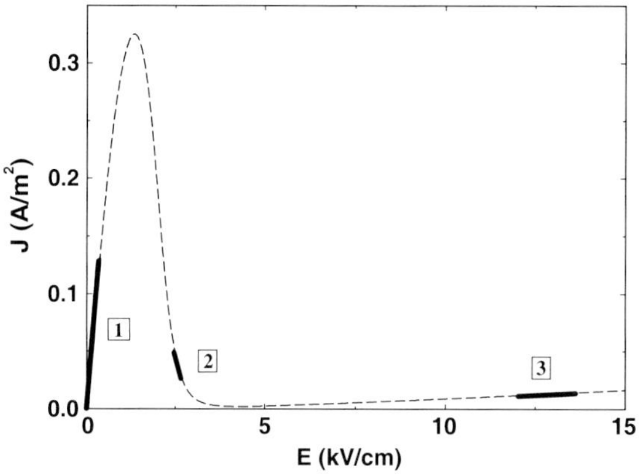

Fig. 9.3 Current density J as a function of the electric field obtained by the numerical solution of Eq. (9.20) (dashed line) together with available experimental results (solid line) for the cases: steady state Ohmic–Ohmic contact (1), high field dipole state for Ohmic–Ohmic contacts (2), and steady state Schottky–Ohmic contact (3) (from Ref. [12]).

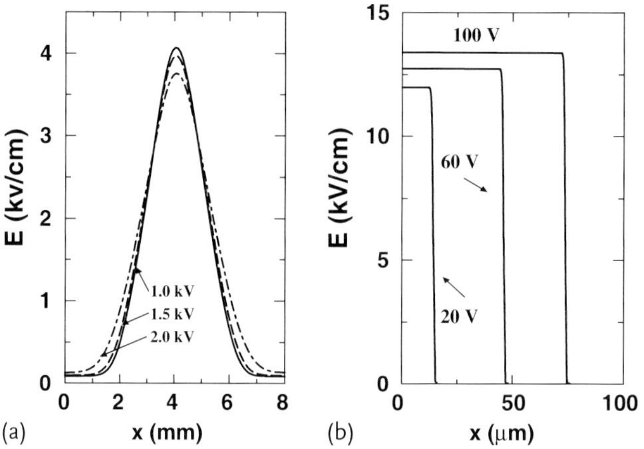

Fig. 9.4 Numerical results obtained in the present analysis for different voltages. In (a), electric field profiles for the high field charge dipole state with Ohmic contacts are shown. In (b), electric field profiles for the high field steady state for Schottky–Ohmic contacts are shown (from [12]).

tion experimentally determined by Piazza *et al.* [13], and it describes a fully developed wave far from the contacts. Piazza *et al.* [13, 14] fitted experimental data for dipole waves by using Gaussian curves, whereas the correct solutions of Eq. (9.20) were used by Bonilla *et al.* [12] to determine the coefficient functions $C_n(\mathcal{E})$ and $X_n(\mathcal{E})$.

9.4 Multidimensional Gunn Effect

Extending the theory of the Gunn effect to samples with point contacts implies considering multidimensional geometries. The simplest case is that of two concentric circular contacts containing bulk n-GaAs (Corbino geometry) in two space dimensions. Equations (6.15), (6.23), (6.18), and (6.19), respectively, in Chapter 6 become

$$\frac{\partial E}{\partial \tau} + v(E)\left[1 + \frac{1}{r}\frac{\partial(rE)}{\partial r}\right] - \delta\frac{\partial}{\partial r}\left[\frac{1}{r}\frac{\partial(rE)}{\partial r}\right] = \frac{J}{r}\,, \tag{9.22}$$

$$\frac{1}{L}\int_{r_c}^{r_a} E\,dr = \phi\,, \tag{9.23}$$

$$E = \rho\left(\frac{J}{r} - \frac{\partial E}{\partial \tau}\right) \quad \text{at} \quad r = r_c,\, r_a\,. \tag{9.24}$$

Here, $2\pi J(t)$ is the total current density through the semiconductor (which is proportional to the current circulating in the external circuit), r_c and r_a, $L = r_a - r_c > 0$ are the radii of cathode and anode and the sample length, respectively. The dimensionless resistivity $\rho > 0$ is chosen so that E/ρ intersects $v(E)\,(1 + E/r_c)$ on the second (decreasing) branch of $v(E)$. Then, there is a Gunn self-oscillation of the current mediated by annular charge dipole waves for appropriate values of ϕ. These waves are periodically generated at the cathode (radius r_c) and move towards the

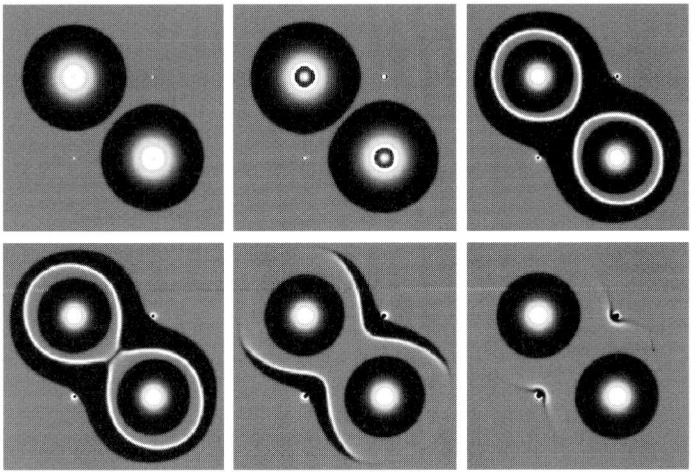

Fig. 9.5 Density plots of the solution of the Kroemer's model (with $v_s = 0$) in a square of side $l = 20$ with four circular contacts forming the vertices of a square of side $d = 4$ located at the center of the sample. Cathodes have potential $\varphi = 0$ and anodes have $\varphi = 10$. Nondimensional units as defined in [17] (from Figure 2 of [17]).

anode contact. There are important differences with the 1D case: the waves may not arrive at the anode (confined Gunn effect), their size diminishes and the current increases as they move, and it is easy to create several dipole domains per cycle of the oscillation. This latter property may give rise to aperiodic and possibly chaotic oscillations for narrow bias intervals. It is possible to extend our asymptotic analysis to this situation and thus to account for many of the oscillation features (see [15]).

In the general (not axisymmetric) case, it is possible to asymptotically reduce the motion of a non-circular dipole wave to the solution of a free boundary problem. Progress can possibly be made by considering simple geometries, for example, a rectangular n-GaAs sample with attached circular contacts. In the case of two contacts, the cathode issues forth dipole waves which are circular near it. They might deform as they have advanced far enough or disappear, as in the Corbino case. This situation is reminiscent of a pacemaker or leading center issuing forth target patterns in excitable media, as shown in Figure 9.5. Thus, it is interesting to consider the interaction of several waves by attaching several cathodes to the semiconductor sample. Further discussion, numerical solutions and analysis can be found in the papers [16–18].

9.5
Fluctuations in Gunn Diodes

Coinciding with the boom of mode-coupling theories, several groups have considered fluctuations in Gunn diodes. A possible formulation is akin to Landau–Lifshitz fluctuating hydrodynamics, that is, white noise sources are added to the electron current density $\boldsymbol{I}$ in Eq. (6.11) of Chapter 6. If current fluctuations are associated to dissipation due to electron diffusion, the electron flux should be $\boldsymbol{I} + \boldsymbol{i}$, where $\boldsymbol{I}$ is given by Eq. (6.11) of Chapter 6 and the fluctuating current $e\boldsymbol{i}$ is a Gaussian white noise satisfying

$$\langle \boldsymbol{i}(\boldsymbol{x}, t)\rangle = 0, \quad \langle i_m(\boldsymbol{x}, t) i_n(\boldsymbol{x}', t')\rangle = 2D[n_0 - n(\boldsymbol{x}, t)]\, \delta_{mn}\, \delta(\boldsymbol{x} - \boldsymbol{x}')\delta(t - t') .$$

Similar terms should be added in the boundary conditions, though we may ignore them in this presentation. The goal of this formulation is to obtain the voltage fluctuations due to the flux $\boldsymbol{i}$ about the steady state or the Gunn oscillation. A straightforward approximation is to linearize the fluctuating equations about the 1D steady state or the Gunn oscillation obtained in Chapter 6. Then, the electric field (for example) is a convolution of the corresponding Green function for this linear problem with the fluctuating flux i. As we know the statistics of i, that of the electric field is immediately obtained. This method has been used by Keizer [19] and Diaz-Guilera and Rubi [20] to predict voltage fluctuations about the stationary state and their enhancement as the critical voltage ϕ_α is approached. The purpose of these theories was to explain measurements by Kabashima *et al.* [21] of voltage fluctuations near the onset of the Gunn instability. No one has extended the theoretical results to fluctuations about the Gunn oscillation proper, in particular near the onset of the instability where fluctuation effects are more important. In principle, the latter is not hard to do by building upon our bifurcation analysis in Section 6.4 of Chapter 6. For Kabashima *et al.*'s sample length, the Hopf bifurcation is subcritical, though close to the length for which the bifurcation changes from super to subcritical. The earliest work by Nakamura [22] tried to do just this analysis. Unfortunately, this work had the flaw that periodic boundary conditions were used. It is easy to show that periodic boundary conditions suppress the Gunn oscillation: a dipole domain may move at constant speed and current through the spatially periodic semiconductor. In fact, let us integrate Ampère's law (6.22) of Chapter 6 using (6.23) of the same chapter to obtain the current density J. The result is

$$J = \frac{1}{L} \int_0^L v(E)\, dx ,$$

provided periodic boundary conditions for E and $\partial E/\partial x$ are used and ϕ is time-independent. If a dipole domain is moving at constant speed c on the sample, $\partial E/\partial \tau = -c\partial E/\partial x$, and therefore $dJ/d\tau = -\frac{c}{L}\int_0^L \frac{\partial v(E)}{\partial x}\, dx = 0$. Then, an oscillation of the current is not possible.

9.6 Dynamics of Dislocations in Mechanical Systems; Nanoarrays

Defects such as dislocations, vacancies and cracks control mechanical properties of materials including crystal plasticity, creep, fatigue, ductility and brittleness, hardness and friction. Crystal growth, radiation damage of materials, and their optical and electronic properties are also strongly affected by defects, particularly dislocations. These phenomena occur over many different scales of length and time and the properties at each scale are influenced by the others. In a certain sense, these multiscale phenomena are reminiscent of fluid turbulence. Aspects of complex physical phenomena such as fluid turbulence may be captured by the dynamics of entangled vortices, whereas dislocations play a paramount role in plasticity and strength of materials. Line vortices are solutions of the Euler equations with singular sources supported on lines and dislocations are solutions of the Navier equations of elasticity with singular sources supported on lines. However, vortices may move in any direction under infinitesimal stress whereas dislocations may move only in glide planes provided they are subject to a stress larger than the *Peierls stress* (defined as the minimal stress necessary for a straight dislocation to move at zero temperature). Both the Euler and Navier equations are approximations of theories that describe the outer structure of vortices and dislocations, respectively. In the case of fluids, the more detailed theory consists of the Navier–Stokes equations, whereas the inner core of dislocations may be described by atomic models or by discrete model equations.

As said, dislocations are defects in crystals. A simple example is obtained by inserting an extra half-plane of atoms in a simple cubic crystal. As shown in Figure 9.6, the added half-plane is the upper part of the y–z plane. The edge of this half-plane (i. e., the z axis) is the *dislocation line* of this *edge dislocation*. Near the dislocation line, the lattice is greatly distorted but, as we move away from it, the planes of atoms fit almost regularly. The distortion is recognized if we form a closed circuit (called Burgers circuit) of lattice points around the dislocation line, as in Figure 9.6. Let the dislocation line point outside the page (the positive z axis) and let the circuit be oriented counter-clockwise, following the right-hand rule [23]. Let r_0 be an arbitrary lattice point which we take as the initial point of the circuit, and let $v_i, i = 1, \ldots, N$ be vectors of length equal to one lattice period comprising the Burgers circuit in the undistorted lattice such that $\sum_{i=1}^{N} v_i = 0$. Each $v_i \in \{(1, 0), (-1, 0), (0, 1), (0, -1)\}$. The points of the Burgers circuit in the distorted lattice are such that r_i is the lattice point closest to $r_{i-1} + v_i$. A circuit that does not enclose a dislocation line ends at the initial point, $r_N = r_0$. A circuit containing a dislocation line is not closed, and we define *the Burgers vector* as [23] $b = r_N - r_0 = \sum_{i=1}^{N} \Delta u_i, \quad \Delta u_i = r_i - (r_{i-1} + v_i)$. In our example, $b = (1, 0, 0)$, one period is in the positive x direction.

An edge dislocation (at low temperature) may move along the *glide plane* formed by its line and its Burgers vector (the plane x–z in Figure 9.6). For edge dislocations, the Burgers vector and the dislocation line are orthogonal. In a finite crystal, the motion of a dislocation to the crystal boundary leaves a permanent deformation.

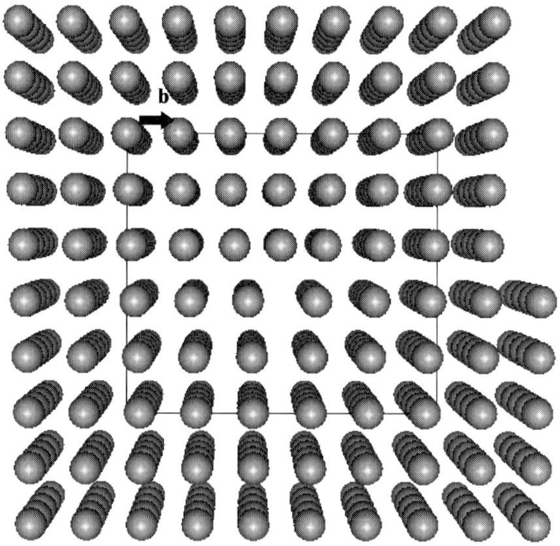

Fig. 9.6 Edge dislocation and a Burgers circuit around it. This is Figure 1 of the paper *Defects, Singularities and Waves*, by L.L. Bonilla and A. Carpio, in the book *Recent Advances in Nonlinear Partial Differential Equations and Applications*, edited by L.L. Bonilla, A. Carpio, J.M. Vega and S. Venakides. *Proceedings of Symposia in Applied Mathematics* vol. 65, pp. 131–150. American Mathematical Society, 2007.

Dislocation dynamics may be understood by simulating the classical motion of all atoms in a crystal (molecular dynamics). While such simulations are used to simulate creation and motion of dislocations, crack propagation, and so on, they have several drawbacks: (i) uncertainty about the potentials, which are derived by fitting empirical data or first-principles descriptions such as density functional theory for some key properties of the material; (ii) huge computational costs which forces simulations to be restricted to small samples during time intervals that may not be enough to display the desired material behavior; (iii) numerical artifacts such as numerical chaos and spurious oscillations have to be eliminated and technical issues as, for example, time discretization for long time computations and elimination of reflected waves from boundary conditions have to be solved; and (iv) qualitative information may be hard to extract from the simulations. Alternative descriptions may be spatially discrete models of dislocations which were analyzed in 1938 when Frenkel and Kontorova (FK) studied a model of interconnected harmonic springs in a periodic potential governed by the equations [24, 25]

$$m\,\frac{d^2 u_n}{dt^2} + \gamma\,\frac{du}{dt} = \kappa\,(u_{n+1} + u_{n-1} - 2u_n) + F - \frac{U\pi}{b}\,\sin\left(\frac{2\pi u_n}{b}\right). \quad (9.25)$$

While this one-dimensional model has several unrealistic features, such as an exponential decay of the distortion far from dislocation cores (instead of the algebraic decay given by elasticity theory), it captures pinning of dislocations by the crystal lattice and the existence of a minimal Peierls stress $F = F_c$ needed to depin them and facilitate their motion. The analysis of dislocation depinning is very similar to the same phenomenon of wave front depinning studied in Chapter 8.

Provided $F < \pi U/b$, a dislocation is a wave front solution joining two lineally stable uniform stationary solutions of Eq. (9.25): $u_n = [b/(2\pi)]\arcsin(Fb/(\pi U))$ and $u_n = [b/(2\pi)]\arcsin(Fb/(\pi U)) + 2\pi$. Firstly, let us consider the case of overdamped dynamics $m = 0$, $\gamma = 1$. In the overdamped case, wave fronts either move if $|F| > F_c > 0$ or are pinned if $|F| \le F_c$ [26]. The depinning transition at F_c was described by Carpio and Bonilla [26, 27] for large and moderate values of $A = U\pi/(b\kappa)$, by King and Chapman [28] in the continuum limit $A \to 0$, and by Fáth [29] when a piecewise linear source term $g(u)$ replaces the sine function in Eq. (9.25). The depinning transition is exactly as explained in Chapter 8 for superlattices. A complete description and analysis can be found in [27].

This picture changes qualitatively if both inertia and damping are included in Eq. (9.25). Static wave fronts are of course the same as for the overdamped case, but the depinning transition changes from the global bifurcation described in Chapter 8 to a subcritical bifurcation. The unstable branch subcritically issuing at $F = F_{cs} = F_c$ eventually coalesces at $F = F_{cd} < F_{cs}$ with a stable branch of moving stable wave front solutions representing dislocations. In the purely conservative case, $m > 0$, $\gamma = 0$, there are simple models with a piecewise linear source that can be analyzed to a great extent using complex variables [30]. Carpio and Bonilla [31] analyzed the piecewise linear model proposed by Atkinson and Cabrera [30] with $\gamma = 0$ ($g(u_n) = u_n \operatorname{sign} u_n$ instead of the sine function in Eq. (9.25)) and showed that the wave fronts representing dislocations join a stable uniform so-

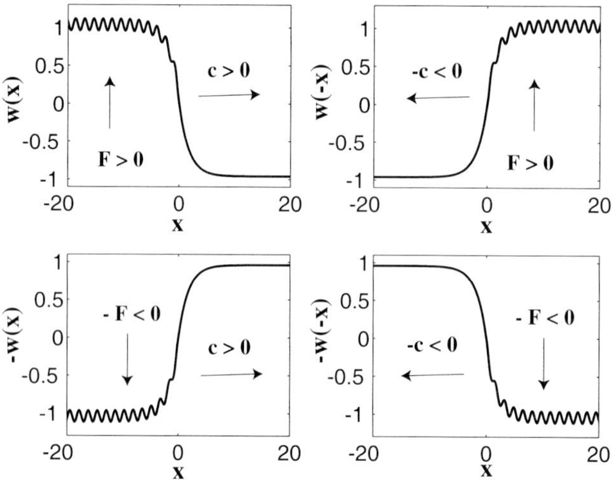

Fig. 9.7 Symmetries in the wavefront solutions for $A = U\pi/(b\kappa) = 0.25$, $\gamma = 0$, $c = 0.5$. (wave front velocity) and $F/\kappa = 0.009$ (from [31]).

lution with a stable oscillation about a stable uniform solution. These oscillatory wave fronts are depicted in Figure 9.7. The description of the corresponding depinning transition of these wave fronts is more complicated: between F_{cd} and $F = F_c$ there are infinitely many turning points in the bifurcation diagram and the bifurcation point at F_c is conjectured to be an accumulation point [31], as shown in Figure 9.8.

More realistic dislocation models are two and three-dimensional and reduce to linear elasticity far from dislocation cores. The main ideas involved in these models can be presented in the case of a two-dimensional simple cubic lattice. In Figure 9.9, we have a single edge dislocation whose Burgers vector is directed along the positive x axis. Then, the extra half-plane of atoms finishing at the dislocation line glides along the x direction, but the atoms do not climb in the y direction. We can capture dislocation motion assuming that the displacement vector is directed along the x direction. As the dislocation moves, atoms such as $u_{i,j}$ and $u_{i,j+1}$ may cease being first neighbors and we should relabel them. This is costly and unnecessary: we may introduce an odd periodic function $g(u)$ (with period equal to the crystal period and such that $g(u) \sim u$ as $u \to 0$) so that the dynamics of lattice points is [32]

$$m\,\ddot{u}_{i,j} + \dot{u}_{i,j} = u_{i+1,j} - 2u_{i,j} + u_{i-1,j} + A[g(u_{i,j+1} - u_{i,j}) + g(u_{i,j-1} - u_{i,j})]. \tag{9.26}$$

Here, $m = MC_{11}/\gamma^2$, $A = C_{44}/C_{11}$, provided γ is a damping coefficient, M the mass of atoms, and the nondimensional time is $MC_{11}t/\gamma \to t$. When the extra half column of atoms corresponding to the edge dislocation moves one period to the right, $u_{i,j+1} - u_{i,j}$ becomes larger than one period for the atoms at the dislocation

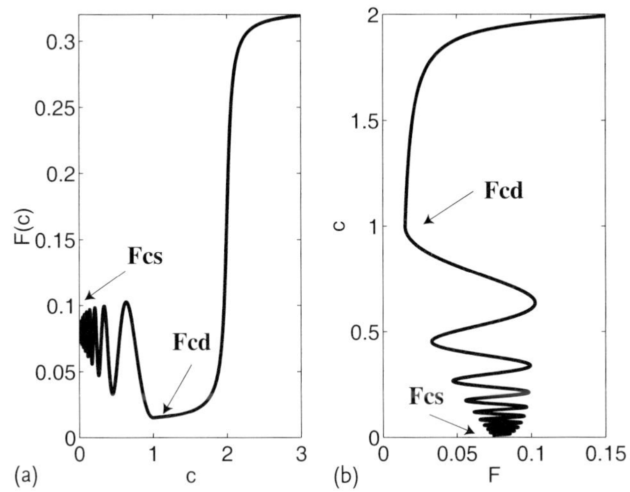

Fig. 9.8 (a) Schematic function $F(c)$ for small $\alpha > 0$ showing infinitely many maxima accumulating at $c = 0$ and $F = F_{cs}$. (b) The bifurcation diagram of wave front velocity versus F: there are infinitely many limit points (saddle-node bifurcations) corresponding to the extrema of $F(c)$ in the interval $F_{cd} < F < F_{cs}$ (from [31]).

core. However, $g(u_{i,j+1} - u_{i,j})$ is quite close to the difference between the new first neighbors at the dislocation core, so we use it instead of relabeling neighbors. The precise form of $g(u)$ can be selected so that the model has the correct Peierls stress; see [33]. These ideas can be easily generalized to the case of a cubic lattice with gliding directions along the x, y and z axes by considering a 3D displacement vector, discretizing the linear elasticity equations and replacing finite differences along the axes by periodic functions thereof [33].

Far from the dislocation core, $g(u_{i,j+1} - u_{i,j}) \approx u_{i,j+1} - u_{i,j} \approx \partial u/\partial y$, and the static solution of Eq. (9.26) approaches the solution of an anisotropic Laplace equation, $\partial^2 u/\partial x^2 + A\,\partial^2 u/\partial y^2 = 0$, which describes scalar elasticity [32]. The solution corresponding to the edge dislocation is the polar angle $\theta(x, Y) \in [0, 2\pi)$, $Y = y/\sqrt{A}$, measured from the positive x axis. Continuum approximations break down near the dislocation core, which should be described by the discrete model Eq. (9.26). This model yields the correct decay for strains and stresses: r^{-1} as $r^2 = x^2 + Y^2 \to \infty$, instead of exponential decay as in the case of the Frenkel–Kontorova model. The profile of an edge dislocation for stresses just above the Peierls stress is depicted in Figure. 9.10. Notice the large plateau similar to the case of a 1D superlattice described in Chapter 8.

Periodized discrete elasticity models of fcc and bcc cubic crystals have been introduced in [33], where interactions of different dislocations and the formation of dislocation dipoles and loops are also discussed. Similar models have been used to study homogeneous nucleation of dislocation dipoles obtained when a crystal is subject to shear stress [34] and the homogeneous nucleation of dislocation loops in

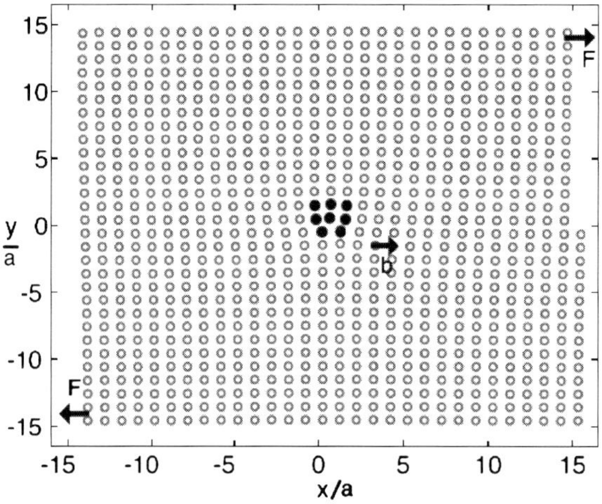

Fig. 9.9 Deformed cubic lattice in the presence of an edge dislocation for a piecewise linear $g(x)$ with $\alpha = 0.24$. We have marked atoms at the dislocation core and depicted the Burgers vector b and a shear force F (from [33]).

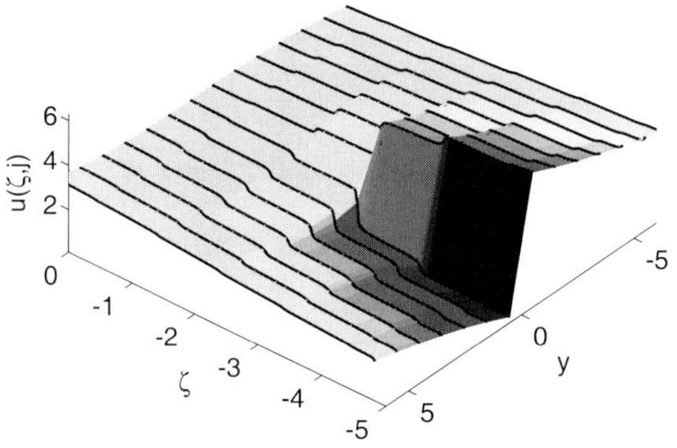

Fig. 9.10 Wave front profiles, $u_{i,j}(t) = u(\zeta, j)$, $\zeta = i - ct$, $c > 0$, near $F = F_{cs}$ for $A = 3$, $m = 0$ and a 25×25 lattice (from [32]).

a nanoindentation test [35]. Defects in the hexagonal lattice of graphene may have far fields corresponding to edge dislocations or dislocation dipoles. These defects and their stability may be found by using periodized discrete stability models introduced by Carpio and Bonilla [36]. These models provide surprises such as the finding that Stone–Wales defects (two pentagon–heptagon pairs such that the heptagons have a common side) are unstable [36, 37], in accordance with experimental results by Meyer *et al.* [38]. Defects in graphene have important consequences for electronic transport and magnetic properties [37].

References

1 Bonilla, L.L., Kochelap, V.A., and Velasco, C.A. (1999) Pattern formation and pattern stability under bistable electro-optical absorption in quantum wells. I. *J. Phys. Condens. Mater.*, **11**, 6395–6411.

2 Schöll, E. (2001) *Nonlinear Spatio-Temporal Dynamics and Chaos in Semiconductors*, Cambridge Univ. Press, Cambridge.

3 Brandl, A., Völcker, M., and Prettl, W. (1989) Fluctuations located at current filament boundaries in n-GaAs. *Solid State Commun.*, **72**, 847–850; Brandl, A., Kröninger, W., Prettl, W., and Obermair, G. (1990) Hall voltage collapse at filamentary current flow causing chaotic fluctuations in n-GaAs. *Phys. Rev. Lett.*, **64**, 212–215.

4 Hess, K., Higman, T.K., Emanuel, M.A., and Coleman, J. (1986) New ultrafast switching mechanism in semiconductor heterostructures. *J. Appl. Phys.*, **60**, 3775–3777.

5 Niedernostheide, F.-J., Ardes, M., Or-Guil, M., and Purwins, H.-G. (1994) Spatio-temporal behavior of localized current filaments in p-n-p-n diodes: numerical calculations and comparison with experimental results. *Phys. Rev. B*, **49**, 7370–7384.

6 Tretiakov, O.A., Gramespacher, T., and Matveev, K.A. (2003) Lifetime of metastable states in resonant tunneling structures. *Phys. Rev. B*, **67**, 073303.

7 Bonilla, L.L., Escobedo, R., and Dell'Acqua, G. (2006) Voltage switching and domain relocation in semiconductor superlattices. *Phys. Rev. B*, **73**, 115341.

8 Amann, A. and Schöll, E. (2005) Coupled lateral and vertical electron dynamics in semiconductor superlattices. *Phys. Rev. B*, **72**, 165319.

9 Xu, H., Amann, A., Schöll, E., and Teitsworth, S.W. (2009) Dynamics of electronic transport in a semiconductor superlattice with a shunting side layer. *Phys. Rev. B*, **79**, 245318. Copyright (2009) by the American Physical Society.

10 Samuilov, V.A. (1995) Nonlinear and chaotic charge transport in semi-insulating semiconductors. *Nonlinear Dynamics and Pattern Formation in Semiconductors and Devices*, Vol. 79 of *Springer Proceedings in Physics* (ed. F.-J. Niedernostheide), Springer-Verlag, Berlin, Heidelberg, p. 220.

11 Sacks, H.K. and Milnes, A.G. (1970) Low frequency oscillations in semi-insulating GaAs. *Int. J. Electron.* **28**, 565–583.

12 Bonilla, L.L., Hernando, P.J., Kindelan, M., and Piazza, F. (1999) Determination of EL2 capture and emission coefficients in semi-insulating n-GaAs. *Appl. Phys. Lett.*, **79**, 988–990.

13 Piazza, F., Christianen, P.C.M., and Maan, J.C. (1996) Electric field dependent EL2 capture coefficient in semi-insulating GaAs obtained from propagating high field domains. *Appl. Phys. Lett.*, **69**, 1909–1911.

14 Piazza, F., Christianen, P.C.M., and Maan, J.C. (1997) Propagating high-electric-field domains in semi-insulating GaAs: experiment and theory. *Phys. Rev. B*, **55**, 15591–15600.

15 Bonilla, L.L., Escobedo, R. and Higuera, F.J. (2002) Axisymmetric pulse recycling and motion in bulk semiconductors. *Phys. Rev. E*, **65**, 016607.

16 Bonilla, L.L. and Escobedo, R. (2001) Two-dimensional oscillatory patterns in semiconductors with point contacts. *Phys. Rev. E*, **64**, 036203.

17 Bonilla, L.L., Escobedo, R., and Higuera, F.J. (2003) Free boundary problems describing two-dimensional pulse recycling and motion in semiconductors. *Phys. Rev. E*, **67**, 036202.

18 Escobedo, R. and Bonilla, L.L. (2002) Wave dynamics in two-dimensional samples of n-GaAs with point contacts. *Chaos Solit. Fract.*, **17**, 283–288.

19 Keizer, J. (1981) Calculation of voltage fluctuations at the Gunn instability. *J. Chem. Phys.*, **74**, 1350–1356.

20 Diaz-Guilera, A. and Rubí, J.M. (1986) On fluctuations about nonequilibrium steady states near Gunn instability. II. Voltage fluctuations. *Physica A*, **135**, 200–212.

21 Kabashima, S., Yamazaki, H., and Kawakubo, T. (1976) Critical fluctuation

near threshold of Gunn instability. *J. Phys. Soc. Japan*, **40**, 921–924.
22 Nakamura, K.-I. (1975) Statistical dynamics of the Gunn instability near threshold. *J. Phys. Soc. Japan*, **38**, 46–50.
23 Bulatov, V.V. and Cai, W. (2006) *Computer Simulations of Dislocations*, Oxford U.P., Oxford.
24 Frenkel, J. and Kontorova, T. (1939) On the theory of plastic deformation and twinning. *J. Phys. Moscow*, **1**, 137–149.
25 Nabarro, F.R.N. (1987) *Theory of Crystal Dislocations*, Dover, New York.
26 Carpio, A. and Bonilla, L.L. (2001) Wave front depinning transition in discrete one-dimensional reaction-diffusion systems. *Phys. Rev. Lett.*, **86**, 6034–6037.
27 Carpio, A. and Bonilla, L.L. (2003) Depinning transitions in discrete reaction-diffusion equations. *SIAM J. Appl. Math.*, **63**, 1056–1082.
28 King, J.R. and Chapman, S.J. (2001) Asymptotics beyond all orders and Stokes lines in nonlinear differential-difference equations. *Eur. J. Appl. Math.*, **12**, 433–463.
29 Fáth, G. (1998) Propagation failure of traveling waves in a discrete bistable medium. *Physica D*, **116**, 176–190.
30 Atkinson, W. and Cabrera, N. (1965) Motion of a Frenkel–Kontorowa dislocation in a 1D crystal. *Phys. Rev.*, **138**, A763–A766.
31 Carpio, A. and Bonilla, L.L. (2003) Oscillatory wave fronts in chains of coupled nonlinear oscillators. *Phys. Rev. E*, **67**, 056621.
32 Carpio, A. and Bonilla, L.L. (2003) Edge dislocations in crystal structures considered as traveling waves of discrete models. *Phys. Rev. Lett.*, **90**, 135502.
33 Carpio, A. and Bonilla, L.L. (2005) Discrete models of dislocations and their motion in cubic crystals. *Phys. Rev. B*, **71**, 134105.
34 Plans, I., Carpio, A., and Bonilla, L.L. (2008) Homogeneous nucleation of dislocations as bifurcations in a periodized discrete elasticity model. *Europhys. Lett.*, **81**, 36001.
35 Plans, I., Carpio, A., and Bonilla, L.L. (2009) Toy nanoindentation model and incipient plasticity. *Chaos Solit. Fract.*, **42**, 1623–1630.
36 Carpio, A. and Bonilla, L.L. (2008) Periodized discrete elasticity models for defects in graphene. *Phys. Rev. B*, **78**, 085406.
37 Carpio, A., Bonilla, L.L., de Juan, F., and Vozmediano, M.A.H. (2008) Dislocations in graphene. *New J. Phys.*, **10**, 053021.
38 Meyer, J.C., Kisielowski, C., Erni, R., Rossell, M.D., Crommie, M.F., and Zettl, A. (2008) Direct imaging of lattice atoms and topological defects in graphene membranes. *Nano Lett.*, **8**(11), 3582–3586.

Index

Nonlinear Wave Methods for Charge Transport. Luis L. Bonilla and Stephen W. Teitsworth

ISBN: 978-3-527-40695-1